Pesticide Chemistry: Crop Protection, Public Health, Environmental Safety

Pesticide Chemistry: Crop Protection, Public Health, Environmental Safety

Contributors

Renata Raina et al.

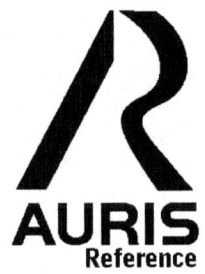

AURIS
Reference

www.aurisreference.com

Pesticide Chemistry: Crop Protection, Public Health, Environmental Safety

Contributors: Renata Raina et al.

Published by Auris Reference Limited

www.aurisreference.com

United Kingdom

Copyright 2016

Printed in 2017 for Sale in the Indian Subcontinent

Pesticide Chemistry: Crop Protection, Public Health, Environmental Safety

ISBN: 978-1-78154-874-5

British Library Cataloguing in Publication Data
A CIP record for this book is available from the British Library

Printed in the United Kingdom

Exclusively distributed by CBS Publishers & Distributors Pvt. Ltd.

Sales & Distribution Rights only for India, Pakistan, Bangladesh, Sri Lanka, Nepal and Bhutan.This book is not to be sold outside these territories.

Contents

List of Abbreviations

API	Atmospheric Pressure Ionization
CID	Collision Induced Dissociation
EI	Electron Impact
FAO	Food and Agricultural Organization
GAP	Good Agricultural Practice
GC	Gas Chromatography
IPM	Integrated Pest Management
LC	Liquid Chromatography
LLE	Liquid-Liquid Extraction
LPME	Liquid-Phase Micro-Extraction
MRL	Maximum Residue Limit
MS	Mass Spectrometry
NCI	Negative Chemical Ionization
OP	Organophosphorus Pesticide
PCI	positive chemical Ionization
PPB	Parts Per Billion
PPM	Parts Per Million
ROS	Reactive Oxygen Species
SBSE	Stir-Bar Sorptive Extraction
SPE	Solid-Phase Extraction
SPME	Solid-Phase Micro-Extraction
SRM	Selected Reaction Monitoring

List of Contributors

Renata Raina
University of Regina, Department of Chemistry & Biochemistry and Trace Analysis Facility
Canada

Mariana Furio Franco Bernardes
Faculdade de Ciências Farmacêuticas de Ribeirão Preto - FCFRP, Universidade de São Paulo-
USP, Brazil

Murilo Pazin
Faculdade de Ciências Farmacêuticas de Ribeirão Preto - FCFRP, Universidade de São Paulo-
USP, Brazil

Lilian Cristina Pereira
Faculdade de Ciências Farmacêuticas de Ribeirão Preto - FCFRP, Universidade de São Paulo-
USP, Brazil

Daniel Junqueira Dorta
Departamento de Química, Faculdade Filosofia, Ciências e Letras de Ribeirão Preto- FFCLRP,
Universidade de São Paulo – USP, Brazil

B. Alewu
Shehu Idris College of Health Science and Technology, Makarfi, Nigeria

C. Nosiri
Shehu Idris College of Health Science and Technology, Makarfi, Nigeria

Harsimran Kaur Gill
University of Florida, Gainesville, FL, USA

Harsh Garg
The University of Sydney, NSW, Australia

Zhi-Jun Zhou
Fudan University, China

Dipsikha Bora
Department of Life Sciences, Dibrugarh University, Dibrugarh, Assam, India

Hiren Gogoi
Department of Life Sciences, Dibrugarh University, Dibrugarh, Assam, India

Bulbuli Khanikor
Department of Zoology, Gauhati University, Guwahati, Assam, India

Simon Koma Okwute
Department of Chemistry, University of Abuja, Gwagwalada, Federal Capital Territory, Abuja,,
Nigeria

Marcela Varona Uribe
Environmental and Occupational Health Group, Research Department, National Institute of Health, Bogotá,, Colombia

Sonia Mireya Díaz
Environmental and Occupational Health Group, Research Department, National Institute of Health, Bogotá,, Colombia

Andrés Monroy
Environmental and Occupational Health Group, Research Department, National Institute of Health, Bogotá,, Colombia

René A. Castro
National Agricultural Inputs Laboratory, Agricultural Protection Associate Management, Analysis and Diagnosis Associate Management, Colombian Agricultural Institute, Mosquera,, Colombia

Edwin Barbosa
National Agricultural Inputs Laboratory, Agricultural Protection Associate Management, Analysis and Diagnosis Associate Management, Colombian Agricultural Institute, Mosquera,, Colombia

Martha Isabel Páez
Environmental Research Group for Metals and Pesticides (GICAMP), Department of Chemistry, University of Valle,, Colombia

Jolanta Stocka
Department of Analytical Chemistry, Chemical Faculty, Gdansk University of Technology, Narutowicza Street 11/12, Gdansk 80-233, Poland

Maciej Tankiewicz
Department of Analytical Chemistry, Chemical Faculty, Gdansk University of Technology, Narutowicza Street 11/12, Gdansk 80-233, Poland

Marek Biziuk
Department of Analytical Chemistry, Chemical Faculty, Gdansk University of Technology, Narutowicza Street 11/12, Gdansk 80-233, Poland

Jacek Namieśnik
Department of Analytical Chemistry, Chemical Faculty, Gdansk University of Technology, Narutowicza Street 11/12, Gdansk 80-233, Poland

Preface

Pesticides are those chemicals that are used to destroy unwanted forms of life or organisms. They include insecticides, rodenticides, herbicides, fungicides, fumigants etc. They are expected to have a selective action or toxicity to animals and man. Pesticides constitute any substance or mixture of substances intended for preventing, destroying, repelling, or mitigating any pest. They can also serve as plant regulators, defoliants, or dessicants. Pesticide Chemistry Crop Protection, Public Health, Environmental Safety provides comprehensive coverage and even captures emerging technologies within the industry. All facets of pesticides are addressed, including agriculture, agrochemicals, and environmental health aspects. First chapter deals with the chemical analysis methods for the main pesticide chemical classes that are most frequently analyzed with gas chromatography (GC) or liquid chromatography (LC) coupled to mass spectrometry (MS). Second chapter explores on impact of pesticides on environmental and human health. The chapter also discusses the pesticides as inducers of oxidative stress and endocrine disruptors action of two important issues. Third chapter describes on human health, and methods of detection of these compounds. Pesticides are those substances which, on entering the body by whatever route, e.g. ingestion, inhalation, or absorption through intact skin, produce harmful effects. The effect may be in the form of damage to the tissues or as a disturbance of the functioning of the body. Fourth chapter details the effect of pesticides on target and non-target organisms including earthworms, predators, pollinators, humans, fishes, amphibians, and birds. Additionally, impact of pesticides on soil, water and air ecosystems is also discussed. Furthermore, an eco-friendly practice (Integrated Pest Management (IPM) approach) has been detailed as a strategy that could minimize the use of pesticides. Fifth chapter presents a study on health problem caused by long-term organo phosphorus pesticides exposure and plant based pesticides are reviewed in sixth chapter. Seventh chapter presents a review on plants as potential sources of pesticidal agents. Eighth chapter determines the biomarkers the inner dosages, exposure and effect caused by the use of organophosphoric (OF), carbamates (C) and organochlorated (OC) pesticides. Green aspects of techniques for the determination of currently used pesticides in environmental samples are described in last chapter.

Chapter 1

CHEMICAL ANALYSIS OF PESTICIDES USING GC/MS, GC/MS/MS, AND LC/MS/MS

Renata Raina
University of Regina, Department of Chemistry & Biochemistry and Trace Analysis Facility Canada

INTRODUCTION

There are well over 500 registered pesticides worldwide for use in agricultural regions and new agrochemicals are introduced to the marketplace continuously. This chapter deals with the chemical analysis methods for the main pesticide chemical classes that are most frequently analyzed with gas chromatography (GC) or liquid chromatography (LC) coupled to mass spectrometry (MS). GC amenable pesticide chemical classes which do not require derivatization include organochlorines (OCs), pyrethroids, organophosphorus pesticides (OPs), triazines, and chloroacetanilides. In addition some transformation products of organochlorines, triazines, and phenylureas are GC amenable and when derivatized some transformation products of OPs, pyrethroids, and phenoxyacid herbicides are also GC amenable. Specific methods have been developed with other injector choices than the standard splitless injection for more thermally labile chemical classes such as trihalomethylthio fungicides to extend the range of GC amenable pesticides. Some chemical classes which are more polar such as phenoxy acid herbicides and carbamates can still be analyzed by GC/MS methods but require derivatization to make them GC amenable. For some other chemical classes a few pesticides have been analyzed by GC/MS usually included in multiresidue methods but these methods have not tackled the entire range of compounds within the chemical class. These include chemical classes such as dicarboximides (vinclozin, iprodione), dinitroaniline (trifluralin, ethalfluralin), dinitrophenol (dinoseb), and dithiocarbamate (triallate). A large number of pesticide classes generally of higher polarity suffer from poor chromatographic performance, poor MS source ionization or stability in GC/MS injectors, on-column, or in MS. For these chemical classes and also to minimize the need for derivatization prior to GC there has been a gradual shift to the development of new methods utilizing LC coupled with tandem mass spectrometry (MS/MS). Tandem mass spectrometry in selected reaction monitoring (SRM) mode is generally now more frequently used for LC rather than selected ion monitoring (SIM) with LC/MS as the ionization process for LC/MS is a softer process (change processes to process) than that of GC/MS ion sources such as EI and CI. For atmospheric pressure ionization (API) sources most frequently used in LC/MS/MS most pesticides have only one ion formed during ionization (the protonated or deprotonated molecular ion or sometimes an adduct ion (eg. sodium or ammonium adduct)) and consequently there is little confirmation ability. Tandem mass spectrometry allows for the controlled collision induced dissociation (CID) of the parent ion making discrimination possible from co-eluting matrix components. No additions (LC/MS/MS or LC/MS methods include phenoxyacid herbicides other pesticides of interest. The main chemical classes of pesticides that have been more recently analyzed by LC/MS/MS or LC/MS methods and include phenoxyacid herbicides and a related nitrile herbicide (bromoxynil) often used in formulations with phenoxyacid herbicides, phenylureas, sulfonyl ureas, carbamates, pyrethroids, azoles, and a more extensive list of dithiocarbamates. Phenylureas, sulfonylureas, and most dithiocarbamates are not GC amenable and many azoles have significantly lower detection limits with LC/MS/MS. Some chemical classes including OPs, pyrethroids, carbamates, phenoxyacid herbicides, and azoles have both GC and LC methods coupled to mass spectrometry that have been developed and will be discussed in more detail in this chapter.

There are a large number of factors that require consideration for the selection of the method for analysis whether that is for an individual pesticide, a chemical class of pesticides, a large number of pesticides of different chemical classes, or for inclusion of their transformation products. These factors include: boiling point or polarity; solubility in desired solvent or mobile phases; stability of pesticides in injector ports, on-column, or in mass spectrometer ion sources; selectivity of columns and chromatographic behavior; interferences in detection; molecular structure or other chemical properties important for both ionization and fragmentation; method detection limit or regulatory requirements; and confirmation ability over linear dynamic range. This chapter does not include a discussion of the sample preparation (pre-concentration or sample clean-up) procedures and does not distinguish methods developed for fruit and vegetables, biological tissues, soil, water, air or other sample matrices. The focus is on issues related to the chromatography-mass spectrometry and instrumental approaches that may be taken advantage of to improve selectivity or sensitivity of analysis.

IDENTIFICATION OF THE PROBLEM

Due to the large number of pesticides under investigation users must firstly decide on whether to choose a GC or LC method coupled to mass spectrometry and if the method can achieve the desired quantitative analysis and confirmation needs. Some laboratories may also be more limited in their choice of instruments or skill of analysts so need to be aware of methods that may be equivalent for those chemical classes that can be analyzed by both GC and LC. Many laboratories are looking towards streamlining sample preparation and analysis needs such as with the Quick Easy Cheap Effective Rugged and Safe (QuEChERSA) pesticide multiresidue methods in combination with GC and LC mass spectrometry methods (Cunha et al., 2007; Payá et al., 2007; Pihlström et al., 2007). Due to the large diversity in sample types and pesticides used or of concern in different regions, multiresidue analysis methods can vary significantly in their choice of target pesticides and transformation products and this makes it challenging for an analyst to select a method for analysis as they may not fully understand the factors that went into the selection of the instrumental parameters and the compromises that were made to resolve matrix effects, chromatographic needs, and detection requirements. This chapter takes a chemical class approach which users can then utilize to select methods with their target pesticide list and can be further built on to include compounds not in these major chemical classes. A main goal is to highlight by chemical class some of the preferences for these methods and the demands or options for improvements. Some of the advances in instrumental approaches to either improve the range of compounds for analysis, reduce background signal, or improve selectivity or sensitivity of the analysis will be highlighted. Due to the increasing need for analysis of transformation products they will be discussed along with their chemical class of parent compounds. Simultaneous analysis of parent pesticides and transformation products is desirable but because of the large diversity in polarity, volatility, stability, and ionization in MS ion sources this is not always feasible. Issues with co-elution of other complex interfering matrix components or other pesticides of interest and their impact on detection and confirmation will also be discussed.

GC/MS, GC/MS/MS, AND LC/MS/MS FOR PESTICIDES AND THEIR TRANSFORMATION PRODUCTS

GC/MS and GC/MS/MS Methods

One of the most important parameters when considering GC/MS methods of analysis particularly when added selectivity or sensitivity are required is the choice of the ionization mode. Sample matrix and sample preparation procedures including clean-up also dictate selection of the ionization method due to presence of co-eluting pesticides or matrix components which can interfere in analysis if they can not be distinguish in the mass spectra. If pesticides are electron-capturing such as those pesticides which contain halogen, NO2, or P ester groups then they will generally give an enhanced response (up to two or three orders of magnitude) with negative chemical ionization (NCI) in comparison to electron impact (EI) or positive chemical ionization (PCI) (Raina and Hall, 2009; Liapis et al., 2003; Bailey and Belzer, 2007; Húšková et al., 2009). The selection of ionization mode often depends upon whether the analysis is targeted for specific chemical classes or is a multiresidue analysis methods for determination of hundreds of pesticides in a sample extract. A comparison of GC/MS or GC/MS/MS with EI to LC/MS/MS has been reviewed for a large number of compounds and suggests for most pesticides other than organochlorines that LC/MS/MS can provide lower detection limits (Alder et al., 2006; Pihlström et al., 2007; Paya et al., 2007; Lambropoulou et al ., 2007). However, lower or comparable detection limits have also been found for chloracetanilides (metolachlor, acetochlor, alachlor) and selected triazines by GC/MS or GC/MS/MS with EI relative to LC/APCI-MS/MS (Dagnac et al., 2005) or LC/ESI-MS/MS (Gomides Freitas et al., 2004). GC/MS of a wider range of triazines has also been done by GC-EI/MS (Nagaraju and Huang, 2007; Zambonin and Palmisano, 2000; Jiang et al., 2005; Gonçalves et al., 2006; Albanis et al., 1998). Chemical ionization is often not considered in comparisons of GC and LC mass spectrometry methods. Reduction of matrix interferences particularly for masses levels of routine analysis particularly for environmental sample analysis there is insufficient concentration to obtain full scan MS spectra of sufficient abundance for library matching when quadrupole or ion trap systems are used.

There is another unique feature of pesticide analysis with mass spectrometry that is often not discussed in detail. Relative to other contaminants, many pesticides including OCs, OPs, pyrethroids, and chloroacetanilides exhibit low intensity for the molecular ion regardless of whether EI or CI is used (Raina and Hall, 2009; Húšková et al., 2009; Yoshida, 2009; Feo et al., 2010; Dagnac et al., 2005). Consequently in SIM mode the quantitative or qualifier ion is rarely selected as the molecular ion. In general >90% of pesticides do not monitor the molecular ion by EI or CI methods as at the working concentration ranges of trace analysis generally the molecular ion is too low in abundance to be observed. The exception are the triazines where the molecular ion is one of the ions monitored but may not be the base peak in the EI mass spectra (Nagaraju et al., 2007; Jiang et al., 2005; Zabonin and Palmisano, 2000). The selection of EI versus NCI or PCI may also be based on instrument design and cost and basic GC/MS instruments often do not include CI capability.

In this section the focus will first be on chemical classes of pesticides where GC/MS methods are superior or equivalent to LC/MS/MS methods and derivatization is not required. The chemical classes that will be discussed include organochlorines (OCs), organophosphorus pesticides (OPs), trihalomethylthio fungicides, pyrethroids, triazines, and chloracetanilides. The ion sources used in LC/MS/MS are not suitable for some of these pesticides including many of the OCs and trihalomethylthio fungicides. OC degradation products have been routinely included in GC/MS methods either with EI or NCI and include OCs such as endosulfan sulphate, DDD, DDE, HCH isomers, endrin ketone, endrin aldehyde, heptachlor epoxide, methoxychlor. Chloroacetanilides are more sensitive with EI than CI modes with GC/MS (Raina and Hall, 2009; Dagnac et al., 2005; Gabaldon et al., 2002) but can be done with similar detection limits with LC/ ESI+ or APCI+ MS/MS (Dagnac et al., 2005). The transformation products of chloroacetanilides are not analyzed by GC/MS, however chloroacetanilide (eg alachlor, propachlor, metalochlor) analysis is frequently included with analysis of OCs by GC/MS. Triazines can be analyzed with comparable GC/EI-MS or LC/MS/MS methods and it depends upon the application needs and availability of instrumentation as to which method is choosen. Transformation products of atrazine: deisopropylatrazine (DIA), desethylatrazine (DEA), didealkylatraizine (DDA) and 3,4-chloroaniline which is a transformation product of phenylureas (linuron and diuron) have also been analyzed by GC/EI-MS or GC/EI-MS/MS methods (Planas et al., 2006; Jiang et al., 2005; Dagnac et al., 2005). GC/MS methods are more suitable to a wider range of OPs than LC/MS/MS as not all OPs are ionized efficiently by API sources (eg. parathion). However, a significant number of OPs which are widely used give significantly lower detection limits with LC/MS/MS (Table 1). In addition GC/MS methods suffer from poor chromatographic performance, low sensitivity, and required derivatization for OP transformation products, whereas LC/MS/MS can be used to simultaneous analyze the OP transformation products including OP oxons with detection limits of 0.06-0.38 µg/L (Raina and Sun, 2008) and OP sulfones and sulfoxides (Chung and Chan, 2010; Jansson et al., 2004; Hiemstra et al., 2007; Economou et al., 2009). For some pyrethroids GC/EI-MS has approximately 100 times higher detection limits than LC/MS/MS (Alder et al., 2006) while for many they are comparable (Yoshida et al., 2009). When NCI is used detection limits for some pyrethroids can be 10-100 times lower than EI (Feo et al., 2010) making GC/MS comparable or better than LC/MS/MS methods. Coupling this with large volume injections can further improve these GC/MS methods if required. In addition, pyrethroid transformation products can be analyzed with GC/EI-MS/MS following 1,1,1,3,3,3- hexafluoroisopropanol (HFIP) derivatization (Arrebola et al., 1999) and there has been no reported LC/MS/MS method.

Table 1. Comparison of GC/MS and LC/MS/MS Limits of Detection for Selected Organophosphorus Pesticides.

OP Pesticide	GC/MS or GC/MS/MS Limit of Detection (μg/L)[Raina and Hall, 2009]	LC/ESI+MS/MS Limit of Detection (μg/L)[Raina and Sun, 2008]
Chlorpyrifos	4.5	0.19
Chlorpyrifos methyl	7.6	0.27
Diazinon	*0.70*	0.08
Malathion	9.5	0.23
Azinphos methyl	>50*	0.32
Azinphos ethyl	>50*	0.47
Dimethoate	NA	0.05
Phorate	7.8	0.37
Fenchlorphos	7.5	16

*note calculated under GC separation conditions for OCs and OPs retention times 25-27 min (higher than most compounds); NA not available. Italics GCEI/MS lowest detection limit otherwise NCI was used.

A comparison of 47 chlorinated organics (including OCs and several chloroacetanilides) and OPs analyzed by GC/MS showed that no one ionization mode could be used to analyze all the pesticides at concentrations <100 ng mL^{-1} for a standard splitless 1 μL injection. In general NCI-SIM provided the lowest method detection limits (MDLs) for the largest number of pesticides along with confirmation at these low levels. When confirmation by NCI-SIM was not sufficient, NCI-SRM could be used and gave additional sensitivity and confirmation ability to ~14% of pesticides studied (Raina and Hall, 2009). Others have also found that GC-MS/MS can provide added selectivity (Zhang and Lee, 2006). Although EISIM is often used for multiresidue GC analysis methods we found that EI-SIM only provided better sensitivity than NCI-SIM or NCI-SRM for 3 of the 19 OPs (aspon, diazinon, sulfotep), and 9 of the 28 OCs or chloroacetanilides studied (alachlor, aldrin, p,p'-DDD, o,p'- DDE. p,p'-DDE, dieldrin, heptachlor, perthane, propachlor) and for other OCs and OPs an additional confirmation approach would be required at these concentrations if EI was used due to low abundance of the confirmation ion (Raina and Hall, 2009). Chloroacetanilides have been previously identified as best analyzed by GC-EI/MS or MS/MS (Galaldon et al., 2002; Dagnac et al., 2005). Others have also found for a range of OCs, OPs, and some pyrethroids that NCI-SIM is up to 100 times more sensitive than EI-SIM (Húšková et al., 2009; Feo et al., 2010). Better S/N ratio with NCI or reduced matrix background interference response was observed particularly at low masses (m/z < 50). NCI provides added selectivity as many interfering matrix components are expected to be hydrocarbons, humic or fulvic acids, or nonhalogenated in nature and thus do not produce a signal with NCI (Bailey, 2005; Bailey and Belzer, 2007). Positive chemical ionization like EI suffers more than NCI from matrix interferences and for these chemical classes of pesticides it is generally less sensitive so it is seldom used for quantitative analysis.

EI full scan mode provides the ability for confirmation with library search matching, however in quantitative analysis generally selected ion monitoring (SIM) is accomplished only with confirmation using an additional one or two ions and the ratio of response of these ions within a specified % relative standard deviation usually determined from standard injections on the day of analysis rather than from libraries. One advantage of GC/MS over LC/MS/MS methods is the lower instrument cost and that pesticides fragment in the EI or CI ion source easily and consequently structural information is available for the pesticide for its confirmation. Fragmentation with EI sources is distinctly different from electrospray ionization used in LC/MS/MS with odd-electron (OE) fragment ions more frequently produced with EI (35% OE ions, 65% EE ions) as compared to 93% even-electron (EE) ions with positive electrospray ionization (Thurman et al., 2007). Chemical ionization is a softer ionization process than EI and the MS spectra generally produce less fragment ions, however for > 90% of pesticides analyzed by GC/MS the two most abundant ions in any ionization mode still generally do not include the molecular ion even when PCI or NCI are used (Raina and Hall, 2009; Húšková et al., 2009; Feo et al., 2010). In addition there may be relatively few fragments available of sufficient abundance for confirmation and consequently often isotope masses of fragment ions are used for confirmation. This has implications on the applicability of GC/MS/MS with our results showing that EI is not suitable for the analysis of OCs or OPs < 100 ng mL^{-1} and NCI-SRM is generally less sensitive than NCI-SIM even though there is reduced background noise (Raina and Hall, 2009). For fruit and vegetable analysis where higher levels of pesticides can be achieved in sample extracts, GC/MS analysis with SRM in EI mode has been used for a similar range of OPs and OCs with preference for these pesticides analyzed by GC/MS/MS over LC/MS/MS (Pihlström et al, 2007). As pesticides easily fragment in the ion sources of GC/EI or NCI-MS, the parent ion selected for collision induced dissociation (CID) is often a fragment ion and this ion must be capable of further fragmentation. In a number of cases for these chemical classes with NCI the presence of higher mass parent ions or the molecular ion improved the potential for lower MDLs with NCI-SRM as compared to EI-SRM. However, most pesticides did not have an abundant molecular ion. Even in NCI-SRM for OCs the SRM transition selected were often $f_1^+ > Cl-$ (m/z=35) with the confirmation SRM utilizing an isotope peak mass (eg $f_1^+ > Cl-$ (m/z=37)) (Raina and Hall, 2009). The fact that the parent ions with GC/MS/MS are often fragment ions makes finding suitable product ions more challenging than with LC/MS/MS ion sources where the parent ion is generally the protonated or deprotonated molecular ion. In the case of HFIP derivatized transformation products of pyrethroids the molecular ion was used for CID and produced better sensitivity and selectivity that GC/EI-MS which observed significant chromatographic resolution problems and reduced MS sensitivity (Arrebola et al., 1999).

There are a number of approaches that can be used to extend the range of pesticides that can be analyzed by GC/MS or to further improve MDLs beyond the most frequently used splitless injections with a hot split/splitless injector. For pesticides such as the trihalomethylthio fungicides that are more thermally labile other injectors including programmable temperature vaporizer (PTV) or cold on-column (COC) injector can be used (Bailey, 2005). Another advantage of these injectors is that they can also be utilized for large volume injections increasing the sample injection size from 1-2 µL to 5-100 µL. With both approaches the sample is injected cold (below or near the boiling point of the solvent). Precolumns have also been utilized with these approaches for focusing and to extend the analytical column lifetime by minimizing build-up of non-volatile matrix components. Both approaches have limitations that are discussed requiring careful consideration.

A standard PTV injector has been used for analysis of chemical classes of pesticides such as OCs and OPs often operated with a solvent vent step where the initial injector temperature is set below the boiling point of solvent (eg 40 °C for toluene) and held at this temperature while the solvent vapour is eliminated via the split exit (Godula et al., 2001). PTV can be used for small injection volumes (2 µL) such as those in neat solvents (eg toluene) (Huskova et al., 2009) or matrix matched standards (Kirchner et al., 2005), and can also be used for large injection volumes of 20 µL (Grob and Li, 1988). More advanced PTV injectors are also available with modifications for dirty matrix injections (DMI) where the GC liner can be replaced for each injection with a robotic autosampler system and contains a small 40 µL DMI microvial. The challenge with larger volume injections with PTV is that the temperature and time the split vent is open must be optimized to remove the solvent without loss of analytes of interest and pesticides with boiling points near the solvent will have a higher potential for loss. An example system that I have used for this approach is a GC Twin-PAL (Leap Technologies, Carrboro, NC) and an Optics 3 PTV inlet (ATAS/GL International BV) with direct thermal desorption (DTD) probe. The crimp top DTD liner is an open liner (80 mm X 5 mm O.D.)

containing a needle guide and the 40 μL DMI microvial held in place at 20 mm from the bottom of the liner by three knobs. The Optics 3 PTV inlet is equipped with DTD probe that allows for interchange of the DTD liners containing the DMI microvial between injections. The inlet has separate gas controls from the GC and a solvent vapour thermal conductivity detector (TCD) sensor and in the example shown below is operated in fixed time mode. During the injection temperature is set below the boiling point of the solvent and there is a high split vent flow (100 mL min^{-1}), after the solvent is vented the injection time starts (Figure 1). For a solvent such as ethylacetate which is often used for extraction procedures in QuEChERS pesticide analysis (Pihlström et al., 2007) a temperature of 70°C can be used and requires a vent time of 330 sec for a 10 μL injection. Increasing the temperature in 10°C increments will reduce vent time required by ~60 sec, however more volatile pesticides such as captan and captafol showed significant loss of signal above 70°C and consequently this temperature and vent time were required for the analysis. When injection size was increased to 20 μL the required vent time increased to 540 seconds and for larger volume samples near the capacity of the microvial the vent time was in excess of 10 minutes which is not practical for analysis. Switching the solvent to a lower boiling solvent such as hexane reduced the temperature to 60°C. For both GC/MS and LC/MS/MS applications there has also been interest in coupling SPE cleanup methods directly with analysis. Table 2 provides the steps required for coupling the LVI-DMI (large volume injection-dirty matrix injection) with the at-line automated SPE approached. The at-line automated SPE LVI-DMI-GC/MS method sequence involving first direct clean-up of a sample with a 96-well plate C-18 SPE format using the Twin-PAL robotic autosampler system for SPE preparation; followed by injection of a portion of the SPE eluted extract directly into DMI liners; and then exchange of the liners in the PTV-DTV probe for sample injection. In this example a 10 μL fraction of each 100 μL fraction eluted from the SPE 96 well plate was analyzed for pesticides. Figure 2 shows that the trihalomethylthio fungicides are eluted with 200 μL of ethylacetate (fractions F2 and F3 of size 100 μL) and illustrates that the at-line SPE approach is capable of replacement of standard off-line SPE procedures. Good linearity from method detection limit (MDL)-500 μg/L (r^2>0.99) was observed with method detection limits of 2.5-5 μg/L similar to that observed for LVI-COC injections (Bailey and Belzer, 2007). The clear advantage of this injection approach over LVI-COC injections is that non-volatile material remains in the injector liner (in the DMI microvial) which is replaced with each injection so there is no build-up of non-volatile material on column reducing maintenance requirements. It is limited in its applicability to pesticides with boiling points near the solvents boiling point as they will be lost during the solvent venting stage so solvent selection is also an important parameter for consideration. PTV inlets may also still cause degradation of pesticides in the injection port as after solvent venting, the injection port temperature is rapidly ramped.

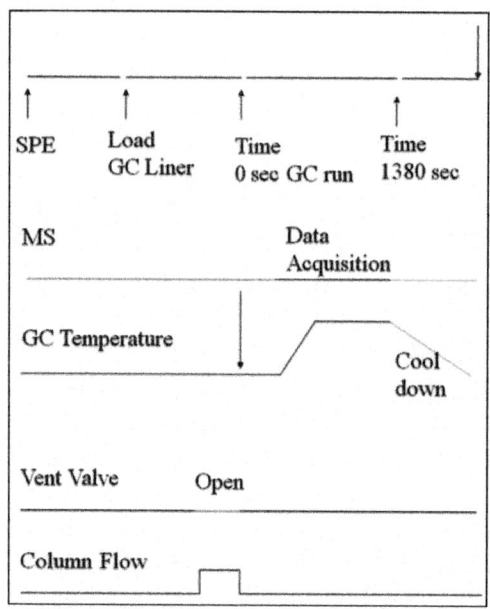

Figure 1. SPE-LVI-DMI-GC/MS run set-up conditions.

Table 2. At-line automated SPE LVI-DMI-GC-MS Method Sequence.

Sequence Step	Conditions
SPE sorbent conditioning with Prep-PAL	1) 500 µL ethyl acetate, apply pressure 2) 500 µL of methanol, apply pressure
Sample Loading with Inject-PAL to SPE 96 well plate	10 µL sample added, rinse syringe
Washing with Prep-PAL	100 µL methanol added, apply pressure
Move 96-well plate with Prep-PAL	Ready for elution step –96 well plate moved forward from over waste to over 96-well collection plate
Elution with Prep-PAL	100 µL ethyl acetate, apply pressure
Addition of IS standard With Inject-PAL	Take 2 µL internal standard solution and mix with 100 µL SPE eluate in SPE collection plate (3-5 strokes)
DMI-Injection –load sample and transfer DTD liner with Inject-PAL	Take 10 µL of SPE eluate from 96-well collection plate and deliver to DTD/DMI liner, move liner into DTD probe, clean syringe

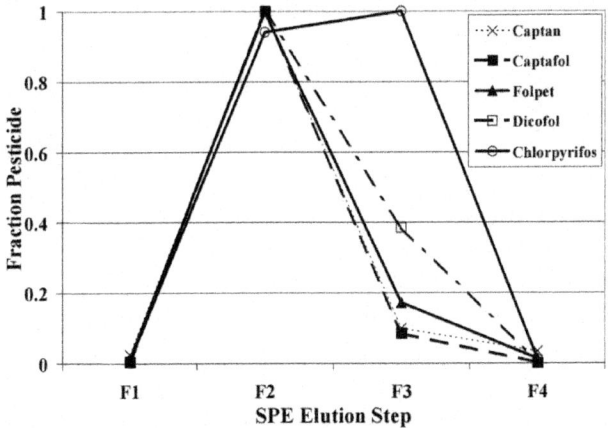

Figure 2. Fraction of Pesticide in Washing and Elutions Steps of At-line SPE Procedure. F1 (washing):100 μL methanol; F2 (elution):100 μL ethylacetate; F3 (elution):100 μL ethylacetate; F4 (elution):100 μL ethylacetate; Sample 10 μL of 0.1 μg mL^{-1} pesticide mixture dissolved in hexane; SPE 96 well plate Bond Elute® C18 100 mg.

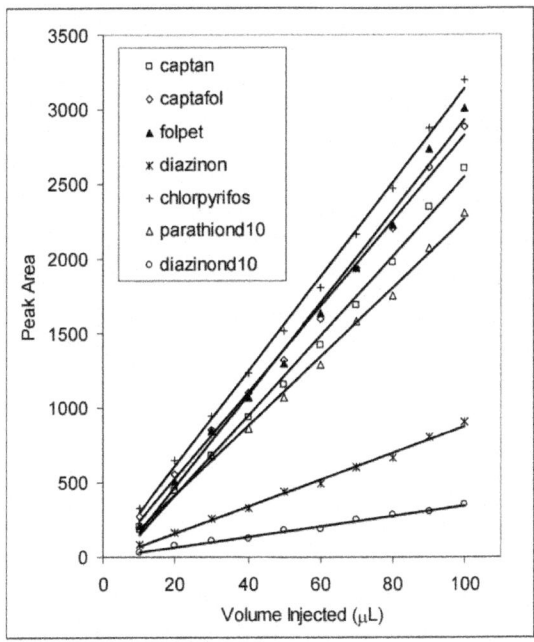

Figure 3. Change in Peak Area with Injected Sample Size for LVI-COC GC/NCI MS. Peak area captan and parathion-d$_{10}$ divided by 3; peak area chlorpyrifos divided by 8 for scaling. Taken from Bailey and Belzer, 2007.

The cold on-column injector is another option for thermally labile pesticides or large volume injections. Figure 3 shows that it can be used for injection sizes up to 100 μL which exceeds the capability of the LVI-DMI injections. The injection size is also more compatible with the needs for at-line SPE approaches. Cold on-column injection reduces the potential for breakdown of pesticides by directly injecting the sample onto typically a wider diameter 1-1.5 m retention gap (0.53 mm i.d.) which is connected to a short pre-column (~0.4 m X 0.25 mm) and then further connected with a T-connector to both the analytical column and a solvent vapour exit valve (50 μm bleed restrictor, Agilent) (Bailey and Belzer, 2007). The oven temperature at the start is set at 60-65°C (hexane as solvent) and the split vent is opened until the solvent is removed which for hexane was 60 seconds. The limitation of this system is that the retention gap and pre-column need periodic replacement due to build-up of non-volatile material from samples and thus there are higher maintenance requirements than standard PTV or LVI-DMI injections. Significant loss in sensitivity or poor chromatographic performance is observed when the retention gap requires replacement. Some of these problems may be alleviated with the availability of high temperature GC columns.

Another key recent advancement in GC/MS analysis that should be considered by users are the use of high temperature columns to extend column lifetime, reduce maintenance needs, to identify high boilers, and reduced column bleed. These columns are available in the full range of polarities from 100% polysiloxane to polyethylene glycol stationary phases and have low column bleed due to the proprietary ESC™ bonding technology. Low and midpolarity columns can be used up to temperatures of 430°C, and higher polarity columns up to 400°C as compared to maximum temperatures of 300-360°C for most standard fused silica GC columns temperatures above which the standard polyimide resin coating pyrolyzes. Zebron™ Inferno™ columns (Phenomenex) utilize a high temperature polyimide coating with the flexibility and robustness of other non-metal columns making it highly compatible for GC/MS analysis. The use of higher temperatures has several advantages even if the pesticides elute prior to these temperatures as it reduces build-up of high boiling point matrix components which can be baked-off at the end of the run.

To extend GC/MS analysis to more polar pesticides often requires preceding or on-column derivatization. One chemical class of pesticides which has been successfully analyzed with derivatization prior to GC/ MS analysis is the phenoxy acid herbicides. Derivatization agents have included pentafluorobenzyl (PFB) bromide, benzyl bromide, trimethylsilyl diazomethane, or alkylchloroformates to produce the corresponding PFB, benzyl, or methyl ester (Nilsson et al., 1998; Rimmer et al., 1996; Henriksen et al., 2001). Methylation with diazomethane or by reaction with 10% sulfuric acid in methanol has also been used (Shin, 2006). The chlorophenols which are transformation products of the phenoxy acid herbicides can also be converted to their carbonates for GC/MS analysis using alkylchloroformates (Henriksen et al., 2001). These approaches can suffer from deteriorating peak shapes over time and reduced column lifetime (Charlton et al., 2009). Carbamates are thermally labile and can breakdown in the injector port or on-column to their corresponding phenols and amines and consequently derivatization using acetylation, silylation, alkylation, or perfluorination is required. On-column derivatization with trimethylphenylammonium hydroxide and trimethylsulfonium hydroxide has been used to give thermally stable products for a variety of carbamates including carbaryl, methiocarb, chlorpropham, propham, and promecarb that can be analyzed by GC-EI/MS (Zhang and Lee, 2006). In more recent years there has been a shift to LC/MS/MS methods (see section 3.2) for both phenoxyacid herbicides and carbamates as these methods do not require the derivatization step and can provide an ability to simultaneous analyze transformation products and often a wider range of pesticides within the same chemical class (Raina and Etter, 2010; Charlton et al., 2009; Chung and Chan, 2010).

To achieve the necessary MDLs required for environmental or food analysis the majority of GC/MS pesticide analysis methods are in SIM mode with either single quadrupole or iontrap systems with ion-traps providing similar or slightly higher MDLs than the more popular quadrupole systems. In addition to the use of tandem mass spectrometry in GC/MS analysis, recent advances in pesticide analysis have included the use of GC/TOF-MS for pesticide analysis to achieve MS scan separation even at these low environmental levels enabling full confirmation ability and added selectivity. In these analysis TOF is generally operated with unit resolution and high scan rates (eg 200-500 scans/sec) to provide for automated mass spectral deconvolution of overlapping signals and library matching (de Koning et al., 2003; Zrostlikova et al., 2003b). GC/TOF-MS can also be operated with high mass resolution (0.02 -0.05 Da) with slower scan rates (2-10 scans/sec). It has had more limited applicability for pesticide analysis (Cajka et al., 2004), however with new designs that include a dynamic range enhancement (DRE) the limitations of saturation at high ion concentrations have been overcome (Leandro et al., 2007). GC/TOF-MS is most often used for fast-eluting peaks and for applications such as comprehensive two-dimensional gas chromatography (GC X GC) analysis of pesticides (Zrostlikova et al., 2003b) but has received much less attention than other GC or LC applications. In these multiresidue analysis applications unit resolution is used with fast scan rates to allow multiresidue screening by GC X GC/MS full scan (50-500 m/z) utilizing spectra library matching in EI mode (Dasgupta et al, 2010). A 5 µL DMI injection has also been used with GC/TOF-MS analysis of pesticides utilizing peak deconvolution and library searching software for isolation of the analyte peaks from matrix components (de Koning et al., 2003). With this smaller DMI injection size and for the list of pesticides under their study the temperature for solvent venting step was set to 50°C with a shorter solvent vent time of 120 sec. Utilizing DMI with GC/TOF-MS is a dual approach of reducing matrix interferences by firstly reducing the amount of matrix introduced into the GC/MS system and secondly utilizing MS spectral library matching ability of TOF-MS. Keeping the upper limit of injector temperature to that just necessary to volatilize analytes also keeps the non-volatile material in the DMI microvial and consequently reduced demands on mass spectral resolution.

LC/MS/MS Methods

LC/MS/MS continues to gain popularity in use for pesticide analysis with most applications focused on non-GC amenable compounds, thermolabile, polar and non-volatile pesticides. Some chemical classes such as phenoxyacids herbicides, triazines, OPs, chloroacetanilides, and pyrethroids can be analyzed by both GC/MS and LC/MS/MS. For phenoxacid herbicides and carbamates LC/MS/MS is regarded as more favourable as it does not require a derivatization step prior to analysis. The use of LC/MS/MS over GC/MS for the chemical classes listed in Table 3 may also be done in order to achieve reduced analysis time by utilizing a multiresidue LC/MS/MS method covering a range of target pesticides from different chemical classes. However the key reason for choosing LC/MS/MS over GC/MS is the need to deal with more polar chemical classes of pesticides and increasingly for the simultaneous analysis of their transformation products.

Transformation products are often more polar and less volatile than their parent compounds and generally have poor chromatographic performance on nonpolar GC columns or are thermolabile. Transformation products many also require derivatization to make them GC amenable for some of these chemical classes as discussed previously. Even for LC/MS/MS methods the large difference in polarity between parent pesticide and transformation product may require different separation conditions or ion source (mode) for adequate sensitivity making development of simultaneous methods challenging.

The use of LC/MS/MS for pesticide residue analysis has focused on systems with atmospheric pressure ionization (API) either atmospheric pressure chemical ionization (APCI) or electrospray ionization (ESI) either in positive or negative mode. Many LC/MS/MS methods are multiresidue analysis methods and have been done for a target list of pesticides requiring analysis for regulatory purposes. Both APCI or ESI have been used for multiresidue methods with ESI+ the most popular as shown in Table 3. Direct comparisons of the sensitivity of APCI and ESI are often not available or not under the same chromatographic conditions. In addition, often regulatory requirements can be met with both approaches with similar MDLs for many pesticides observed under optimal conditions (Titato et al., 2007; Thurman et al., 2001). The design and operational parameters of individual API ion sources can also lead to varying results between the sensitivity of ESI versus APCI and consequently should be evaluated for the system under use and expected flow rate conditions. Table 3 shows that flow rate conditions for the separation are an important consideration as ESI is generally most sensitive at lower flow rates typically near 0.2 mL/min and consequently it may be desire to utilize smaller particle size (2-3μm) LC/MS columns however the reduction in sample loading capacity should also be consider (Asperger et al., 2001; Titato et al.,2007). If using higher flow rate conditions for the separation on columns (5μm, 150 to 250 mm X 4.6 mm) then the flow is generally split prior to MS (Banerjee et al., 2009; Crescenzi et al., 1995; Di Corcia et al., 2000). APCI is most often operated under high flow rates 1-2 mL/min (Table 3) and optimal flow varies with chemical class (Asperger et al., 2001; Titato et al., 2007). OPs are distinctly different and require lower flow rates for optimal sensitivity with APCI (Asperger et al., 2001; Jansson et al., 2004; Titato et al, 2004). Even if sensitivity is better with the more popular ESI methods there may be preference to use APCI for some chemical classes (OPs, chloracetanildes, pyrethroids, phenoxyacid herbicides, carbamates) to take advantage of other factors which include the following: (1) APCI is generally less prone to sodium adduct formation that ESI; (2) APCI can be less prone to matrix impacts as compared to ESI (Souverian et al., 2004); and (3) in some cases the SRM transition can differ from ESI so that co-eluting peaks can be isolated with MS/MS thereby reducing chromatographic resolution needs (see Table 3).

As a general rule the choice of positive or negative mode depends upon polarity and acidity of analytes and sample matrix impacts. In general, ESI- is more sensitive for phenoxyacid herbicides and their transformation products (Raina and Etter, 2010; Koppen et al., 1998; Dijkman et al., 2001) and chloroacetanilide transformation products; ESI+ for sulfonylureas, phenylureas, N-methylcarbamates, organophosphorus pesticides (Cessna et al., 2006; Degenhardt et al., 2010; Hernandez et al., 2006; Steen et al., 1999; Raina and Sun, 2008); APCI+ for triazines (Dagnac et al., 2005; Jeannot et al., 2000); and APCI+ or ESI+ for chloroacetanilides (Dagnac et al., 2005; Ferrer et al., 2007; Banerjee et al., 2009). It should be noted that phenoxyacid herbicides have been analyzed with APCI- (Puig et al., 1997); sulfonylureas, phenylureas, carbamates, OPs with APCI+ (see Table 3); triazines with ESI+

(Dagnac et al., 2005; Jeannot et al., 2005); and for methods where acidic pesticides are analyzed separately from neutrals the sulfonylureas along with phenoxyacid herbicides are analyzed together with ESI- (Dijkman et al., 2001; Koppen et al., 1998; Di Corcia et al., 2000). Similarly triazines and atrazine metabolites may be done with ESI+ rather than APCI+ due to the diversity and sensitivity of ESI+ for other chemical classes that are analyzed simultaneously with only a small loss in sensitivity.

Table 3. LC/MS/MS Separation Conditions by Chemical Class

Column, flow rate, ion source	Mobile phase organic modifier (MeCN - acetonitrile; MeOH-methanol) and additive	Reference
Chloroacetanilides and transformation products* - LC/ESI+MS/MS unless specified		
Zorbax Eclipse SB-C18 (1.8 µm, 150 X 4.6 mm) 0.6 mL/ min	0.1 % formic acid, MeCN 10-100%	Ferrer et al., 2007
Omnisper C-18, (3 µm 150 X3 mm) 0.4 mL/min		
Purosphere STAR RP-18e (5 µm, 150 X 4.6 mm) 1.0 mL/ min split to MS	MeCN 85% to 40%	Dagnac et al., 2005
Gromsil C18 (3 µm, 150 X 2.0 mm) 0.15 mL/min, ESI-	5 mM ammonium formate, 34%-90% MeOH	Banerjee et al., 2009
	0.6% formic acid, MeOH 40-95%	Gomides Freitas, 2004*
Dithiocarbamates (anionic¹ and neutral²) and transformation products² -LC/MS		
ZIC-pHILIC column (5 µm 150 X4.6 mm) 0.7 mL/min with 50% split to LC/MS, ESI-	10 mM ammonia, 10 to 40% MeCN	Crnogorac et al ., 2007¹
C8 (5 µm 150 X4.6 mm) 0.8 mL/min, ESI+ and APCI+	10-90% MeOH	Blasco and Pico, 2004²
OPs -LC/MS/MS ESI+ unless specified		
C6phenyl Gemini (3 µm, 150 X 2.0 mm) 0.2 mL/ min	0.1% formic acid, 2 mM ammonium acetate, MeOH 40-95%	Raina and Sun, 2008
C12 (4 µm, 150 X 2.0 mm) 0.25 mL/min	5 mM ammonium formate, MeOH 5%-90%	Chung and Chan, 2010
Chromolith SpeedROD RP-18e (50 X 4.6 mm) varied 0.2-1.2 mL/min ESI+; APCI+ 0.2-2.8 mL/min	MeOH 97%	Asperger et al., 2001
Zorbax Eclipse SB-C18 (1.8 µm, 150 X 4.6 mm) 0.6 mL/ min	0.1 % formic acid, MeCN 10-100%	Ferrer et al., 2007
Atlantis C18 (5 µm, 100 X 2.1 mm) 0.2 mL/ min		
Xterra MS C18 (3.5 µm, 150 X 2.1 mm) 0.2 mL/ min	0.01% formic acid gradient MeOH 5%-90%	Hernandez et al., 2006
Genesis C18 (4 µm, 100 X 3 mm) 0.3 mL/min	0.1% formic acid, MeCN 10-90%	Botitsi et al., 2007
Genesis C18 (4 µm, 100 X 3.0 mm) 0.2 mL/ min, APCI+ and ESI+	10 mM ammonium formate pH 4, MeOH	Pihlstrom et al., 2007
Supelcosil LC18 (5µm, 150 mm X 2.1 mm) 0.6 mL/min for APCI+; ODS (5µm, 150 mm X 2.1 mm) 0.1 mL/min	10 mM ammonium formate, pH 4, MeOH 0-90%, flush with MeCN 80% each run	Jansson et al. , 2004
Aqua C18 (3 µm, 150 mm X 2 mm), 0.3 mL/min	0.05% TFA, MeCN 70% isocratic	Titato et al ., 2007
Synergie RP (4 µm, 50 mm X 2.0 mm), 0.6 mL/min	5 mM ammonium formate, MeOH 0-90%	Paya et al. , 2007
Alltima C18 (5 µm, 150 X 3.2 mm) 0.3 mL/ min	5 mM ammonium acetate MeOH 20-80%	Muller et al ., 2007
XTerra MS C18 (3.5 µm, 150 X 2.1 mm) 0.2 mL/ min	5 mM ammonium formate, 25-95% MeOH 0.1% formic acid, MeCN 10%-90%	Hiemstra et al. 2007
Purosphere STAR RP-18e (5 µm, 150 X 4.6 mm) 1.0 mL/ min split to MS	5 mM ammonium formate 34%-90% MeOH	Economou et al., 2009 Banerjee et al 2009
OP transformation products (oxons¹, sulfoxides and sulfones²) LC-ESI+/MS/MS		
C6phenyl Gemini (3 µm, 150 X 2.0 mm) 0.2 mL/ min	0.1% formic acid, 2 mM ammonium acetate, MeOH 40-95%	Raina and Sun, 2008¹

Column, flow rate, ion source	Mobile phase organic modifier (MeCN – acetonitrile; MeOH–methanol) and additive	Reference
C12 (4 µm, 150 X 2.0 mm) 0.25 mL/min	5 mM ammonium formate, MeOH 5%-90%	Chung and Chan, 2010[2]
Genesis C18 (4 µm, 100 X 3.0 mm) 0.2 mL/ min	10 mM ammonium formate, pH 4, MeOH 0-90%, flush with MeCN 80% each run	Jansson et al., 2004[2]
Alltima C18 (5 µm, 150 X 3.2 mm) 0.3 mL/ min	5 mM ammonium formate, MeOH 25-95%	Hiemstra et al., 2007[2]
XTerra MS C18 (3.5 µm, 150 X 2.1 mm) 0.2 mL/ min	0.1% formic acid, MeCN 10%-90%	Economou et al., 2009[2]
Pyrethroids LC-ESI+/MS/MS		
Genesis C18 (4 µm, 100 X 3 mm) 0.3 mL/min	MeOH, 10 mM ammonium formate pH 4, gradient not specified	Pihlstrom et al., 2007
Waters Symmetry (5 µm, 250 X 4.6 mm) 1.0 mL/min	50 mM ammonium formate, formic acid pH 3.5, 70-100% acetontrile	Martinez et al., 2006
Waters Symmetry(5 µm, 250 X 4.6 mm) 1.0 mL/min	50 mM ammonium formate, formic acid pH 3.5, 70-100% acetontrile	Gil-Garcia et al., 2006
Zorbax C18 (5 µm, 250 X 4.6 mm) 0.6 mL/min	MeOH 77%-100%	Chen et al., 2007
Pyrethroid Transformation Products –only GC/MS/MS – Arrebola et al., 1999		
Phenoxyacid herbicides LC/MS/MS ESI- unless specified		
Hypersil-BDS C18, (5µm, 250 X 2.0 mm) 0.2 mL/min	A Water:MeOH:acetic acid; B MeOH:water 90:810:1 for A; 900:1 B, 0-50% B	Koppen et al., 1998
Zorbax Eclipse XDB-C18 (1.8 µm, 50 X 4.6 mm) 0.15 mL/ min	2 mM ammonium acetate, MeOH 65-90%	Raina and Etter, 2010
C-18 (5 µm, 50 to 100 mm X 2.1 mm in general) 0.2 mL/min	0.1% formic acid, Gradient ranges on column, MeCN 0-65%or higher starting	Dijkman et al., 2001
LiChrocart C-18, (5µm, 250 X 4.6 mm) 0.9 mL/min, APCI-	ammonium formate, 5 mM formic acid (pH 3), 40%	Santos et al., 2000
Alltima C18, (5 µm 250X4.6 mm) 0.8 mL/min split 3:1 to MS	MeCN	
Alltima C18, (5 µm, 250 X 4.6 mm) 1.0 mL/min split 30/970 to MS/UV	50-95% MeOH, 1% v/v acetic acid	Baglio et al., 1999
Hypersil C18 (5µm, 250 X 4.6 mm) 1 mL/min 50% to MS	0.1 mM K₂HPO₄ 0.2 mM TBAF, MeOH 30-75%	Crescenzi et al. ,1995
	Formic acid and ammonia –pH varied 2.9-8.4, 20-100% MeOH	Di Corcia et al., 2000
Phenoxyacid herbicide transformation products (chlorophenols) and nitro substituted phenols, phenols LC/MS/MS		
Hypersil green ENV (C-18) (150 X 4.6 mm) 1mL/min, APCI-	1% acetic acid, 25-100% gradient with 1:1 MeOH/MeCN	Puig et al. , 1997
Zorbax Eclipse XDB-C18 (1.8 µm, 50 X 4.6 mm) 0.15 mL/ min, ESI-	MeOH 2 mM ammonium acetate	Raina and Etter, 2010

Column, flow rate, ion source	Mobile phase organic modifier (MeCN – acetonitrile; MeOH–methanol) and additive	Reference
	65-90% + postcolumn addition of ammonia in MeOH	
Sulfonyl ureas LC/MS/MS ESI+ unless specified		
C6-phenyl (3 µm, 150 mm X 2.0 mm), 0.2 mL/min	0.1% formic acid, 2 mM ammonium acetate, 35% MeCN	Degenhardt et al., 2010
Zorbax Eclipse Plus C-18 (3.5 µm 150 mm X 2.1 mm)	MeCN 0.1 % formic acid, MeCN 45-60%	Fang et al., 2010
Zorbax Eclipse SB-C18 (1.8 µm, 150 X 4.6 mm) 0.6 mL/ min	0.1 % formic acid, MeCN 10-100%	Ferrer et al ., 2007
Varied 5 µm C-18 from 50 to 100 mm X 2.1 mm in general, ESI+ and ESI-	0.1% formic acid, Gradient ranges on column, 0-65%MeCN or higher starting %	Dijkman et al., 2001,
Hypersil-BDS C18, (5µm, 250X2.0 mm i.d.) 0.2 mL/min, ESI+ and ESI-	MeOH:acetic acid:water ;MeOH /water	Koppen et al., 1998
Hypersil C18 (5µm, 250 X 4.6 mm) 1 mL/min 50% to MS, ESI+ and ESI-	90:810:1 for A; 900:1 B, 0-50% B gradient	
	formic acid and ammonia –pH varied 2.9-8.4, MeOH 20-100%	Di Corcia et al ., 2000
Sulfonyl urea transformation products LC-ESI+ or APCI+/MS/MS		
Hypersil BDS C18 (5µm, 250X2.0 mm i.d) 0.2 mL/min, ESI+ and APCI+	confirmation only with LC/MS/MS MeOH 10-100%	Bossi et al., 1999
Triazines LC/ESI+ MS/MS unless specified		
Omnisper C-18, (3 µm 150 X3 mm) 0.4 mL/min, APCI+	MeCN (85-40%)	Dagnac et al ., 2005
Vydac C18 (5µm, 250 mm X 4.6 mm) 1.0 mL/min	10 mM ammonium acetate (pH 4.5) MeOH 45-90%, or MeCN 27-78%	Steen et al.,1999
Zorbax Eclipse SB-C18 (1.8 µm, 150 X 4.6 mm) 0.6 mL/ min	0.1 % formic acid, MeCN 10-100%	Ferrer et al .,2007
Chromolith SpeedROD RP-18e (50 X 4.6 mm) varied 0.2-1.2 mL/min ESI+, 0.2-2.8 mL APCI+	MeOH 97%	Asperger et al., 2001
Genesis C18 (4 µm, 100 X 3 mm) 0.3 mL/min		
Supelcosil LC18 (5µm, 150 mm X 2.1 id) 0.6 mL/min for APCI+;	MeOH, 10 mM ammonium formate pH 4, gradient not specified,	Pihlstrom et al., 2007
ODS (5µm, 150 mm X 2.1 id) ESI+ 0.1 mL/min	0.05% TFA, MeCN 70%	Titato et al .,2007
Alltima C18, (5 µm, 250X4.6 mm 1.0 mL/min), APCI+		
Hypersil ODS (5 µm, 250X4.6 mm 1.0 mL/min, APCI+	50-95% MeOH	Baglio et al., 1999
Synergie RP (4 µm, 50 mm X 2.00 mm), 0.6 mL/min	MeCN 15-60%	Jeannot et al., 2000
Uptispher ODB, (3 µm, 50 mm X 2 mm), 0.2 mL/min	5 mM ammonium acetate MeOH 20-80%	Muller et al., 2007
Purosphere STAR RP-18e (5 µm, 150 X 4.6 mm) 1.0 mL/ min	0.5% acetic acid, MeCN 10%-100%	Bichon et al ., 2006

Column, flow rate, ion source	Mobile phase organic modifier (MeCN – acetonitrile; MeOH–methanol) and additive	Reference
split to MS	5 mM ammonium formate 34%-90% MeOH	Banerjee et al., 2009
Triazine Transformation Products LC/ESI+ MS/MS unless specified		
Omnisper C-18 (3 µm, 150 X3 mm) 0.4 mL/min, APCI+	MeCN 85% to 40%	Dagnac et al., 2005
Hypersil ODS (5 µm, 250X4.6 mm 1.0 mL/min, APCI+	MeCN 15-60%	Jeannot et al., 2000
Vydac C18 (5µm, 250 mm X 4.6 mm) 1.0 mL/min	10 mM ammonium acetate (pH 4.5) MeOH 45-90%, or MeCN 27-78%	Steen et al., 1999
Uptispher ODB, 3 µm, 50 mm X 2 mm, id, 0.2 mL/min	0.5% acetic acid, MeCN 10%-100%	Bichon et al., 2006
Carbamates and transformation products* LC/ESI+MS/MS unless specified		
Xterra MS C18 (5µm, 100 mm 2.1 mm) 0.2 mL/min	0-75% MeOH with 0.01 % formic acid	Goto et al., 2006*
Zorbax Eclipse SB-C18 (1.8 µm, 150 X 4.6 mm) 0.6 mL/ min	0.1 % formic acid, MeCN 10-100%	Ferrer et al., 2007*
Atlantis C18 (5 µm, 100 X 2.1 mm) 0.2 mL/ min		
Xterra MS C18 (3.5 µm, 150 X 2.1 mm) 0.2 mL/ min	0.01%formic acid MeOH 5%-90%	Hernandez et al, 2006
C12 (4 µm, 150 X 2.0 mm) 0.25 mL/min	0.1% formic acid, MeCN 10-90%	Botitsi et al., 2007*
Supelcosil LC18 (5µm, 150 mm X 2.1 id) 0.6 mL/min for APCI+;	5 mM ammonium formate, MeOH 5%-90%	Chung and Chan, 2010
ODS (5µm, 150 mm X 2.1 mm) 0.1 mL/min ESI+	0.05% TFA, MeCN 70%	Titato et al ., 2007
Genesis C18 (4 µm, 100 X 3 mm) 0.3 mL/min		
Genesis C18 (4 µm, 100 X 3.0 mm) 0.2 mL/ min, APCI+	10 mM ammonium formate pH 4, MeOH gradient not specified	Pihlstrom et al., 2007
Aqua C18 (3µm, 150 mm X 2. mm), 0.3 mL/min	10 mM ammonium formate, pH 4, MeOH 0-90%,	Jansson et al., 2004*
Alltima C18 (5 µm 250mm X 4.6 mm) 1.0 mL/min	flush with MeCN 80% each run	
Synergie RP (4 µm, 50 mm X 2.00 mm), 0.6 mL/min	5 mM ammonium formate, MeOH 0-90%	Paya et al., 2007
Polaris C18 3 µm, 150 X 2.0 mm id, 0.2 mL/min	50-95% MeOH	Baglio et al., 1999
	5 mM ammonium acetate MeOH 20-80%	Muller et al., 2007
Luna C18 (5 µm, 150 X 4.6 mm), 0.4 mL/min	2 mM ammonium formate, pH 2.8 MeOH 20-85%	Martinez Vidal et al., 2005
Alltima C18 (5 µm, 150 X 3.2 mm) 0.3 mL/ min	10 mM Ammonium formate MeOH 35-90%	
XTerra MS C18 (3.5 µm, 150 X 2.1 mm) 0.2 mL/ min	5 mM ammonium formate, MeOH 25-95% 0.1%	Pico and Kozmutza, 2007*
Purosphere STAR RP-18e (5 µm, 150 X 4.6 mm) 1.0 mL/ min	formic acid, MeCN 10%-90%	Hiemstra et al., 2007
split to MS, ESI+	5 mM ammonium formate 34%-90% MeOH	Economou et al., 2009
		Banerjee et al., 2009
Phenylureas LC/ESI+/MS/MS unless specified		
Microsphere 3 µm LC-LC (50 X4.6 mm – 100 X 4.6) 1.0 mL/min split to MS to 0.5 mL/min, APCI+	MeOH 10-60%	Van der Heeft et al., 2000

Column, flow rate, ion source	Mobile phase organic modifier (MeCN – acetonitrile; MeOH-methanol) and additive	Reference
Omnisper C-18, (3 µm 150 X3 mm) 0.4 mL/min	MeCN (85-40%)	Dagnac et al., 2005
Vydac C18 (5µm, 250 mm X 4.6 mm) 1.0 mL/min	10 mM ammonium acetate (pH 4.5)	Steen et al., 1999
Zorbax Eclipse SB-C18 (1.8 µm, 150 X 4.6 mm) 0.6 mL/ min	MeOH 45-90% or MeCN 27-78%	
Atlantis C18 (5 µm, 100 X 2.1 mm) 0.2 mL/ min	0.1 % formic acid, MeCN 10-100%	Ferrer et al., 2007
Supelcosil LC18 (5µm, 150 mm X 2.1 mm) 0.6 mL/min for APCI+;		
ODS (5µm, 150 mm X 2.1 mm) ESI+ 0.1 mL/min	0.01%formic acid gradient MeOH 5%-90%	Hernandez et al., 2006
Genesis C18 4 µm, 100 X 3 mm) 0.3 mL/min, ESI+	0.05% TFA MeCN 70%	Titato et al ., 2007
Aqua C18 (3µm, 150 mm X 2. mm), 0.3 mL/min, ESI+	MeOH, 10 mM ammonium formate pH 4, gradient	Pihlstrom et al., 2007
Alltima C18 (5 µm 250 mm X 4.6 mm) 1.0 mL/min, APCI+	not specified	
Hypersil ODS (5 µm 250 mm X 4.6 mm) 1.0 mL/min, APCI+	5 mM ammonium formate, MeOH 0-90%	Paya et al., 2007
Polaris C18 (3 µm, 150 mm X 2.0 mm), 0.2 mL/min, ESI+	50-95% MeOH	Baglio et al., 1999
	MeCN 15-60%	Jeannot et al., 2000
Uptispher ODB (3 µm, 50 mm X 2 mm), 0.2 mL/min, ESI+	2 mM ammonium formate, pH 2.8 MeOH 20-85%	Vidal et al., 2005
Genesis C18 (4 µm, 100 X 3.0 mm) 0.2 mL/ min, ESI+	0.5% acetic acid, MeCN 10%-100%	
		Bichon et al., 2006
Alltima C18 (5 µm, 150 X 3.2 mm) 0.3 mL/ min, ESI+	10 mM ammonium formate, pH 4, MeOH 0-90%,	
Discovery C18 ((5.0 µm, 150 X 3 mm) 0.5 mL/ min, ESI+ and ESI-	flush with MeCN 80% each run	Jansson et al., 2004
	5 mM ammonium formate, 25-95% MeOH	
	MeOH 20-100%	Hiemstra et al., 2007
		Zrostlikova et al. 2003
Phenylurea Transformation Products		
Omnisper C-18, (3 µm 150 X3 mm) 0.4 mL/min, ESI+	MeCN (reverse 85% decreased to 40%, water 15% to 100%)	Dagnac et al ., 2005
Vydac C18 (5µm, 250 mm X 4.6 id) 1.0 mL/min, ESI+ and ESI-	10 mM ammonium acetate (pH 4.5)	Steen et al ., 1999
Chromolith SpeedROD RP-18e (50 X 4.6 mm) varied 0.2-1.2 mL/min ESI+, 0.2-2.8 mL APCI+	MeOH 45-90% or MeCN 27-78%	
	97% MeOH, (note lower flows when % MeOH	Asperger et al., 2001
Uptispher ODB, 3 µm, 50 mm X 2 mm, id, 0.2 mL/min, ESI+	decreased	
Genesis C18 (4 µm, 100 X 3.0 mm) 0.2 mL/ min, ESI+	0.5% acetic acid, MeCN 10%-100%	Bichon et al., 2006
	10 mM ammonium formate, pH 4, MeOH 0-90%, flush with MeCN 80% each run	Jansson et al., 2004

Column, flow rate, ion source	Mobile phase organic modifier (MeCN – acetonitrile; MeOH–methanol) and additive	Reference
Azoles (Triazoles and benzimidazoles) and triazole transformation products* LC/MS/MS ESI+ unless specified		
Synergi Hydro HP (4 μm, 150 X 2 mm) 0.25 mL/ min	MeCN 0.1 % formic acid 50%-100%	Trosken et al., 2005
Symmetry C18 (5.0 μm, 250 X 4.6 mm) 0.3 mL/ min	0.2% formic acid, MeOH 50% to 82%	Schermerhorn et al., 2005*
Discovery C18 (5.0 μm, 150 X 3 mm) 0.5 mL/ min	MeOH 20-100%	Zrostlikova et al., 2003
Zorbax Eclipse SB-C18 (1.8 μm, 150 X 4.6 mm) 0.6 mL/ min	0.1 % formic acid, MeOH 10-100%	Ferrer et al., 2007
Supelcosil LC18 (5μm, 150 mm X 2.1 mm) 0.6 mL/min for APCI+;		
ODS (5μm, 150 mm X 2.1 mm) 0.1 mL/min, ESI+	0.05% TFA, MeCN 70%	Titato et al., 2007
Hypersil ODS (5 μm, 250mm X 4.6 mm) 1.0 mL/min		
Genesis C18 (4 μm, 100 X 3 mm) 0.3 mL/min	MeCN 15-60%	Jeannot et al ., 2000
Aqua C18 (3μm, 150 mm X 2.0 mm), 0.3 mL/min	10 mM ammonium formate pH 4, MeOH	Pihlstrom et al., 2007,
Synergie RP (4 μm, 50 mm X 2.00 mm), 0.6 mL/min	5 mM ammonium formate, MeOH 0-90%	Paya et al. , 2007
Alltime C18 (5 μm, 150 X 3.2 mm) 0.3 mL/ min	5 mM ammonium acetate MeOH 20-80%	Muller et al., 2007
Atlantis C18 (5 μm, 100 X 2.1 mm) 0.2 mL/ min	5 mM ammonium formate, 25-95% MeOH	Hiemstra et al., 2007
XTerra MS C18 (3.5 μm, 150 X 2.1 mm) 0.2 mL/ min	0.01%formic acid, MeOH 5%-90%	Hernandez et al., 2006
Purosphere STAR RP-18e (5 μm, 150 X 4.6 mm) 1.0 mL/ min	0.1% formic acid, MeCN 10%-90%	Economou et al., 2009
split to MS	5 mM ammonium formate 34%-90% MeOH	Banerjee et al, 2009

Common co-elution problems that must be resolved prior to detection exist for a number pesticides either within the same chemical class or for multi-class residue methods. Table 3 shows the vast majority of LC/MS/MS methods utilize C18 columns in order to achieve the desired selectivity for the separation. A few separations have taken advantage of different selectivity from ZIC-pHILIC, C6-phenyl, C8, or C12 (Raina and Sun, 2008; Degenhardt et al., 2010; Crnogorac et al., 2007; Blasco and Pico, 2004). In general methods provide the necessary resolution for compounds with the same SRM transitions. For chloracetanilides care must be taken with ESI+ as acetochlor, metalochlor, and alachlor can co-elute and acetochlor and alachlor which both have molecular mass of 269.5 g/mol have the same precursor ions with ESI+ (m/z 270 or 292) from [M+H]+ and [M+Na]+ (Dagnac et al., 2005; Ferrer et al., 2007). This can be resolved by adequate chromatographic resolution prior to detection or switching to APCI + where in addition to monitoring [M+H]+ for both the 224 and 256 precursor ion can be monitored for acetochlor, while for alachlor 162 and 238 can be monitored (Dagnac et al., 2005). Phenoxyacid herbicides and sulfonylureas also require adequate chromatographic resolution. Niocsulfuron, ethametsulfuron-methyl, bensulfuronmethyl have a similar parent ion (411.2, 411.8, 411.5) but can be separated using C6-phenyl column (Degenhardt et al., 2010). With ESI- the phenoxy acid herbicides MCPA, mecoprop, MCPB have a common confirmation SRM transition of 140.9 > 105.2 which is also the quantitative SRM for degradation product, chloromethylphenol. In addition, 2,4-D, dichlorprop, and 2,4-DB also have the sample confirmation SRM of 160.9 > 124.7 which is also the quantitative SRM for dichlorophenol (Raina and Etter, 2010). Separated of these phenyoxyacid herbicides can be achieved using a short C18 column with methanol gradient and mobile phase containing 2 mM ammonium acetate (Raina and Etter, 2010). The list of pesticides that are required for screening or quantitative analysis and those potentially present in samples should determine your requirements for detection and separation.

Degradation products by LC/MS/MS such as more chlorophenols (degradation products of phenoxyacid herbicides) and nitrophenols are move sensitive with ESI- , while other phenols are more sensitive with APCI (Reesmtsma et al., 2003; Raina and Etter, 2010). However these chloro and nitrophenols have also been analyzed successfully by APCI (Silgoner et al., 1997). OP degradation products including 3,5,6-trichloro-2-pyridinol, diethyl phosphate, 2- isopropyl-6-methyl-4-pyrimidinol, malathion monocarboxylic acid, and OPoxons (Raina and Sun, 2008) as well as OP sulfones and sulfoxides (Jansson et al, 2004; Chung and Chang, 2010, Hiemstra et al., 2007; Economou et al, 2009) are more sensitive with LC-ESI+/MS/MS and some OPs observe a drastic loss in sensitivity with an APCI source at high flow rates (Asperger et al., 2001). Degradation products of triazines and phenylureas have been done by LC-APCI+/MS/MS but for triazines lower MDLs can be achieved with GC-EI/MS/MS (Dagnac et al., 2005; Goncalves et al., 2006). Diuron and its degradates 3,4- dichlorophenyurea and 2,4-dichlorophenylurea also observed better sensitivity in methanol compared to acetonitrile mobile phases and switching the organic modifier has greater impact for these degradation products when using ESI+ than ESI- (Steen et al., 1999). Chloroacetanilide metabolites have been analyzed by LC-ESI-/MS/MS (Gomides Freitas et al., 2004). LC-ESI+/MS/MS has also been used for analysis of a number of carbamate and azole degradation products the most common of which for carbamates are aldicarb sulfone and sulfoxide, and 3-hydroxycarbofuran (Goto et al., 2006; Ferrer et al., 2007; Botitsi et al., 2007; Jansson et al., 2004; Pico and Kuzmutza, 2007; Schermerhorn et al., 2005). A more extensive list of dithiocarbamates and their transformation products have only been analyzed using LC/MS methods (Crnogorac et al., 2007; Blasco and Pico, 2004) with triallate routinely analyzed by GC/MS methods. Transformation products of pyrethroids currently have no LC/MS or LC/MS/MS.

Selection of the organic modifier (generally methanol or acetonitrile), and the presence or absence of formic or acetic acid, and salts can greatly impact the ionization of a pesticide and its sensitivity. Some pesticides have the potential to form sodium or ammonium adducts in positive ion mode, or acetate or formate adducts in negative ion mode with an API source. The formation of adducts decreases the abundance of the protonated or deprotonated molecular ion and there is greater potential for adduct formation with ESI than APCI. In general positive ion mode is more prone to adduct formation than negative ion mode. Methanol mobile phases have a higher degree of adduct formation particularly for sodium adducts relative to acetonitrile although many pesticides see better ionization in methonal than acetonitrile. The formation of sodium adducts in mobile phases with methanol can be reduced or suppressed by the addition of ammonium or hydrogen ions. The most common additives for this purpose are ammonium acetate (2-10 mM), ammonia, acetic acid (1%), formic acid (0.05-.2 v/v%), or trifluoroacetic acid (TFA, 0.05 v/v%) (see Table 3). The impact of adjustment of the pH of the mobile phase on chromatographic resolution for closely eluting pesticides with the same SRM transitions or parent ions should be considered along with the impact of changing pH on sensitivity particularly for acidic pesticides (Raina and Etter, 2010). In practice a balance must be met between separation needs and MS sensitivity for the range of pesticides and transformation products under study and can vary significantly even for those of the same chemical class. Table 3 shows that there is a large diversity in additives and organic solvent used even within the same chemical class.

For OPs ESI+ is superior and generally [M+H]+ is observed as the parent ion even in mobile phases only containing methanol. The presence of both ammonium and formic acid was shown to give optimal sensitivity or OPs, OP oxons, and other OP transformation products (Raina and Sun, 2008). For OP sulfone and sulfoxide transformation products sodium adducts can form in mobile phases containing only methanol or methanol with formic acid. Switching either to acetonitrile with formic acid or addition of a salt such as ammonium formate (or ammonium acetate) suppresses the formation of adducts (Hiemstra et al., 2007; Economou et al., 2009; Jansson et al., 2004; Raina and Sun, 2008; Muller et al., 2007).

Individual phenylureas and carbamates are also more likely to form sodium adducts as compared to triazines. Aldicarb, 3-hydroxycarbofuran, aldicarb sulfone, and aldicarb sulfoxide form sodium adducts in a methanol mobile phase with 0.01% formic acid (Hernandez et al., 2006). Switching to acetonitrile reduces adduct formation for the sulfone and sulfoxide of aldicarb but aldicarb is still present as sodium adduct (Botitsi et al., 2007). Addition of ammonium formate (Pihlström et al., 2007; Pico and Kozmutza, 2007) or ammonium acetate leads to suppression or reduction in the formation of the sodium adduct. Depending upon the ammonium ion concentration aldicarb may form the ammonium adduct (Pico and Kozmutza, 2007) as well as oxamyl (Pihlström et al., 2007). In a mobile phase of methanol with 5 mM ammonium formate, methiocarb sulfone and ethiocarb sulfone both form the ammonium adduct as well as the protonated molecular ion with SRM transitions of 275→122 and 258→122 for methiocarb sulfone and 275 →107 and 258→107 for ethiocarb sulfone (Hiemstra and de Kok, 2007). Aldicarb does not form [M+H]+ under most mobile phase conditions so either [M+Na]+ (213→116 or 213→89) or [M+NH4]+ (208 →116 or 208→89) are used for SRM transition under the proper mobile phase conditions (Table 3). Other carbamates generally form [M+H]+ regardless of mobile phase composition.

With ESI+ the most commonly monitored phenylureas produce [M+H]+ under varying mobile phase conditions and the mass selected may be utilizing 35 or 37 Cl isotopes. A few phenylureas such as isoproturon and chlortoluron can form sodium adducts and consequently ammonium formate or ammonium acetate may be added to the mobile phase (Table 3). Often these additives are required more for other chemical classes that are analyzed along with the phenylureas such as carbamates. The more commonly analyzed phenylureas such as diuron and linuron do not require addition of additives. Phenylureas sensitivity is impacted by the organic solvent selected with methanol having significant improved response in ESI+ for phenylureas and their degradation products (Steen et al., 1999). In addition, the use of a higher percentage of methanol to achieve the desired separation conditions also improves sensitivity for both for ESI+ and ESI- as sensitivity improves with the percentage of organic modifier. The reduction in signal intensity for the degradation products when switching organic modifier to acetonitrile was not as great in ESI- (Steen et al., 1999). For chloracetanilides and phenylureas using APCI also reduces potential for sodium adduct formation (Dagnac et al., 2005).

Sulfonylureas also observe predominately [M+H]+ but can form sodium adducts as has been observed for sulfometuron-methyl with ESI+ with acetonitrile-aqueous 0.1%formic acid mobile phase (Dijkman et al., 2001). Consequently separation conditions generally contain both 0.1% formic acid and 2 mM ammonium acetate (Degenhardt et al., 2010) and if an organic modifier is used it is generally acetonitrile as shown in Table 3. In negative ion mode adduct formation is not observed. For phenoxy acid herbicides the presence of formic acid will result in decreased abundance of [M-H]- (Raina and Etter, 2010) while for many neutral pesticide chemical classes it will improve the ionization so typically is added at ~0.1- 0.2 v/v%. For phenoxyacid herbicides methanol is generally choosen as the organic modifier as it improves the efficiency of ionization and the sensitivity improvement relative to acetonitrile mobile phases ranges from 3-5 orders of magnitude (Raina and Etter, 2010). OPs and their degradation product signal intensity is also much better with methanol (Raina and Sun, 2008) and it has been shown as the % of methanol increases the signal intensity improves as was also observed for phenylureas (Steen et al., 1999). This suggests an advantage in using gradient elutions with methanol rather than acetonitrile for these pesticides as a higher percentage of organic solvent will be required to achieve the same chromatographic resolution during the separation.

Users must also be aware of particularly for gradient elution whether the pesticides of interest are soluble over the range of mobile phase conditions for the separation. For some chemical classes or multi-residue analysis gradient elution programs will start at a very low percentage of methanol or acetonitrile and peak broadening or distortion and even carryover and increasing MS background signal may be observed. Reduced sensitivity and reproducibility over time may become apparent due to low solubility of some of the analytes in mobile phases of high aqueous content. For these challenging chemical classes which are more prone to build-up a flushing step with high concentration of acetonitrile is used prior to re-equilibration of the column to reduce carry-over issues.

A number of LC/ESI+MS methods for pyrethroids have been developed (Chen and Chen, 2007; Gil-Garcia et al., 2006; Martinez et al., 2006) which have comparable MDLs to GC/EI MS methods (Yoshida, 2009). For halogenated pyrethroids GC/NCI-MS provides the best sensitivity (Feo et al., 2010). These methods have largely focused on LC/MS where either the protonated molecular ion or ammonia adduct are predominately observed in mobile phases containing ammonium acetate or formate (Table 3). There is little structural information available and only a few multi-residue LC/MS/MS methods contain selected pyrethroids (Pihlstrom et al., 2007). In addition, the only available methods for analysis of pyrethroid metabolites currently require derivatization with GC/MS analysis.

There are a number of approaches that have been used to further improve sensitivity of LC/MS/MS methods. When separation needs do not permit changes in mobile phase composition to improve MS sensitivity then alternatively post-column reagents may be added using an additional pump (Raina and Etter, 2010; Carabias-Martinez et al., 2004) at lower flow rates (eg 50 µL/min) such that the total flow is still optimal for the ESI or APCI used for the analysis. Bases have been used as post-column reagents to enhance ionization including ammonia, trimethylamine, tris(hydroxymethyl) aminomethane, and 1,8- diazabicyclo-(5,4,0) undec-7-en (Raina and Etter, 2010; Carabais-Martinez et al., 2004; Marchese et al., 2002; Gomides Freitas et al., 2004). This approach has been used to improve the sensitivity of transformation products of phenoxyacid herbicides with ammonia in methanol (Raina and Etter, 2010). Reagent addition should consider the change in solvent composition as this may also alter sensitivity with most pesticides observing enhanced sensitivity with higher percentages of organic modifiers such as methanol.

Similar to GC/MS methods large volume injections have also been used for LC/MS/MS applications although not specifically for pesticides. Direct on-column loop injection of 2 mL of water samples to a standard C18 column with LC/APCI+MS/MS achieved sub-µg/L range detection (Speksnijder et al., 2010). For urine samples an on-line LC-MS approach was used where the sample is pumped into the LC system and diluted through a mixing Tee with ammonium acetate after which it is loaded onto a restricted access material (RAM) precolumn while the analytical column equilibrates. The analytes of interest are then backflushed to transfer them to the analytical column followed by a typical gradient elution (Liu et al., 2008). The use of the RAM pre-column enables matrix removal of proteins as it retains only low molecular weight analytes. Matrix effects with API sources can lead to suppression or enhancement of analyte response due to co-eluting matrix constituents (Niessen et al., 2006). The choice of solvent used in extraction procedures can reduce matrix impacts with ethylacetate or acetonitrile often preferred for QuEChERS methods. If matrix suppression/enhancement can not be eliminated by sample preparation procedures prior to LC/MS/MS then deuterium or carbon-C13 labeled internal calibrations should be used for stable isotope dilution. If sufficient levels are available sample dilution or infinite dilution (matrix-free solution) can be utilized with typical dilution factors of 0.05 or 0.025 (Kruve et al., 2009). UV detection can also be utilized to identify co-eluting matrix issues requiring improvements in chromatography which can be achieved either with other column choices or comprehesive LCXLC (Hajšlová and Zrostíková, 2003). LC/TOF-MS has also gained considerable interest for confirmation of pesticides or transformation products with exact mass measurements for those applications where LC/MS/MS may not have the required sensitivity from the confirmation SRM transition (Portolés et al., 2009; Kuster et al., 2009). Pesticides such as aldicarb, diuron, linuron, aldicarb sulfone and sulfoxide can be distinguished and quantified by exact mass measurements with mean error of 2.3 ppm (Maizels and Budde, 2001).

CONCLUSION

GC/MS, GC/MS/MS, LC/MS/MS, and in some cases LC/MS methods are required to cover the full range of pesticide chemical classes and their transformation products. No one method can meet the needs of all the current pesticide chemical classes. There are also a number of chemical classes including phenoxyacid herbicides, pyrethroids, triazines, acetanilides, and azoles with both GC and LC methods coupled to mass spectrometry which may meet the needs of users. GC/MS or GC/MS/MS multiresidue methods with NCI are recommended for use with OCs, most OC transformation products, trihalomethylthio fungicides, and if a wide range of OPs require analysis for the best sensitivity and selectivity. Pyrethroids are recommended to be done with GC/EI-MS methods for those which are not chlorinated and for their derivatized degradation products. NCI may be also used for added sensitivity and selectivity for those pyrethroids that are halogenated. When developing GC/MS multiresidue methods it is also essential to have a GC/EI-MS method which is more suitable for acetanilides, triazines, atrazine transformation products, some OCs and a few OP pesticides. For some pesticides or transformation products the second confirmation ion or SRM transition may not be sensitive enough and both NCI SIM and NCI SRM methods or NCI and EI SIM methods may be required. If sufficient sample concentration is available then EI SRM methods may also be useful for a wide range of these pesticides. If sample concentrations are lower but added confirmation beyond these methods is required then GC/TOF-MS or GCXGC/TOF-MS is an alternative. In general it is not recommended to analyze azoles with GC/MS methods due to often higher detection limits as compared to LC/MS/MS methods and thermal instability. As a strategy LC/MS/MS methods should be utilized for carbamates, phenylureas, sulfonyl ureas, azoles, and transformation products from these chemical classes. If a laboratory desires to minimize the need for derivatization then phenoxyacid herbicides and their degradation products can also be accomplished by LC/MS/MS and postcolumn reagent addition can be used if added sensitivity is required for the chlorophenol transformation products. If a multiresdiue LC/MS/MS method is developed for these pesticide classes then the inclusion of chloroacetanilides, triazines, and transformation products from these chemical classes should also be considered. As with GC/MS methods, one ionization method can not be utilized for all chemical classes for LC/MS/MS methods and matrix impacts should be assessed to determine if there is an advantage in utilizing an alternative ionization mode to minimize impacts from interferences. For example one may consider analyzing sulfonylureas and acetanilide transformation products with phenoxyacid herbicides with LC/ESI-MS/MS, while analyzing phenylureas, carbamates, azoles, chloroacetanilides, triazines, OPs and transformation products of OPs, carbamates, and some sulfonylureas by LC/ESI+MS/MS, or LC/APCI+MS/MS method for triazines, phenylurea and their transformation products for best sensitivity. LC/APCI+/MS/MS can also be used to resolve matrix issues or co-elution problems for OPs, chloroacetanilides, phenylureas, carbamates, triazines, sulfonyl ureas (if phenoxyacid analysis is not required). Inclusion of transformation products will likely be the largest factor in selection of multiple methods with ESI and APCI in positive and negative mode as some transformation products require alternative ionization methods from that used for parent pesticides for adequate sensitivity. In addition the dithiocarbamates with the exception of triallate which can be included with standard GC/MS methods should be done with separate LC/MS methods either for anionic or neutral dithiocarbamates and their transformation products. Although their analysis requires a separate LC/MS method this is a significant improvement over prior GC methods that were not specific to individual dithiocarbamates and were laborious. If no GC/MS methods are necessary then pyrethroids may be included in LC methods. Currently the advantages of new column choices, GCXGC or LCXLC, large volume injections, on-column clean-up, post-column reagent addition, and TOF-MS are underutilized to resolve matrix and confirmation needs and should be considered in future method development.

REFERENCES

1.   Albanis, T.A.;hela, D.G.; Sakellarides, T.M.; Konstantinou I.K. (1998). *J. Chromatogr. A,* 823, 59-71.

2.   Alder, L; Gruelich, K; Kempe, G.; Vieth, B. (2006). *Mass Spectrometry Reviews,* 25, 838-865.

3.   Arrebola, F.J.; Martinez Vidal, J.L; Fernandez-Gutierrez, A.; Akhtar, MIL (1999). *Anal. Chico. Ada,* 401, 45-54.

4.   Asperger, A.; *Efer,* J.; Koal, T.; Engewald, W. (2001). *J. Chromatogr. A,* 937, 65-72.

5.   Baglio, D.; Kotzias, D.; Larsen, B.R (1999). *J. Chromatogr. A,* 854, 207-220.

6.   Bailey (Raina), R (2005). *J. Environ. Monitor.,* 7,1054-1058

7.   Bailey (Raina), R.; Belzer, W. (2007). *J. Agric. Food Chem.,* 55, 1150-1155.

8.   Banerjee, K.; Oulkar, D.P.; Pahl, 5.13.; Jadhav, M.R; Dasgupta, S.; Patil, S.H.; Bal, S.; Adsule, P.C. (2009). *J. Agric. Food Chem,* 57, 4068-4078.

9.   Bichon, E.; Dupuis, M; Le Bizec, B.; Andre, F. (2006). *J. Chromatogr. A,* 838, 96-106.

10.   Boni, R; Vejrup, K; Jacobsen, C.S. (1999). *J. Chromatogr. A,* 855, 575-582.

11.   Botitsi, H.; Economou,   Tsipi, D. (2007). *Anal. Bioanal. Chem.,* 389,1685-1695.

12.   Cajka, T.; Hajslova, J.; Lacina, O.; Mastovska, K.; Lehotay, S.J. (2004). *J. Chromatogr. A,* 1058, 251-261.

13.   Carabis-Martinez, R.; Rodriguez-Gonzalo, E.; Revilla-Ruiz, P. (2004). *J. Chromatogr. A,* 1056, 131-138.

14.   Cessna, A. J.; Donald, D.B.; Bailey, J.; Waiser, M.; Headley, J.V. (2006). *J. Environ. Qual.,* 35, 2395-2401.

15.   Charlton, A.J.A., Stuckey, V.; Sykes, M.D. (2009). *Bull Erwin's'. Cantata. Taricol.,* 82, 711-715

16.   Chen, T.; Chen, G. (2007). *Rapid Commun. Mass Spectrom.,* 21(12),1848-1854.

17. Chung, S.W.C.; Chan, B.T.P. (2010). *J. Chromatogr. A*, 1217, 4815-4824.

18. Corcia, A. D.; Nazzari, M.; Rao, R; Samperi, R., Sebastiani, E. (2000). *J. Chromatogr. A*, 878, 87-98.

19. Crescenzi, C.; Corcia, A.D.; Marchese, S.; Samperi, R (1995). *Anal. Chem.*, 67,1968-1975.

20. Crnogorac, G.; Schwack, W. (2007). *Rapid Commun. Mass Spectrom.*, 21, 4009-4016.

21. Cunha, S.C.; Lehotay, S.J.; Mastovska, K; Fernandes, J.0.; Beatriz, M; Oliveira, P.P. (2007). *J.*

22. *Sep. Sci.*, 30, 620-632.

23. Dagnac, T,; Bristeau, S.; Jeannot, R; Mouvet, C.; Baran, N. (2005). *J. Clunnurtogr. A*, 1067, 225233.

24. Dasgupta, S.; Banerjee, K; Patil, S.11.; Chaste, M.; Dhumal, K.N.; Assule, P.G. (2010). *J. Chromatogr. A*, 1217, 3881-3889.

25. Degenhardt, D.; Cessna, A.J.; Raina, R; Pennock, D.J.; Farenhorst, A. (2010). J. *Environ. Sci. Health, Part B*, 45,11-24.

26. de Koning, S.; Lack G.; Linkerhagner, M.; Loscher, R; Tablack, P.14.; Brinkman, U. A. Tit (2003). *J. Chromatogr. A*, 1008, 247-252

27. Dijkman, E.; Mooibroek, D.; Hoogerbrugge, R; Hogendoom, E; Sancho, J.-V.; Pow, O.; Hernandez, F. (2001). *J. Chromatogr. A*, 926,113-125.

28. Economou, A; Botitsi, H.; Antoniou, S.; Tsipi, D. (2009). *J. Chromatogr. A*, 1216, 5856-5867.

29. Feo, M.L; Eljarrat, E.; Barcel6, D. (2010). *J. Chromatogr. A*, 1217, 2248-2253.

30. Ferrer, I.; Thurman, EM.; Zweigenbaum, J.A_ (2007). *Rapid Common. Mass Spectrom.*, 21, 3869-3882

31. Gabalden, J. A.; Cascales J.M.; Maquieira, A.; Puchades, R (2002). *J. Chromatogr. A*, 963, 125- 136.

32. Godula, M.; Hajalova, J.; Maatouslca, K.; Ktivilmkova, J. (2001). *J. Sep. Sci.*, 24, 355-366.

33. Gonsalves, C.; Carvalho, J.J.; Azenha, M.A.; Alpendurada, M.F. (2006). *J. Chromatogr. A*, 1110, 6-14.

34. Gil-Garcia, M.D.; Barranco-Martinez, D.; Martinez-Calera, M.; Parrilla-Vazquez, P. (2006). *Rapid Commun. Mass Spectrom., 20* (16), 2395-2403.

35. Gomides Freitas, L; Getz, C.W.; Ruff, M.; Singer, H.P.; Muller, S.R (2001). *J. Chromatogr. A*, 1028,277-286.

36. Goto, T.; Ito, Y.; Yamada, S.; Matsumoto, H.; Oka, H.; Nagase, H. (2006).

Anal. Chim. Acta, 555, 225-231

37. Grob, K.; Li, Z. (1988). *J. High Resolut. Chrom.*, 11, 626-632

38. Hajalova, J.; Zrostilcova, J. (2003). *J. Chromatogr. A*, 1000, 181-197.

39. Henriksen, T.; Svensmark, B.; Lindhardt, B.; Juhler, RK (2001). *Chemosphere*, 44,1531-1539.

40. Hernandez, F.; Pow, 0.J.; Sancho, J.V.; Barreda, M.; Pitarch, E. (2006). *J. Chromatogr. A*, 1109, 242-252

41. Hiemstra, M.; de Kok, A. (2007). *J. Chromatogr.* A, 1154, 3-25.

42. Hogenboom, AC.; Niessen, W.IVLA.; Brinkman, U.A. Th. (2001) *J Sep Sri.*, 24, 331-354.

43. Httakova, R; Matisova, E.; gvorc, L.; Moak, J.; Kirchner, M. (2009). *J. Chromatogr.* A, 4927-4932.

44. Jansson, C.; Pihlstrom, T.; Osterdahl, B.-G.; Markides, K.E. (2004). *J. Chromatogr. A*, 2004, 1023, 93-104.

45. Jeannot, R; Sabik, H.; Sauvard, E; Genin, E. (2000). *J. Chromatogr. A*, 879, 51-71.

46. Jiang, H.; Adams, C.D.; Koffskey, W. (2005). *J. Chromatogr. A*, 1064, 219-226

47. Kirchner, M.; Matisova, E.; Otrekal, R; Hercegova, A.; de Zeeuw, J. (2005). *J. Chromatogr. A*, 1084, 63-70.

48. Koppen, B.; Spliid, N.H. (1998).*J. Chromatogr. A*, 803,157-168.

49. Kuster, M.; de Alda, M.L.; Barcel6, D. (2009). *J. Chromatogr. A*, 1216, 520-529.

50. Lambropoulou, D.A.; Albartis, T.A. (2007). *Anal. Normal. Chem.*, 389,1663-1683.

51. Leandro, CC.; Hancock, P.; Fussell, RJ.; Keely B.J. (2007).*J. Chromatogr. A*, 1166, 152-162

52. Liapis, KS.; Aplada-Sarlis, P.; Kyriakidis, N.V. (2003). *J. Chromatogr. A*, 996,181-187.

53. Maizels, M.; Budde, W.L. (2001). *Anal. Chem.*, 73, 5436-5440.

54. *Marchese,* S. Perret, D.; Gentili, A; D'Ascenzo, G.; Faberi, A. (2002). *Rapid Commun. Mass Spedrom.,16,* 134-141.

55. Martinez Vidal, J.L; Garrido Frenich, A.; Lopez Lopez, T.; Martinez Salvador, I.; Hajjaj el Hassani, L.; Hassan Benajiba, M. (2005). *Chromatogmphia*, 61(3/4),127-131.

56. Martinez, D.B.; Parrilla Vazquez, P.; Martinez Galera M.; Gil Garcia,

M.D. (2006) *Chromatographia,* 63 (9/10), 487-491.

57. Muller, A.; Flottmann, D.; Schulz, W.; Seitz, W.; Weber, W.H. (2007). *Clean,* 35(4), 329-338.

58. Nagaraju, D.; Huang, S.-D. (2007). *J. Chmmatogr.* A, 1161, 89-97.

59. Nilsson, T; Baglio, D.; Galdo-Miguez, I.; Madsen J.O.; Facchette, S. (1998). *J. Chromatogr. A,* 826, 211-216.

60. Paya, P.; Anastassiades, M.; Mack, D.; Sigalova, I.; Tasdelen, B.; Oliva, J.; Barba, A. (2007). *Anal. Bioanal. Chem.,* 389,1697-1714.

61. Pico, Y.; Kozmutaz, C. (2007). *Anal. Bioanal. Client.,* 389,1805-1814.

62. Pililstrom, T.; Blomkvist, G.; Friman, P.; Pagard, U.; Osterdahl, B.-G. (2007). *Anal. &c anal. Chem.,* 389,1773-1789.

63. Planas, C.; Puig, A.; Rivera, J.; Caixach, J. (2006). *J. Chromatogr. A,* 1131, 242-251

64. Portolts, T.; Ilguiez, M.; Sancho, J.V.; Lopez, F.J.; Hernandez, F. (2009). *J. Agric. Food Chem.,* 57, 4079-4090.

65. Puig, D.; Silgoner, I.; Grasserbauer, M.; Barce16, D. (1997). *Anal. Chem.,* 69, 2756-2761.

66. Raina, R; Sun, L (2008). *J. Environ. Sci. Health, Part B,* 43, 323-332

67. Raina, R; Hall, P. (2009). *Anal. Chem. Insights,* 3, 111-125.

68. Raina, R; Etter, M.L (2010). *Anal. Chem. Insights,* 5,1-14.

69. Reesmtsma, T. (2003). J. *Chromatogr. A,* 1000, 2003, 477-501.

70. Rimmer, D.A.; Johnson, RD., Brown, RH. (1996). *J. Chromatogr. A,* 755, 245-250.

71. Santos, T.C. R; Rocha, J.C.; Barce16, D. (2000). *J. Chromatogr. A,* 879, 3-12

72. Schermerhom, P.G.; Golden, P.E.; Krynitsky, A. J.; Leimkuehler, W.M. (2005). *J. AOAC International,* 88 (5),1491-1502

73. Shin, H.S. (2006). *Chromatogmphia,* 63, 579-583.

74. Souverian, S; Rudaz, S.; Veuthey, (2004). *J. Chromatogr. A,* 1058, 61-66.

75. Steen, RJ.C.A.; Hogenboom, A.C.; Leonards, P.E.G.; Peerboom, RA.L.; Cofino, W.P.; Brinkman U.A.Th. (1999). *J. Chromatogr. A,* 857,157-166.

76. Thurman, EM.; *Ferrer,* I.; 13acele, D. (2001). *Anal. Chem.,* 73, 5441-5449.

77. Thurman, E.M.; Ferrer, I.; Pozo, 0.J.; Sancho, J.V.; Hernandez, F. (2007). *Rapid Commun. Mass Spedrom.,* 21, 3855-3868.

78. Titato, G.M.; Bicudo, RC.; Lamas, F.M. (2007). *J.* Mass *Spectrom.,* 4Z 1348-1357.

79. Trosken, E.R; Bittner, N.; Wad, W. (2005). *J. Chromatogr. A,* 1083, 113-119.

80. Van der Heeft, E.; Dijkman, E; Baumann, RA.; Hogendoorn, E.k (2000). *J. Chromatogr. A,* 879, 39-50.

81. Yoshida, T. (2009). *J. Chromatogr. A,* 1716, 50695076.

82. Zambonin, C.G.; Palmisano, F. (2000). *J. Chromatogr. A,* 874, 247-255.

83. Zang, J.; Lee, H.K (2006). *J. Chromatogr. A,* 1117, 31-37.

84. Zrostlilcova, J.; Hajalova, J.; Kovalczuk, T.; gtepan, R; Poustka, J. (2003a). *J. AOAC International,* 86 (3), 612-621

85. Zrostlikova, J.; Hajalova, J.; eajka, T. (2003b). *J. Chromatogr. A,* 1019, 173-186.

Chapter 2

IMPACT OF PESTICIDES ON ENVIRONMENTAL AND HUMAN HEALTH

Mariana Furio Franco Bernardes[1], Murilo Pazin[1], Lilian Cristina Pereira[1], and Daniel Junqueira Dorta[2]

[1]Faculdade de Ciências Farmacêuticas de Ribeirão Preto - FCFRP, Universidade de São Paulo- USP, Brazil

[2]Departamento de Química, Faculdade Filosofia, Ciências e Letras de Ribeirão Preto- FFCLRP, Universidade de São Paulo – USP, Brazil

INTRODUCTION

Pesticides constitute any substance or mixture of substances intended for preventing, destroying, repelling, or mitigating any pest. They can also serve as plant regulators, defoliants, or dessicants [1].

Chemicals have long been used to control pests. Sumerians already employed sulfur compounds to control insects and mites 4500 years ago. Pyrethrum, a compound derived from the dried flowers of *Chrysanthemum cinerariaefolium*, has been applied as an insecticide for over 2000 years. Salt or sea water has been used to control weeds. Inorganic substances, such as sodium chlorate and sulfuric acid, or organic chemicals derived from natural sources were widely employed in pest control until the 1940s [2].

During World War II (1939-1945), the development of pesticides increased, because it was urgent to enhance food production and to find potential chemical warfare agents [3]. Consequently, the1940s witnessed a marked growth in synthetic pesticides like DDT, aldrin, dieldrin, endrin, parathion, and 2,4-D. In the 1950s, the application of pesticides in agriculture was considered advantageous, and no concern about the potential risks of these chemicals to the environment and the human health existed [2].

In 1962, Rachel Carson published the book "Silent Spring", in which she mentioned problems that could arise from the indiscriminate use of pesticides.

This book inspired widespread concern about the impact of pesticides on the human health and the environment. In 1967, Ratcliffe [4] noted increased incidence of raptor nests with broken eggs in the United Kingdom. This author showed that the sharp decline in eggshell thickness coincided with the beginning of the widespread use of DDT in agriculture (1945–1946). In the 1970s, pest resistance emerged which, combined with influence of the book "Silent Spring", and accumulated evidence on the effects of pesticides, culminated in banning of the use of DDT in the United States in 1972. Thereafter, other countries discontinued the use of DDT, as well [5].

The 1970s and 1980s saw the introduction of more selective pesticides. In the 1990s, research activities concentrated on finding new members of existing pesticides that were even more selective. Besides, pesticides with new chemical groups emerged. During this period, safer chemicals arose. In addition, Integrated Pest Management (IPM) systems, came into play – these systems used crop production methods that attracted predators or parasites that attacked pests and timed pesticide applications to coincide with the most susceptible period of the pest's life cycle, thereby reducing the amount of applied pesticides [2].

However, IPM or related methods did not eliminate the need for pesticides. These chemicals ensure the production of adequate quantities of high quality pest-free crops, which is important for food supply, prevents human diseases transmitted by insect or rodent vectors, and positively impacts public health [6].

The best pesticide policies need to reconcile environmental concerns with economic realities – pest management is mandatory, and farmers must survive economically. A number of studies have described the problems that not using pesticides would cause. Without pesticides, food production would be lower, and larger cultivated farm areas would be necessary to produce the same amount of food, which would impact the wildlife habitat. More frequent cultivation of the fields would be increase soil loss due to erosion, too. Knutson et al. [7] have pictured the U.S. society without pesticides: agricultural production would decrease, food prices would rise, farmers would be less competitive in global markets, and U.S. exports would drop, leading to many job losses [8].

Despite their benefits, pesticides can be hazardous to both humans and the environment. Countless chemicals are environmentally stable, prone to bioaccumulation, and toxic [6]. Because some pesticides can persist in the environment, they can remain there for years. Environmental contamination or occupational use can expose the general population to pesticides residues, including physical and biological degradation products present in the air, water, and food [9].

Less than 1% of the total amount of pesticides applied for weed and pest control reach the target pests. A large quantity of pesticides is lost via spray drift, off-target deposition, run-off, and photodegradation, for instance, which can have undesirable effects on some species, communities, or ecosystems as a whole, as well as on the humans [10]. Another relevant factor is that low concentrations of many chemicals may not elicit acute detectable effects in organisms, but they may induce other damage, like genetic disorders and physiological alterations, which reduce life span in the long run [11].

There are various ways to group pesticides, including classification based on the pests they control. Some example, insecticides combat insect growth or survival, herbicides act against plants, weeds, and grasses, rodenticides tight against rats and other rodents, avicides act against bird populations, fungicides attack fungi, and nematicides combat nematodes [12]. The global pesticide market divided according to the type of pesticide is as follows: 42.48% herbicides, 25.57% insecticides, 24.19% fungicides, and 7.76% other types of pesticides [13].

Pesticides grouping can also rely on their chemical structure. Organophosphorus (chlorpyrifos and diazinon), carbamates (carbaryl and aldicarb), organochlorine (DDT and aldrin), pyrethrins and pyrethroids (cyfluthrin and cypermethrin), benzoic acids (dicamba), triazines (atrazine and simazine), phenoxyacetic derivatives (2,4-D), dipyridyl derivatives (diquat and paraquat), glycine derivatives (glyphosate), and dithiocarbamates (maneb and ziram) [12].

Pesticides that bear similar chemical structures exhibit similar mechanism of toxicity and physicochemical properties, as well as comparable fate and transport properties. This chapter will deal with pesticides according to their chemical group. Pesticides belonging to different chemical classes but which have similar toxic effects, such as the ability to induce oxidative stress and act as endocrine disrupters will be treated as well.

PHYSICAL AND CHEMICAL PROPERTIES AND STAGES OF INTOXICATION

Organophosphorus

Organic compounds containing phosphorus, the so called organophosphorus compounds (OP), have found application as pesticides and war gases since their synthesis, in 1937 [14]. OP contain carbon and derive from phosphorous acid. Their primary structure may vary depending on whether they bear sulfur (S) or oxygen (O) double binds. X is a group of the general structure that separates when the compound binds to acetylcholinesterase (AChE). On the basis of the variations in their general structure, it is possible to subdivide these compounds into phosphates, phosphorothioates, phosphoramidates, and phosphonate, for example. The structural difference between these compounds results in peculiar characteristics regarding OP metabolism and toxicity. Some representatives of this class of pesticides are Diazinon, Malathion, and Paration [15].

The skin, conjunctiva, gastrointestinal tract, and lungs rapidly absorb most OP compounds. Cytochrome P_{450} isozymes metabolize these chemicals in the liver, which sometimes generates metabolites that are more toxic than the parent compounds [16]. One example is the oxon form, which may bind to cholinesterase or undergo hydrolysis to a dialkyl phosphate and a hydrolyzed organic moiety specific to the pesticide [14]. Most OP are polar and hence water soluble. Their metabolites arise 12 to 48 h after exposure. However, a few compounds, such as dichlofenthion, possess high partition coefficients, which culminates in long-lasting symptoms [17].

These pesticides can reversibly or irreversibly stablish covalent bonds with the serine residue in the active site of acetyl cholinesterase, to prevent the natural function of this enzyme in the catabolism of neurotransmitters [14]. The formation of complexes between acetylcholinesterase enzymes and organophosphates leads to phosphorylation and deactivation, and the neurotransmitter acetylcholine consequently accumulates in the synaptic cleft. The accumulation of large amounts of acetylcholine stimulates and exhausts cholinergic synapses due to the excessive cholinergic activity produced by these agents [18]. The cholinesterase-bound phosphate group can lose the o,o-dialkyl groups or undergo hydrolysis, to regenerate the active enzyme. This process occurs not only in insects, but also in humans and the wildlife [14].

The main symptoms of pesticides intoxication can be differentiated into syndromes like the muscarinic syndrome, in which the action of acetylcholine on the smooth muscle, heart, and exocrine glands increases bronchial secretion, tearing, and sweating; disrupts the gastrointestinal tone to cause nausea, vomiting, and diarrhea; and elicits urinary incontinence, bronchospasm, miosis, and bradycardia. Another example is the nicotine syndrome, in which acetylcholine accumulates at the motor nerve endings in the autonomic ganglia and causes tremors, spasms, hypertonicity, hyperreflexia, paralysis, or muscle weakness and stimulates the sympathetic autonomic ganglia, to promote tachycardia, pallor, hyperglycemia, and hypertension. Additional effects on the central nervous system (CNS) include anxiety, headache, dizziness, ataxia, sleep and memory disorders seizures, tremors, respiratory depression, and coma. Some OP still have teratogenic potential and mutagenic effects.

Laboratory diagnosis of this syndrome involves determination of cholinesterase activity [19].

To treat poisoning with OP, it is necessary to maintain vital functions and assess cholinesterase levels in the red cells and pseudocholinesterase levels in the plasma, before therapy. It is important to avoid the use of parasympathomimetic agents, which may increase the anticholinesterase activity. Treatment should start with atropine, which acts as a competitive muscarinic anticholinergic agent, together with pralidoxime, until complete control of the symptoms. After atropinization, administration of furosemide prevents pulmonary congestion, whereas administration of benzodiazepines controls seizures [20].

Carbamates

Carbamates insecticides produce clinical signs and symptoms of cholinergic excess that resemble the signs elicited by organophosphate toxicity, except that the effects are more reversible and less severe [14]. The mechanisms underlying carbamates poisoning involve carbamylation of the active site of acetylcholinesterase, which inactivate this essential enzyme in the nervous system of humans and other animal species [21]. The reaction of carbamates with acetylcholinesterase is similar to the reaction of OP with the same enzyme. However, reactivation of the carbamylated enzyme by hydrolysis is faster as compared with reactivation of the phosphorylated enzyme, with reversal of inhibition typically occurring half an hour or less after exposure [22]. Nevertheless, reports on cases of neuropathy after poisoning exist [23].

Organisms readily absorb carbamates through the lungs, gastrointestinal tract, and skin. Fortunately, carbamates poorly penetrate the blood-brain barrier. Therefore, they affect brain cholinesterases activity minimally and promote fewer CNS symptoms as compared with organophosphates. In addition, the spontaneous in vivo hydrolysis of the carbamate-cholinesterase complex contributes to less severe and less enduring symptoms.

The main symptoms of carbamates intoxication are miosis, salivation, sweating, tearing, rhinorrhea, behavioral change, abdominal pain, vomiting, diarrhea, urinary incontinence, bronchospasm, dyspnea, hypoxemia, bradycardia, bronchial secretions, pulmonary edema, respiratory failure, drop in body temperature, incoordination, lip tingling, tremors, and seizures. Less common symptoms include muscle spasms, twitching, muscle weakness (including respiratory muscles), paralysis, tachycardia, and hypertension [24].

The treatment of carbamates intoxication includes maintenance of vital functions. It is crucial to avoid the use of parasympathomimetic agents, because they may increase the anticholinesterase activity. Treatment should start with atropine, followed by administration of furosemide, only if necessary. If poisoning is due to pure carbamates only, it is not necessary to administrate pralidoxime, except in cases that these carbamates are associated with OPs [15].

Organochlorines

Organochlorine is used mainly as insecticides. Human body burden due to organochlorine pesticides results from the universal presence of these contaminants in the environment. This constitutes a major public health concern; indeed, organochlorines have been linked with cancer, asthma, diabetes, and growth disorders in children [25]. Organochlorine pesticides include cyclodienes, hexachlorocyclohexane isomers, and DDT and its analogues (e.g., DDE, methoxyclor, and dicofol) [14].

Exposure to organochlorines occurs via ingestion of contaminated food or water, inhalation of vapor, and absorption through the skin. Occupational and other domiciliary exposures are also possible. Dietary exposure results in bioaccumulation of these chemicals in the human body [26].

Organochlorines have similar structure – they all contain a cyclodiene ring. The lungs, gastrointestinal tract, and skin can absorb all these compounds. In addition, although the organism absorbs approximately 10% of the applied dose, lipid solvents increase dermal penetration [15], thereby raising the risk of intoxication in the case of workers who apply these products in crops without proper protective equipment.

The accumulation of organochlorine compounds is a result of their chemical structure and their physical properties such as polarity and solubility. These fat-soluble compounds persist in both the body and the environment. Consequently, researchers and regulatory agencies have banned several organochlorines [14].

The main symptoms of organichlorines intoxication are dizziness, headache, anorexia, nausea, vomiting, malaise, dermatitis, diarrhea, apprehension, excitement, irritability, gait disorders, excessive sweating, altered reflexes, muscle weakness, tremors, spasms, mental confusion, anxiety, seizures, coma, and death. The carcinogenicity of this class of compound is assigned to polychlorocyclodiene compounds that form epoxides during their biotransformation. Because organochlorines have long half-life, these levels in the serum constitute a marker of exposure to these pesticides [15].

To treat organochlorines intoxication, it is necessary to maintain the vital functions, administer diazepam and phenobarbital by slow injection, to control seizures, and to monitor the airways closely. Lorazepam constitutes an alternative to diazepam. Ion exchange resins can also be administered orally. Arrhythmias that damage the myocardium rarely occur. Lidocaine is the treatment of choice [27].

Pyrethrins and Pyrethroids

Pyrethrins and pyrethroids function mainly as iseticides. Pyrethrins are natural compounds originating from the plant *Chrysanthemum cinerariaefolium*. They comprise active agents (pyrethrins I-VI), but pyrethrins I and II are the most active. These compounds decompose rapidly in the presence of light, but synthetic production of pyrethroids around 1950 overcame some disadvantages of natural pyrethrins [15].

Crude pyrethrum is a dermal and respiratory allergen, probably due it is to non-insecticidal ingredients. Contact dermatitis and allergic respiratory reactions (rhinitis and asthma) have occurred after exposure to this compound [28].

Both pyrethrins and pyrethroids bear an acid moiety, a central ester bond, and an alcohol moiety in their structure. This class of compounds typically exists as stereoisomers (*trans* and *cis*) for a total of eight different stereoenantiomers. In adittion, they comprise two main groups, Type I and Type II, which bear a cyano group in the alpha position or not, respectively [29].

After absorption, rapid pyrethroids distribution occurs in the organism. Therein, these compounds undergo biotransformation via two mechanisms: hydrolysis of the ester linkage by carboxylesterases and oxidation of the alcohol moiety by cytochromes P_{450} [30]. Pyrethroids exert the same mechanism of action in insects and mammals. Both pyrethrins and pyrethroids have insecticide potential because they can disrupt the muscular system and alter the normal functioning of voltage-dependent sodium channels. Sodium channels play an important role in the cell-to-cell communication, which is vital for the function of more excitable cells involved in the action potential that the excitable cells can propagate in the CNS. Pyrethroids bind to the α-subunit of the sodium channel that is left open for a longer time, to increase membrane permeability to sodium. Consequently, these compounds cause paralysis, especially in flying insects, known as knockdown. The specific interaction of pyrethroids with the sodium channel shows both the activation and inactivation properties of the sodium channel, making the hyperexcited cells [31]. After interaction of moderate levels of pyrethroids with the sodium channel, the cell can continue to operate in an abnormal state of hyperexcitability. The amplitude of the sodium current remains unchanged until the level of hyperexcitability overwhelms the maintenance of the activity of the sodium channel. This culminates in depolarization and blocks conduction of the action potential until the situation in the cell becomes unsustainable [31].

The toxicodynamics of pyrethroids may also include other mechanisms such as antagonism of gamma-aminobutyric acid (GABA), stimulation of chloride channels modulated by protein kinase, modulation of nicotinic cholinergic transmission, increased release of noradrenaline, and deregulation of calcium homeostasis. Authors have also proposed that pyrethroids act on the voltage-sensitive chloride channels as well as on the voltage-dependent calcium channels [31].

Diagnosis can be difficult because acute pyrethroid poisoning can be mistaken for OP intoxication. Pyrethroid poisoning symptoms are: tremors, spasms, incoordination, prostration, drooling, irregular movements of the limbs, tonic and clonic convulsions, and hypersensitivity to stimuli. It can also cause skin irritation and tingling due to hyperactivity of cutaneous sensory nerve fibers. Eye miosis also occurs due to exposure [32].

Because exposure to pyrethroids does not usually prompt systemic effects, most patients only require decontamination of the skin and eyes, besides basic maintenance of the vital functions. Paresthesia usually subsides within 12-24 h, which dismisses direct treatment. If severe skin irritation occurs, application of DL-α-tocopherol acetate (Vitamin E) should alternate this problem. Gastric lavage is discarded in case of ingestion, because solvents present in many formulations may increase the risk of aspiration pneumonia. Ingestion of a potentially toxic amount requires administration of activated charcoal within one hour of the event [32].

Triazines

Triazines are effective and inexpensive compounds that have found application as herbicides. They combat a wide spectrum of weeds by inhibiting photosynthesis and the electron transport chain in plants. Physiological and molecular changes due to accumulation of these compounds in organisms remain unclear. Human exposure to triazines has been associated with carcinogenicity and endocrine disruption, but these effects are still debatable [33]. The chemical structures of triazine herbicides correspond to permutations of the alkyl substituted 2,4-diamines of chlorotriazine [14].

After absorption, these compounds undergo conjugation with glutathione or simply dealkylation. The chlorine group of the triazine structure is replaced with the free-SH group of glutathione, the terminal peptide is cleaved, and the cysteine moiety is N-acetylated. The mercapturate residues and the dealkylation metabolites are subsequently excreted in the urine [14]. Triazines have low acute oral and dermal toxicity. Chronic toxicity studies have primarily indicated reduced body weight gain [16].

Atrazine is the often most studied triazine herbicide. Authors have investigated their carcinogenic potential in mice and rats. Tumor incidence did not augment in mice, whereas atrazine appeared to increase the incidence of mammary carcinoma in Sprague-Dawley rats [34, 35].

Reports of human poisoning by this class of compounds are rare. When they happen, irritation at the site of contamination such as the skin, eyes, nose, and TGI occurs. Triazines may be carcinogenic and teratogenic, but there is still no evidence that this is really the case. Contamination with atrazine may also cause sensory motor polyneuropathy [15, 33].

Because exposure to triazines usually causes local irritation, in most cases it is only necessary to decontaminate the site exposed to the substance, besides offering basic life support [15].

Phenoxy Derivatives

The structures of phenoxy derivatives bear an aliphatic carboxylic acid moiety attached to a chloride or methyl-substituted aromatic ring. The commonest phenoxy herbicides are 2,4-dichlorophenoxyacetic acid (2,4-D) and 2,4,5-trichlorophenoxyacetic acid (2,4,5-T). A combination of these two herbicides in equal proportions affords Agent Orange, a product applied in the jungles of Vietnam, Laos, and Cambodia during the Vietnam War. Manufacture of phenoxy herbicides often requires co-formulation with ioxynil and/or bromoxynil, which are generally more toxic than the herbicides. Moreover, other more toxic substances can emerge during the fabrication of some of these herbicides at excessively high temperatures, such as at chlorinated dibenzo dioxin and chlorinated dibenzo furan [36]. Because 2,4,5-T contains the highly toxic and persistent 2,3,7,8-tetrachlorodibenzeno-p-dioxin along with other chlorinated dioxins and furans, regulatory agencies have banned it for most applications [14].

Phenoxy salts and esters rapidly dissociate or hydrolyze in vivo, so the toxicity of the derivative will depend mainly on the acid form of the pesticide. Individuals and species vary substantially in terms of phenoxy herbicides elimination. The biological half-life of herbicides in humans reportedly varies from 12 to 72 h [36], but long half-lives occur at large doses and after prolonged exposure [28].

The gastrointestinal tract absorbs phenoxy derivatives. The lungs absorb them less, their cutaneous absorption is minimal, and fat does not store them. Phenoxy derivatives exhibit a variety of mechanisms of toxicity including dose-dependent cell membrane damage, uncoupling of oxidative phosphorylation, and disruption of acetylcoenzyme A metabolism [36]. Phenoxy acids and esters are moderately irritating to the skin, eyes, and the respiratory, gastrointestinal, and mucous membranes. Their toxicity on the CNS is dose-dependent. These derivatives disrupt the blood-brain barrier and the neuronal membrane transport mechanisms, and damage to the intracellular membrane results in uncoupling of oxidative phosphorylation [36]. In addition, prolonged inhalation of these herbicides may cause burning sensation in the nasopharynx and dizziness. Some recent studies have examined female exposure to herbicides and assessed effects such as spontaneous abortion, birth defects, and infertility, among others [28].

Intoxication by this class of compounds is uncommon, but when they occur they can cause serious sequelae. The main symptoms are nausea, dizziness, vomiting, burning in the mouth, constipation, abdominal pain, numbness, diarrhea, gastrointestinal bleeding, gastrointestinal fluid loss, vasodilation and/ or direct toxicity due to grafting hypotension, ECG alterations like ventricular or supraventricular tachycardia and, on rare occasions, sinus bradycardia. In more severe cases, agitation, confusion, weakness, paralysis, coma, and death by ventricular fibrillation can occur, and chances of survival are small. Other disrupted functions comprise changes in the NCS [36]. Some compounds of this class (e.g., 2,4,5-T) can also produce carcinogenic and teratogenic effects as well as hepatotoxicity. As for metabolic acidosis, clinical signs such as hyperthermia (due to uncoupling of oxidative phosphorylation), renal failure, increased aspartate aminotransferase and alanine and lactate dehydrogenase, thrombocytopenia, hemolytic anemia, and hypocalcemia activities can arise [36].

In general, treatment of phenoxy derivatives poisoning includes maintenance of the vital functions. If the poisoning is due to ingestion, administration of activated charcoal is necessary for adsorption of the compounds, provided that intoxication occurred within an hour. Systemic poisoning calls for hemodialysis, but other effective purification methods exist, like alkalinization of the urine flow and increase of urine volume to facilitate excretion. To control seizures, administration of benzodiazepines is mandatory [36].

Dipyridyl Derivatives

The dipyridyl compounds paraquat and diquat are non-selective contact herbicides that have found wide application in agriculture and industries. They help to control weeds. However, these compounds are highly toxic and managing poisoning with these substances requires a great skill and knowledge of proper management procedures [28].

Paraquat (1,1'-dimethyl-4,4'-dipyridylium) is a dipyridylium quaternary ammonium compound related to diquat and morfamquat. The latter product is the least toxic but also the least effective herbicide [15]. Their biotransformation produces free radicals, with consequent lipid peroxidation and cell injury [37].

Paraquat causes aggressive tissue damage in the lungs, kidney, and liver. The major target organ of paraquat poisoning is the lung, which consists of the most lethal and the least treatable manifestation of toxicity. Reactive oxygen species (ROS) play a crucial role in paraquat induced pulmonary injury, characterized by edema hemorrhage and hypoxemia, as well as infiltration of inflammatory cells [28,38].

The other representative of this class is diquat (1,1'-ethylene-2,2'-bipyridilium), which causes fewer poisoning events than paraquat, the reason why reports on human toxicity and animal experimental data are less extensive for diquat than paraquat. The mechanisms of paraquat and diquat toxicity are similar: radicals destroy lipid membranes. After absorption, diquat does not selectively concentrate in the lung tissue, but it exerts severe toxic effects on the CNS, an event that is not typical in the case of paraquat [28]. The kidney is the main excretory pathway for absorbed diquat. Renal damage is therefore an important feature of diquat poisoning [15, 28].

A very interesting action against poisoning by diquat and paraquat is the addition of an emetic agent in their formulations, wherein the additive acts rapidly in the body and causes the individual to regurgitate the pesticide before it performs its toxic action [38 - 40]. The main poisoning symptoms are dehydration resulting from vomiting. The high oxidative stress elicited by these herbicides causes necrosis in the gastrointestinal tract, kidney tubules, liver, and lung; in the latter case, respiratory failure and pulmonary fibrosis may occur. Ingestion of large amounts of these compounds leads to death within two to three weeks, a result of acute renal failure, hepatitis, and especially respiratory failure caused by pulmonary inflammation and fibrosis. In addition to the systemic effects, these compounds are very harmful to the skin and may cause severe burns [38, 41].

The treatment of poisoning with dipyridyl derivatives includes maintenance of the vital functions, minimization of the absorption of the compound more cathartic (activated charcoal), acceleration of excretion (forced diuresis, hemodialysis, or hemoperfusion), abatement of the effects on the affected tissue, and fluid replacement. Topical lesions should be treated with topical silver sulfadiazine, combined with systemic antibiotics [41]. An addition method to recognize paraquat poisoning is to test the urine with sodium dithionite [42].

Glycine Derivatives

Two representatives of this class are glyphosate (N-phosphonomethyl glycine) and glufosinate (N-phosphonomethyl homoalanine), marketed primarily as the isopropylamine salt (glyphosate) or ammonium salt (glufosinate). Both substances are broad-spectrum nonselective systemic herbicides with application in for post-emergent control of annual and perennial plants. Although both compounds contain a P=O moiety, they are not organophosphates, but organophosphonates, and they do not inhibit AChE [36].

Glyphosate, which contains phosphorus, is a herbicide used in 75% of all the genetically modified crops (GMCs), which tolerate high concentrations of this compound [36, 43]. Glyphosate inhibits an enzyme in the biosynthesis of tryptophan, phenylalanine, and tyrosine, present in plants, fungi, and bacteria, but not in animals or humans. [44]. However, according to literature reports, glyphosate can enter living organisms, including humans, where it exerts various toxic effects [45].

One pathway of glyphosate metabolism involves formation of aminomethylphosphonic acid (AMPA) by action of glyphosate oxidoreductase; AMPA is also the metabolite that emerges in humans [36]. Knowledge of the toxicokinetics of glyphosate derives mainly from animal studies and the similar patterns of absorption, metabolism, and eliminations in humans [46]. Rats absorb only 30% glyphosate after oral administration [36]. Glyphosate plasma concentrations peak at 1-2 h, and declined thereafter, with distributions to the intestine, colon, kidney, and bones [47].

The mechanisms of toxicity of glyphosate formulations are complicated [36]. The most widely used glyphosate product is Roundup®, formulated as a concentrate containing 41% glyphosate [16]. Some in vitro studies have suggested that, at high concentrations of glyphosate, its metabolites and impurities may reduce acetylcholinesterase (AChE) activity [48], although no evidence for significant AChE inhibition in mammals in vivo exists [36]. A study published in the *Archives of Toxicology* by Koller and colleagues showed increased in nuclear aberrations after exposure to glyphosate concentrations between 10 and 20 mg/L, which indicated DNA damage [49]. In adition, in vitro tests using isolated rat liver mitochondria showed that glyphosate uncoupled the electron transport chain [50].

Glufosinate inhibits the synthesis of glutamine in plants, and plant death occurs as a consequence of the increased ammonia levels [16]. Glufosinate supress the activity of glutamine synthetase and glutamate decarboxylase, reducing glutamic acid levels and elicits various types of moderate-to-severe CNS toxicities [51]. Given the differences in the biochemical and metabolic pathways of plants and mammals, glufosinate ammonium formulations are minimally toxic to humans [52]. However, ingestion of the undiluted form can cause grave outcomes such as seizures, respiratory arrest, coma, and disturbance of consciousness, which appear after a latent period of 4 – 60 h [53]. No work has reported that this compound induces genotoxic or carcinogenic effects or that impacts reproduction and fertilization [16].

The effects of this class of compounds range from irritation upon local contact (skin, GI), to hypotension, development of acute renal failure with oliguria, and severe hypoxia and death [54].

The treatment of glycine derivatives poisoning includes maintenance of the vital functions. Hemodialysis is crucial to reduce the amount of toxins normally excreted by the kidney, thereby preventing the impacts on this organ [54].

Dithiocarbamates

Dithiocarbamates comprise two groups: [1] dimethyldithiocarbamate and [2] ethylenebisdithiocarbamate, depending on which metal cation is present in the chemical structure. The nomenclature of various compounds of this class is related to the association of the metal cations; e. g., maneb (manganese), and zineb and ziram (zinc) [16, 50].

The slow absorption of these compounds means that they have low acute oral and dermal toxicity. On the other hand, chronic exposure to dithiocarbamates leads to adverse effects due to contact with dithiocarbamate acid or metal ligand [16].

The metabolite that arises from dithiocarbamates biotransformation is ethylenethiourea (ETU), which induces thyroid cancer and modifies thyroid hormones. Moreover, dithiocarbamates and disulfiram have similar structures, and both can inhibit acetaldehyde dehydrogenase, the enzyme that converts acetaldehyde into acetic acid [55].

Although these products are little toxicity to humans, they are potential precursors of ethylenethiourea, which has carcinogenic and teratogenic action.

There is no specific treatment for poisoning with this class of compounds, so only maintenance of vital functions and minimization of their absorption (activated charcoal) are necessary.

Others

Others classes of pesticides exist, including the chloroacetanilide commonly used in agriculture. A number of chloroacetanilides, like alachlor, acetochlor, metolachlor, and propachlor are carcinogenic [56]. The metabolism of chloroacetanilides most likely proceeds via conjugation with glutathione, as judged from the amount of glutathione-related metabolites in the urine of rats treated with these herbicides. [57]. However, the predicted differences between humans and rats in terms of disposition together with the lower rates of alachlor metabolism in human nasal microsomes have led scientists to question the human relevance of chloroacetanilide olfactory carcinogenicity [58].

Benzimidazoles are another important class of pesticides. They are commonly used as veterinary medicines (anthelmintics) and pesticides. They inhibit microtubule formation when they bind to free β-tubulin monomers at the colchicine-binding site [59].

Regarding new technologies, nanopesticides or nanoplant protection products represent an emerging technological development. In relation to pesticide use, these technologies could offer a range of benefits including increased efficacy and durability, and they use of smaller amounts of active ingredients [60]. Nanopesticides "involve either very small particles of a pesticide active ingredient (ai) or other small engineered structures with useful pesticidal properties" [61]. Nanoformulations combine several surfactants, polymers (organic), and metal nanoparticles (inorganic) in the nanometer size range [62].

Table 1. Summary of the physicochemical properties and phase of the intoxication of different classes of pesticides

	Physical and Chemical Properties	Exposition	Toxicokinetics	Toxicodynamics	Signs and Symptoms	Treatment
Organophosphorus	Organic compounds containing phosphorus[15]. The properties vary with the size and structure. In general are more soluble in organic solvents [65].	Skin, conjunctiva, gastrointestinal tract, and lungs[16].	Rapidly absorbed and metabolized by P450 isozymes in oxom form, more toxic than the parent compounds[16].	Covalent bonds with the serine residue in the active site of acetyl cholinesterase (reversibly or irreversibly)[14].	Muscarinic syndrome and nicotine syndrome, resulting of excess acetylcholine in the synaptic cleft[19].	Maintenance of vital functions and cholinesterase levels. It is important to avoid the use of parasympathomimetic agents[20].

Carba-mates	The carba-mate is an ester deriva-tive[14]. A wide range of melt-ing points (50 to 150°C) is found for these com-pounds and the majority have low va-por pressures and poor volatiliry at usual tem-peratures[21].	Lungs, gastroin-testinal tract, and skin[22].	Readily ab-sorbed by or-ganisms with exception the blood-brain barrier[22].	Carba-mylation of the active site of acetylcholin-esterase[22].	Miosis, salivation, sweating, tearing, rhinorrhea, behavioral change, abdomi-nal pain, vomiting, diarrhea[24].	Main-tenance of vital functions and cho-linesterase levels. It is important to avoid the use of parasympa-thomimetic agents[24].
Organo-chlorines	They all contain a cy-clodiene ring. Fat-soluble compounds persist in both the body and the environ-ment[15]. The majority of organochlo-rines are sparingly soluble and semivolatile[66].	Lungs, gastroin-testinal tract, and skin[26].	The organ-ism absorbs approximate-ly 10% of the applied dose, but the lipid solvents increase the accumula-tion[15].	Endocrine disrupters and growth disorders in children[25].	Dizziness, headache, anorexia, nausea, vomiting, malaise, dermatitis, diarrhea, muscle weakness, tremors, spasms, mental confusion, anxiety[15].	Mainte-nance of vital func-tions and administer diazepam and pheno-barbital to control sei-zures, and to monitor the airways closely[27].
Pyrethrins and Pyre-throids	Both bear an acid moiety, a central ester bond, and an alcohol moi-ety in their structure[29]. Generally, have been low vapor pressures, low Henry's law constants, and large octanol/water coeffi-cients (Kow), and are not very soluble in water[67,68].	Skin, lungs and gastroin-testinal[28].	After absorption, are rapidly distributed in the organism and undergo biotransfor-mation by hydrolysis or oxidation by P450 isozymes[30].	They can dis-rupt the mus-cular system and alter the normal functioning of voltage-dependent sodium chan-nels. This interaction shows the hyperexcited cells[31].	Tremors, spasms, incoor-dination, prostra-tion, drooling, irregular move-ments of the limbs, tonic and clonic convul-sions, and hypersen-sitivity to stimuli[32].	Decon-tamination of the skin and eyes, besides basic main-tenance of the vital functions[32].

Triazines	Permutations of the alkyl substituted 2,4-diamines of chlorotri-azine[14]. The retention in soils can varies as a function of the alkyl chain-length, such as the melting point varies between (133 – 177 °C) [69].	Skin, eyes, nose, and gastroin-testinal[16].	Undergo conjuga-tion with glutathione or dealkyl-ation[14].	Mechanism not defined[34, 35].	Irritation at the site of con-tamination. Carcino-genic and teratogenic evidenc-es[15, 33].	It is neces-sary to decontami-nate the site exposed to the sub-stance[15].
Phenoxy Deriva-tives	An aliphatic carboxylic acid moiety attached to a chloride or methyl-substituted aromatic ring[36]. It's adsorption coefficient (Koc) varied by four-fold, from 76 to 315 L kg(-1) [70] and the main com-pound has melting point is around 140 °C [71].	Gastro-intestinal and Lungs[36].	They rapidly dissociate or hydrolyze in vivo, and fat does not store them[36].	Cell membrane damage, uncoupling of oxidative phosphory-lation and and esters are irritating to the skin, eyes, and the respiratory, gastrointes-tinal, and mucous membranes[36, 28].	Nausea, dizziness, vomiting. As for metabolic acidosis, clinical signs such as hyper-thermia (due to uncoupling of oxida-tive phos-phoryla-tion), renal failure, increased aspartate amino-transferase and ala-nine and lactate dehydro-genase[36].	Mainte-nance of the vital functions, decrease the adsorption of the com-pounds[36].

Dipyridyl Deriva-tives	Are a di-pyridylium quaternary ammonium[28]. Diquat for example it is practically nonvolatile with a vapour pressure of <0,013 mPa and very soluble in water[21].	Skin, eyes, lungs, and gas-trointesti-nal[28, 38].	Their bio-transforma-tion produces free radicals, with conse-quent lipid peroxida-tion and cell injury[37].	Tissue damage in the lungs, kidney, and liver as consequence to lipid peroxida-tion[37, 38].	Dehydra-tion result-ing from vomiting. The high oxida-tive stress causes necrosis in the gastro-intestinal tract, kidney tubules, liver, and lung[38, 41].	Minimiza-tion of the absorption of the com-pound more cathartic, acceleration of excre-tion, abate-ment of the effects on the affected tissue, and fluid replacement[41, 42].
Glycine Deriva-tives	Marketed primarily as the isopro-pylamine salt (glyphosate) or ammo-nium salt (glufosinate). Although compounds contain a P=O moiety they do not inhibit AChE such as organo-phosphates[36]. Glyphosate is a relatively strong acid with a pH of 2 in 1% aque-ous solution [21].	Skin, gastroin-testinal[36].	Formation of aminometh-ylphosphonic acid (AMPA) by action of glyphosate oxidoreduc-tase[36].	DNA dam-age and uncoupling the electron transport chain [49,50].	Seizures, respira-tory arrest, coma, and distur-bance of conscious-ness and irritation upon local contact[33].	Mainte-nance of the vital func-tions[54].

| Dithio-carba-mates | Maneb and Zineb, for example, are identical in structure with exception the cátion. Maneb is moderately water soluble and stable under normal conditions, while Zineb are slightly soluble in water and unstable in light [21]. | They show slow absorp-tion by oral and dermal contact [16]. | Biotransfor-mation of dithiocarba-mates form ethylenethio-urea (ETU)[55]. | Your meta-bolic induces thyroid can-cer, modifies thyroid hor-mones and can inhibit acetaldehyde dehydroge-nase [55]. | Carcino-genic and teratogenic action and thyroid problems [55]. | There is no specific treatment for poison-ing [55]. |

Recently, some studies have reported on the nanomaterial-induced perturbation of different cell death pathways. In the majority of the cases, the key to understanding the toxicity of nanomaterials is that their smaller size as compared with cells and cellular organelles allows them to penetrate these basic biological structures and disrupt their normal function [63]. Thus, advances in research into the mechanism of action of nanopesticides will allow better prediction of the consequences of human exposure to these materials.

All these compounds are among more than 1000 active ingredients that are marketed as insecticide, herbicide, and fungicide. However, with the news pest resistance and need to hygienic controls the quantities of the formulations have been increased constantly [64].

Ass seen above, pesticides currently used over the world are numerous and have various chemical and physico-chemical properties [21]. Nevertheless, is already known that long-term contact to pesticides can harm human life and can disturb the function of different organs in the body, including nervous, endocrine, immune, reproductive, renal, cardiovascular, respiratory systems, and chronic diseases, including cancer, Parkinson, Alzheimer, multiple sclerosis, diabetes [64].

PESTICIDES AND OXIDATIVE STRESS

Interest in the toxicological aspects of oxidative stress has grown in recent years. Many researchers have focused on the mechanistic aspect of oxidative damage and cellular responses in biological systems, mainly in the case of pesticides, because oxidative stress is a condition that stems from exposure to various classes of these compounds.

Oxidative stress occurs when the rate of reactive species production exceeds the rate of reactive species decomposition in antioxidant systems, which culminates in increased oxidative damage in different cellular targets [72]. Reactive species comprise substances that do not necessarily have unpaired electrons but are very reactive due to their instability [73]. Free radicals are atoms, molecules, or ions with unpaired electrons on an otherwise open shell configuration are examples of reactive species. Their electrons are usually highly reactive [74].

Oxygen and nitrogen free radicals play an essential role in the physiological control of cell function in biological systems. Living cells continuously produce these radicals [73]. In aerobic organisms, several basic cellular metabolic processes induce production of reactive oxygen species (ROS) within cells. Cellular respiration involves reduction of molecular oxygen (O_2) to water during oxidative phosphorylation in the electron transport chain, to generate reactive, partially reduced intermediates such as the superoxide anion radical ($O2^-$), hydrogen peroxide (H_2O_2), and the hydroxyl radical ($HO\cdot$) [75]. Around 5% of these ROS originate from electron transport chain processes and can damage cellular components [76]. Moreover, several enzymes produce ROS, thereby constituting a second source of ROS synthesis in cells [77].

Regulated ROS production is higher in organisms, and maintenance of redox homeostasis is essential for the physiological health [78]. Living organisms have developed adequate enzymatic and nonenzymatic antioxidant mechanisms to protect cellular components from oxidative damage [79].

Exposure to some xenobiotics, especially toxic chemical pollutants, such as pesticides, may produce an imbalance between endogenous antioxidants and ROS, with subsequent decrease in antioxidant defenses to trigger oxidative stress in biological systems, damage to tissues, inflammation, degenerative diseases, and aging [79]. As mentioned previously, many classes of pesticides induce oxidative stress, through several different mechanisms. They may affect the redox cycle by donating electrons to or withdrawing electrons from cell components. During metabolism, they may deplete glutathione (endogenous antioxidant) or even inactivate other endogenous antioxidants [74]. In short, oxidative stress can take place either through overproduction of free radicals or alteration in antioxidant mechanisms [80]. Increased concentration of plasma and red blood cell thiobarbituric acid reactive substances (TBARs), changes in the antioxidant status, and altered activities of cellular enzymes such as superoxide dismutase (SOD) and catalase (CAT) indicated higher oxidative stress in pesticides sprayers. Hence, many researchers have associated exposure to pesticides with oxidative stress [81].

Several works have described oxidative stress induction after exposure to organophosphorus insecticides. The stress is a result of intracellular Ca^{+2} influx, which leads to cholinergic hyperactivity and activates proteolytic enzymes and nitric oxide synthase, which in turn generates free radicals [74]. Fenitrothion, a phosphorothioate, has been linked to histopathological effects on the liver and kidneys and cytotoxic effects on the lungs. These effects originate from ROS generation via pesticide metabolism by P450 or via high-energy consumption coupled with inhibition of oxidative phosphorylation [82]. Moreover, hydrocarbon insecticides chlorinated like DDT can induce oxidative stress after metabolic activation by CYP_{450} [80].

Synthetic pyrethroids are less persistent and less toxic to mammals and birds. Deltamethrin is one of the pyrethroids that has found wide acceptability. Nevertheless, this pyrethroid has effects on the nervous, respiratory, and hematological systems in fish, and it displays tumorigenicity in rodents [80]. All these effects are due to oxidative stress; they impact various antioxidants [83].

A classic example of oxidative stress induction among pesticides is the action of dipyridyls such as paraquat. This compound enters the redox cycle and constantly generates ROS such as the superoxide anion and the hydroxyl and peroxyl radicals [74]. ROS play a crucial role in the development of paraquat-induced pulmonary injury [38]. The basic mechanism of oxidative stress in this class is simple: the dipyridyl initiates a cyclic oxidation/reduction process. First, they undergo one-electron reduction by NADPH to form free radicals. The latter donate their electron to O_2, to give a superoxide radical. Upon NADPH exhaustion, to superoxides react to produce hydroxyl free radicals and other reactive species that lead to oxidative stress and consequent cell death [80]. The free radicals react with lipids in cell membranes, to start a destructive process known as lipid peroxidation. The lung is the organ that is mostly involved in this case [38]. Other compounds, like dithiocarbamates mainly inhibit antioxidant enzymes, such as SOD and catalase [84].

Pesticides can induce free radical formation by altering the way that cell organelles, like mitochondria operate. Rotenone, an insecticide of the class of rotenoids, strengthens the case for complex I inhibition – rotenone specifically binds to complex I, to inhibit the electron flow through the respiratory chain. Deficiencies in the mitochondrial respiratory chain diminish ATP synthesis and produce ROS, which culminates in oxidative stress, mitochondrial depolarization, and initiation of cell death processes [16,75].

In this context, many studies in human or animals have evidenced that pesticides exert their toxic action in the body via oxidative stress induction both upon acute and chronic exposure.

PESTICIDES AND ENDOCRINE DISRUPTION

The endocrine system refers to glands located in several areas of the body. Glands release some hormones that enter the circulation and act on specific "target" organs. If an event disrupts the endocrine system, some organs will not receive the correct amount of hormones and might not function properly or even function wrongly. In this context, low levels of some pesticides in the environment can impair the endocrine system [85].

Besides their primary action as pesticides, organophosphorus, carbamates, and organochlorines can act as endocrine disruptors and affect the function of hormones by blocking, mimicking, displacing, or acting to subvert their natural roles in living species. DDT and its metabolites are among the most famous endocrine disruptors. DDT was widely used in the 1950s and 1960s, and it is still allowed in some countries. Its proven estrogenic action can affect the reproductive system of mammals and birds [86].

In vitro and in vivo studies have shown that pyretroids also act as endocrine disruptors, but their effects only arise at relatively high levels [87]. Atrazine, a triazine herbicide, may also exert endocrine-disrupting effects on amphibians [5].

PESTICIDES AND HUMAN HEALTH

Many workers and residents, especially in the rural sector, are in contact with pesticides on a daily basis, so they are at high risk of poisoning by these compounds. This exposure can cause neuropsychiatric sequelae (mood disorders, depression, and anxiety), because many pesticides underlie changes in the function (e.g., cholinergic crisis) of the central, peripheral, and autonomic nervous system, which are often followed by suicide attempts. In addition to being causative agents of neuropsychiatric disorders that might culminate in suicide, these effects may lead to the use of pesticides as a weapon [88].

According to data released by the World Health Organization (WHO) [89], suicide by pesticides is common in many Asian and Latin American countries. Pesticides are often poorly controlled and widely available, particularly in countries of low and middle income [42]. The first epidemiological reports of suicides involving pesticides appeared in the beginning of the 1990s. Currently, homicides and suicides involving pesticides have raised the concern of many organizations and governments as, depression and suicide clearly correlate with high exposure to pesticides. This concern has motivated and still motivates many studies into how and why exposure to pesticide occurs; researchers have also caught methods to solve this serious social problem [88].

Detoxification measures after poisoning are crucial, no matter whether exposure was intentional, accidental, or occupational. Recognition of poisoning is easy when the patient knows which pesticide he/she was exposed to or when symptoms are typical. However, poisoning may be unclear if the patient has generalized symptoms. Therefore, along with the procedures to terminate contamination, an investigation with family members and the people present at the time of contamination, and information on patient care should exist. These individual will be questioned, about the way in which the patient was exposed to the contaminant and about the possibility of simultaneous intoxication with other poisons [27]. Along with these recognition steps the analytical detection of pesticides is mandatory.

Decontamination methods must be combined with care and maintenance of vital signs and administration of antidotes. It is important to bear in mind that new cases of contamination may appear. Furthermore, professionals as well as other patients staying in the same ward as the contaminated individuals must wear protective equipment until decontamination and treatment are complete [27].

Methods exist to decontaminate patients poisoned via gastrointestinal tract. Gastric lavage is extremely invasive and aggressive to the body, so it is indicated only in potentially fatal cases. The cathartic method, which elicits bowel movement to force excretion of the pesticide, is not suitable when poisoning induces diarrhea. Administration of adsorbents is an alternative – adsorbents can bind to the toxic agent, to form a stable compound. This compound is not absorbed by the gastrointestinal tract and is subsequently excreted with the feces. This method is commonly performed in conjunction with the cathartic method. The most usual adsorbent is activated charcoal, but it does not adsorb all pesticides. Finally, the syrup ipeac, a medicinal plant, can help to induce vomit. However, this procedure is contraindicated in the case of ingestion of hydrocarbons or corrosive substances [27, 28].

In the case of dermal exposure, it is necessary to start the decontamination process by placing the patient under a shower and using soap and water to remove the chemicals from the skin, hair, nails, ear canals, and other possibly contaminated body parts. If contact occurs by the ocular route, it is essential to rinse the eyes with plenty of clean water. All the materials and clothes used by the patient at the time of intoxication, like clothes and shoes, should be removed. In cases of large contamination, it is crucial to consider the need to decontaminate all the people who work in the emergency system [27, 28].

Because hundreds of pesticides compositions exist, we will focus on the clinical profile and treatment of pesticides that cause major poisoning, in terms of quantity and severity of cases. In general, treatment aims to override the mechanism of action of the toxic pesticides, and many possibilities exist (Table 2).

Table 2. Methods used to override the mechanism of toxic action of pesticides [39].

Pharmacological antagonism - competes with pesticides for the target site
Physiological antagonism - reversal of a physiological effect of the pesticide
Changing distribution to tissues – e.g., competition with membrane pumps
Modification of biochemical pathways - interferes with the biochemical response of the pesticide
Chelation of a pesticide to disable it
Treatment of pathological response to tissue injury caused by pesticides

An example of suicide attempt has been the case of a man aged 22 who tried to kill himself by drinking a solution of paraquat (50 mL). He underwent gastric lavage and received activated charcoal. Later, he was discharged. However, the treatment did not suffice – four days later, the man returned to the hospital with sore throat, dysphagia, retrosternal pain, hemoptysis, and blistering and ulceration of the mouth and tongue. Biochemical tests revealed elevated creatinine levels, leukocytosis, hyponatremia, and metabolic acidosis. Because the effect had become systemic, the patient had to undergo hemodialysis and immunosuppressive therapy (cyclophosphamide, methylprednisolone, and dexamethasone). The patient did not improve and presented hemoptysis. Examination of the thoracic region detected localized alveolar infiltrate, pulmonary opacities, pneumomediastinum, pneumothorax, and subcutaneous emphysema. The patient's condition worsened, and he -underwent the same immunosuppressive therapy again. The patient recovered gradually; he was discharged after four weeks. After four months, he was working again. His lungs did not return to perfect condition – the man still this place crackles in the lower lung fields, universally distributed wheezing and pleural friction in the right hemithorax, and dyspnea after physical exertion [40].

An example of homicide involving pesticides is the case of a 52-year-old entrepreneur that was killed by injections of poison in his abdomen, conducted by their business rivals. Soon after he was attacked, the man was taken to a private clinic to receive primary treatment, and later he was taken to a hospital, where hours later he was pronounced dead. The body was sent to the morgue for post-mortem examination. Necropsy revealed distended abdomen and two punctures by needles in this region; necrotic changes appeared in the tissue around these two holes. Analysis of the organs revealed congested and edematous brain and lungs, as well as congested stomach with hemorrhagic spots. The toxicological analysis report described the presence of organochlorine pesticides in the region of the piercings and all viscera. This suggested that the man died due to cerebral and pulmonary edema after organochlorine poisoning [90].

Apart from intentional exposure to pesticides, cases of accidental poisoning occur frequently. A Latin American man (66 years old), who had a history of diabetes mellitus (type 2), hypertension, and alcohol abuse, was admitted to the emergency department unconscious, reaching a score of 5 in the Glasgow Coma Scale; he also presented hypotension (blood pressure 87/45 mmHg), sweating, and hypoxia. On the basis of reports by his wife, she had accidentally mixed Roundap in his alcohol, and he had ingested between 350 and 500 mL of rum Roundup. About two hours after ingestion, she found him with altered mental status, non-bilious vomiting, and difficulty to wake, but he did not present bleeding. Biochemical analysis revealed high hypoxia and lactic acidosis as well as AG and high osmolar gap. First care included intubation, ventilation, and fluid bolus with 2 L of normal saline and 1 L of sodium bicarbonate. His condition worsened, and he rapidly went into shock (blood pressure 66/43 mmHg), with acute renal failure, hyperkalemia, leukocytosis, and worsening lactic acidosis. On the basis of these results, health professionals administered high dose of Levophed (Hospira, Lake Forest, Illinois) and vasopressin to provide pressure support and continuous veno-venous hemofiltration. After 24 h, the patient's conditions improved. Treatment was discontinued, and renal and cerebral functions were fully recovered [91].

Finally, cases of poisoning due to occupational exposure exist. Some pesticides can cause topical damage when they come in contact with the skin, as in the case of two farm workers admitted to the hospital in great pain due to extensive chemical burns in the perineal and scrotal regions, caused by Ducatalon (a dipyridyl herbicide containing a mixture of diquat and paraquat). The men suffered burns due to a leak in the equipment they used to spray the herbicide. Lesions reduced upon topical treatment with silver sulfadiazine associated with systemic administration of antibiotics. Fortunately, in a few days, the damaged skin recovered without scars. After replacement of the faulty equipment, no more injuries occurred [41].

PESTICIDES AND ENVIRONMENTAL HEALTH

Pesticides reach the environment primarily during preparation and application. Application can take place via different techniques, depending on factors such as the formulation type, the controlled pest and, the application timing. In agriculture, it is possible to apply pesticides to the crop or to the soil. Liquids sprays are commonly used in crops; for example, boom sprayers, tunnel sprayers, or aerial application. Systemic pesticides can also be employed. As for soils, pesticides can be applied as granules, injected as a fumigant, or sprayed onto the soil surface, which is possibly followed by pesticide incorporation into the soil top layer. Seeds are sometimes treated with pesticides prior to planting. [92].

After application, pesticides can be taken up by target organisms, degraded, or transported to the groundwater; they can also enter the surface water bodies, volatilize to atmosphere, or reach non-target organisms by ingestion, for example. The physical and chemical properties of the pesticide, soil, site conditions, and management practices influence the behavior and fate of pesticides [93].

Concerning the physical and chemical properties of pesticides, their solubility determines their transport in surface runoff and their leaching to groundwater. The higher the solubility, the greater the carrying and leaching. The partition coefficient also affects the behavior of pesticides, and many chemicals do not leach because soil particles adsorb them.

Adsorption depends on the chemical and also on the soil type. The volatility of pesticides indicates their tendency to become a gas; the higher the volatility (high vapor pressure), the larger their loss to the atmosphere. Environmental conditions such as temperature and humidity impact volatility, which can occur from soil, plants, or surface water, and may continue for several days or weeks after pesticide application. In the atmosphere, the chemicals can be transported over long distances. Subsequent atmospheric deposition can contribute to surface water pollution. Finally, the degradation of pesticides that also determines the behavior and fate of these compounds in the environment. Degradation (their brake down into other chemical forms) can occur by photodecomposition, microorganisms, and a variety of chemical and physical reactions. Pesticides with low biodegradation are called persistent, they can remain in the environment for a long time [94, 95].

Soil properties can also affect the movement of pesticides. In relation to the soil texture, coarse-textured sands and gravels have high infiltration capacities, and water tends to percolate through the soil and reach groundwater. Fine-textured soils such as clays generally have low infiltration capacities, so water tends to run off, reaching streams and lakes. Moreover, soil containing more clay in its composition bears larger surface area to adsorb pesticides. Regarding permeability, highly permeable soils allow water to more easily. This water may contain dissolved pesticides, which will reach groundwater. Texture influences soil permeability. Ultimately, soils with high organic matter content can adsorb pesticides and retain water with dissolved chemicals. Moreover, these soils possess a larger population of microorganisms that can degrade the pesticides [93, 94].

The site conditions that can determine pesticide behavior in the environment are depth until the groundwater, geological conditions, topography, and climate. In the case of shallow groundwater, the soil filters smaller amount of water with chemicals and adsorbs and degrades lower quantities of pesticides, so contamination is a major concern. Regarding the geological conditions, the presence of wells, sinkholes, and highly permeable materials, such as gravel deposits, facilitates groundwater contamination. On the other hand, the existence of drainage ditches, streams, ponds, and lakes increases the probability that rainfall or irrigation runoff will contaminate surface water. In relation to topography, flat landscapes, areas with closed drainage systems where water drains toward the center of a basin, and especially sinkhole areas, are more susceptible to groundwater contamination. As for climate, large rainfall or irrigation may culminate in large amounts of water percolating through the soil, to reach groundwater. Rainfall can also carry pesticides to surface waters, contaminating rivers, lakes, and seas, and taking these chemicals to distant places [94].

Finally, management practices can affect the movement of pesticides. With respect to the application methods, pesticides injected or incorporated into the soil are more available for leaching and reaching groundwater, whereas pesticides sprayed onto crops are more susceptible to volatilization and surface runoff, reaching surface waters and the atmosphere. Concerning the application rates and timing, the use of larger amounts of a pesticide during are rainfall or irrigation facilitates the assess of the chemical to groundwater. With respect to handling practices, correct storage and disposal of the pesticides containers impact environmental contamination [94].

The fact that a contaminant is present in the environment does not necessarily mean that it will reach an organism. The contaminant and the organism must overlap in time and space for exposure to occur. Contact can be dermal or oral or even via inhalation, gills, and, more rarely, injection [5].

Once pesticides reach non-target organisms, they may undergo biotransformation via reactions like hydrolysis, oxidation, reduction, or conjugation catalyzed by liver enzymes. Biotransformation is an effort of the organism to detoxify and eliminate xenobiotics, but this process can also produce metabolites that are more toxic than their parent compound, a phenomenon called bioactivation. An example of bioactivation is the biotransformation of DDT, which is not highly toxic to birds, into DDE, which causes thinning of eggshells because it disrupts calcium metabolism [5].

In organisms, the absorption of a pesticide with high lipid solubility and low elimination rate can lead to bioaccumulation of this chemical in the fatty tissue, and the final concentration of the chemical in the organism will be higher than its concentration in the environment [96]. When the bioaccumulated chemical passes from lower to higher trophic levels through the food chain, successively greater pesticide concentrations emerge in animals of higher trophic level. This phenomena is called biomagnification. The offspring of top predators can also become contaminated, mainly in the case of marine mammals, because they can consume milk with extremely high fat and pesticides content [5].

Application of pesticide involves not only the active ingredient but also the whole formulation. Therefore, the environment and the human are exposed to both the active and inert ingredients. Although inert ingredients have no pesticidal activity, facilitate application of the pesticides – they enhance the active compound penetration into the target organism as well as the toxic action. Hence, the inert ingredients raise the formulation toxicity even in non-target organisms [35]. One example is the formulation of glyphosate, which is an active ingredient. It contributes a little to the total toxicity of the formulated product, particularly in the case of aquatic organisms, which are more sensitive to surface-active substances [97].

The categorization of pesticides commonly relies on their persistence in the environment. Organochlorine pesticides are persistent, whereas organophosphates, carbamates, phenoxyacid derivatives, chloroacetanilides, pyrethroids, and others are non-persistent. Compared with persistent pesticides, non-persistent chemicals have much shorter environmental half-lives and do not tend to bioaccumulate. Nevertheless, because of the heavy agricultural use of these chemicals, exists concern about their presence in the environment [14].

The non-persistent pesticides organophosphorus and carbamates act on acetylcholinesterase. The presence of this enzyme in insects, birds, fish, and all mammals allows these pesticides to reach both target and non-target organisms. [98]. Pesticides such as organophosphorus and carbamates can affect numerous teleost behaviors [99]. The pesticides that inhibit acetylcholinesterase are polar and water soluble. Moreover, their metabolism in the body is fast, and their degradation in the environment is relatively rapid. Therefore, organophosphorus and carbamates do not tend to bioaccumulate in aquatic species. However, the accumulation of these compounds in fish and invertebrates was reported long ago [100].

Organophosphorus compounds do not persist in the environment. However, their large-scale use and their decomposition rates in the environment cause these compounds to accumulate in soils, from where they subsequently enter groundwater and rivers [101]. A recent study detected the organothiophosphate insecticide chlorpyrifos in air and seawater in the Arctic, which demonstrated the long-range transport of this chemical [102]. Diazinon, another organophosphorus compound, frequently occurs in point sources (wastewater treatment plant effluent) and non-point sources (storm water runoff) in urban and agricultural areas. This pesticide is extremely toxic to birds and the aquatic life [103].

Organophosphorus compounds are acutely toxic, broad-spectrum pesticides. In the environment, secondary poisoning can occur when predators consume animals poisoned by these chemicals. Examples of contamination by organophosphorus are numerous. In Argentina in 1995–1996, approximately 6000 wintering Swainson's hawks (*Buteo swainsoni*) became poisoned after they fed on grasshoppers sprayed with the organophosphorus insecticide monocrotophos [5].

An example of carbamate contamination occurred with the pesticide, aldicarb, which polluted groundwater in the United States. Other carbamates such as carbaryl and its degradation product 1-naphthol have emerged in surface waters. The metabolite 1-naphtol is more toxic than its parent compound, and it has arisen in India [104].Methomyl, carbaryl and carbofuran, commonly used carbamates, have appeared in the aquatic environment [105].Carbofuran has commonly been associated with wildlife pesticide poisoning events when applied in the granular form. Apparently, birds mistake them for seeds [5].

Organochlorines have long environmental half-lives and tend to bioaccumulate and biomagnify in organisms. A series of evaporation and deposition steps as well as migration of animals containing bioaccumulated organochlorines can transport these compounds through the environment, carrying it to animals in higher levels of the food chain. These persistent chemicals thus occur thousands of miles away from their origin [14]. The properties of organochlorines like aldrin and dieldrin result in direct mortality of predatory birds, such as sparrow hawks and kestrels [5]. These chemicals have intensive use in agricultural and industrial activities, so they emerge across the world, including the deserted plateau and the polar zone [106]. The organochlorine chlorothalonil is a fungicide that has arisen in seawater and air in the Arctic as well as in snow cores in Arctic Canada. Endolsulfan, an organochlorine insecticide, has appeared in animals from Greenland like marine fish and mammals [102].

Despite the ban on many organochlorine compounds in the 1970s, some countries still fabricate and use chemicals such as DDT to control vector disease [98]. Other countries have replaced organochlorines with the less persistent and more effective organophosphorus compounds [107].

Pyrethrins and Pyrethroids are non-persistent pesticides used worldwide as insecticides in agriculture, forestry, households, public health and stored products [108]. Therefore, urban and peri-urban populations are potentially chronically exposed to these compounds [87].

Pyrethrins and Pyrethroids act on sodium channels in the nervous system of numerous phyla, such as arthropods and chordates [87]. Pyrethrins and Pyrethroids present low acute toxicity to mammals and birds and constitute one of the safest insecticides to man. However, at low concentrations these chemicals are acutely toxic to a wide range of aquatic organisms and insects [108].

Pyrethrins are natural compounds extracted from chrysanthemum flowers; pyrethroids are synthetic compounds whose structure resembles the structure of pyrethrins [87]. Light degrades these chemicals. Modification of pyrethroids over the years has enhanced their insecticidal activity and persistence in the environment [109]. Compared with pyrethrins, pyrethroids are more stable under light [108], which incurs increased environmental risks associated with their use [5]. Pyrethrins and Pyrethroids display high selectivity and easy degradability in the environment as compared with other pesticides, been a favored replacement for organophosphorus compounds [110].

Pyrethroids strongly adsorb to soil particles, but they can move in runoff with soil particles and reach sediments, consequently entering aquatic ecosystems and affecting aquatic organisms like invertebrates and fish [108]. Fish are highly sensitive to pyrethrin and pyrethroid products, and contamination of lakes, streams, ponds, or any aquatic habitat is a concern [109]. Moreover, some formulations contain additional insecticides, insect repellents, and solvents such as alcohol and petroleum, which increase pesticide toxicity [109].

Triazines basically consist of herbicide compounds, are relatively persistent and migrate easily through the soil into surface and ground waters [111]. In soil, they undergo degradation mainly in a microbial action, but the role of photodegradation is still significant [112]. Residues of triazines have emerged in soil, surface waters, and groundwater in areas where the application of agrochemicals has taken place [111].

Herbicides are often benign with regard to impacts on animals; however, these compounds can have toxic effects at concentrations found in the environment [5]. Furthermore, indiscriminate use of this herbicide, careless handling, accidental spillage, or discharge of untreated effluents into natural water ways can harm the fish population and other aquatic organisms and may contribute to long-term effects in the environment. Atrazine, a triazine herbicide, is one of the most often detected pesticides in streams, rivers, ponds, reservoirs, and groundwater [113].

Phenoxy derivatives basically consist of compounds with herbicide action. They are soluble in water and can pollute surface and ground waters. Phenoxy derivatives display moderate toxicity, but some chlorinated metabolites can be toxic to human and aquatic organisms [114]. In addition, the metabolites may have mutagenic and carcinogenic properties. 2,4-D and MCPA, which are also phenoxy herbicides, can undergo degradation by biotic and abiotic mechanisms. However, these processes may not suffice to reduce the concentrations of chlorinated phenoxy derivatives on many sites [115].

Regarding dipyridyl derivatives, the best-known compounds are diquat and paraquat, developed as herbicides and desiccants. Diquat is water soluble and persistent in the aquatic system. However, it can bind to soil, which reduces its mobility in the environment. Although herbicides are usually little toxic to animals, diquat is toxic to some aquatic organisms [116]. Soil adsorbs paraquat, which presents its leaching to ground water; soil microorganisms and photolysis degrade this herbicide [117]. The herbicide glyophosate bears glycine, which adsorbs to soil, undergoes degradation by bacteria, and has low potential for runoff. However, is it highly water soluble and emerges in surface waters. Glyphosate is little toxic to mammals, but the surfactants present in some formulations rise the toxicity of this chemical. Hence, some formulations, mainly those intended for aquatic vegetation control, can kill amphibians [5]. Many authors have demonstrated that glyphosate formulations can cause genetic damage in fish [97].

Dithiocarbamates (DTC) function mainly as fungicides that protect crops, but they also work as rodent repellents [118]. The intensive use of dithiocarbamates in agriculture often contaminates water bodies [119]. Ziram, one of the best-known dithiocarbamates, is toxic to aquatic organisms [120].

Other examples of chemical classes of pesticides exist. Alachlor and metolachlor belong to the group of chloroacetanilides. These herbicides and their degradation products have arisen in surface and groundwater [121]. Diuron, a urea derivative, can pollute freshwaters by leaching through the soil. It has appeared in marinas and coastal areas [122]. Additionally, trifluralin, a dinitroanilin, has emerged in Arctic air and seawater [102].

Therefore, a huge amount and variety of pesticides exist in the environment. Many chemicals that exist at low concentrations may not cause acute detectable effects in organisms, but they may induce other kinds of damage, like genetic disorders and physiological alterations that, in the long run, reduce the organisms life span [11].

METHODS TO DETECT PESTICIDES

A wide range of methodologies exist to identify possible exposure to pesticides. When identification is necessary due to poisoning of a patient attended in the clinic, the general procedures include anamnesis, physical examination, evaluation of clinical signs, and diagnostic and toxicological analysis. If the investigation aims to qualify and/or quantify a possible pesticide, it is generally necessary to collect a sample and analyze it for the presence of pesticides and/ or metabolites in biological samples (blood, liver, stomach contents) and/or the environment (air, water, ground). Selection of the test will depend on the purpose of the analysis. It is also essential to consider the financial costs of a method. Simpler tests are still important, – apart from been inexpensive, many offer high sensitivity, specificity, precision, and accuracy, all of which are factors that are crucial for reliable analysis [123, 124].

Prior to analyzes pesticides samples analysts have to go through similar steps: definitions of the analytical problem (target analyte and its properties), choice of detection methods (immunoassays spectrometry), sampling (how to collect and store the sample), sample preparation (solubilization, extraction, concentration, and separation), calibration (qualification and/or quantification of the analyte), calculation and evaluation of the results, and actions to complete the analysis [125].

Sample storage for long periods should ensure that no sample degradation or external contamination occurs. Well-sealed containers stored under refrigeration and protected from light are mandatory. To avoid any type of external interference during analysis, none of the employed materials should modify or degrade the pesticide in the sample. The analysis of pesticides, mainly in water, ambient air, and soil sediments, often requires a purification step to clean the sample and pre-concentrate the analytes, to improve the quality of the analytical results. The extraction process is a key analytical step – it extracts the desirable compounds for further separation and characterization. Liquid-liquid extraction, and pre-concentration procedures, such as solid-phase extraction and solid-phase microextraction, are the most commonly used methods, but other extraction methods are also applicable depending on the objective [126]. Extraction of residues from the sample matrix demands appropriate solvents for maximum extraction efficiency and minimal co-extraction of interfering substances. The extraction solvents must be highly pure. Blank tests help to prove that the matrix does not interfere in the analyzes. After extraction, a purification step removes the interfering substance with minimal loss of the analyte. The final solution should include an appropriate solvent for analyte determination by the selected method [127].

Below is a didactic description of the main separations and detection methods.

Physicochemical Methods

Gas chromatography (GC), Liquid Chromatography (LC), and Capillary Electrophoresis (CE) constitute physicochemical separation methods.

When the analyzed pesticide is volatile or semi-volatile, GC still is the method of choice: it offers higher resolution and lower detection limits. GC is usually associated with multiple detectors whose choice will depend on the characteristics of the target analytes. GC is based on sample volatilization and introduction into a chromatographic column coated or packed with a solid or liquid stationary phase. A gaseous mobile phase elutes the analyte; this phase is inert, and does not interact with the analyte. The carrier gases should be pure and chemically inert, too, and the choice will depend on the detector. The commonest carrier gases are helium, argon, nitrogen, carbon dioxide, and hydrogen [128].

LC has emerged as a great separation tool. It allows for effective separation of nonvolatile and thermally unstable pesticides that are incompatible with GC. During LC, extracts pass through multiple adsorbent columns that can discriminate between the components of the matrix and target analyte. The degree of selectivity will vary according to the adsorbent present in the column (alumina, silica gel, or Florisil), mesh size, and activity levels. Columns can be used separately or in combination [129].

CE is a powerful tool to separate and identify a wide range of molecules. EC provides high resolution, and large separation efficiency. It requires small sample size and low solvent consumption analyzes is faster and operational coats are low [130].

An ideal detector should ensure adequate sensitivity, good stability and reproducibility, and linear response to various concentrations of the analytes. It should also operate in a wide range of temperature, have reduced response time (independent of the flow), and be easy to handle. The detector response should be equivalent for all the analytes or selective to certain classes of compounds. Ultimately, the detector should not destroy the sample. Unfortunately, a switch that exhibits all these characteristics does not exist, so it is necessary to select the detector according to the desired goal [128].

Several types of detectors are commercially available. They can come coupled to the separation device. These detectors use photometric or fluorimetric methods, thermal conductivity, diode array detection, electrons capture, atomic absorption, or pesticides mass/charge evaluation. The latter method is currently in evidence due because it is highly sensitive, offers autonomy, and performs a variety of functions. Electron capture and mass spectrometry are the most often used to detect pesticides.

The electron capture detector (ECD) is usually employed to search for organic pesticides, because it is highly sensitive and selective toward molecules containing electronegative functional groups. It also detects masses in the order of pictograms and can analyze traces of pesticides. However, ED cannot detect compounds with low electron affinity. Its excellent properties are useful for analysis of pesticides in both the environmental area and hospitals. A detector called µECD is also available in the market. It is advantageous over ECD in term of sensitivity, stability, and robustness [131].

Mass spectrometry (MS) is based on the ionization, acceleration, and separation of the generated molecules and ions according their mass/charge (m/z) ratio. This Provides a typical spectrum that gives the relative mass abundance of the different ionic species as a function of m/z so, which permits unambiguous identification of molecules. Mass spectrometry is a confirmation technique that is less subject to misunderstanding. Nevertheless, it has a drawback – it destroys the analyte [132].

As mentioned previously, the choice of method will depend on the case. LC-MS and GC-MS are the methods that generally separate and detect pesticides most suitable. These methods play a very important role in the analysis of pesticides and related compounds and are applicable in several areas like environmental analysis, food safety, and occupational toxicology, among others. Because they can serve various purposes, these methods also help to detect compounds in different samples, such as water, soil, sediment, sludge, vegetables and fruits, and animals and humans tissues and fluids [124,126]. Obviously, method will based on the needs and characteristics of the target pesticide, and each sample will have their own features, which will depend on their physicochemical properties.

Biological

Chemical analysis of isolated compounds is commonly used to monitor environmental pollution, but such analyses can be limited and expensive and cannot indicate the biological effects. In contrast, biological tests indicate the toxicity of a ride range of compounds or environmental samples, and are therefore essential to determine the environmental impacts of the presence of these chemicals [133]. Immunoassays and biosensors are methods related to the biological factor. Immunoassays are a powerful tool in clinical laboratories and one of the most widely applied analytical techniques. The reagents kits and the equipment necessary to perform immunoassays are commercially available and rely on fluorescent, chemiluminescent or other detection methods. Immunoassays can detect a wide range of compounds including drugs, proteins, and hormones; they can also identify and quantify the presence of pesticides residues in various samples such as natural water, food, and blood, among others [129].

Regarding biosensors, organisms such as *Drosophila melanogaster* fly species may aid the detection of pesticides in food samples and other matrixes such as water, soil, plants, and animal tissue. This test model is advantageous, because these insects have low tolerance to toxic substances with insecticidal character, besides being experimental models of easy creation, manipulation, and maintenance. In addition, they require few financial resources and can remain under laboratory conditions. However, this method only serves to detect the presence of pesticides, but it cannot identify the detected compound. Therefore, after using this probe, the analyst has to employ a chromatographic, for example, to identify the group of pesticides in that sample [123].

SUMMARY OF IMPORTANT POINTS AND PERSPECTIVES

The chapter begins with an introduction about pesticides, citing the Second World War and the publication of the book "Silent Spring" by Rachel Carson. Even in the introduction, it is mentioned the Integrated Pest Management (IPM) and the risks and benefits of pesticides use.

Subsequently, the chapter presents the topic "physicochemical properties and stages of intoxication." This topic cites the physicochemical properties, the exposure, toxicokinetic, toxicodynamic and clinical phase of organophosphorous, carbamates, organochlorines, pyrethrins and pyrethroids, triazines, phenoxy derivatives, dipyridyl derivatives, glycine derivatives, dithiocarbamates, and others. In the latter group, the nanopesticides are mentioned.

The chapter also discusses the pesticides as inducers of oxidative stress and endocrine disruptors action of two important issues. Beyond, adress three topics differences: pestidas and human health, pesticides and environmental health, and methods of detection of these compounds. In the first, there are examples of intoxication from occupational, accidental and intentional exposure, besides decontamination methods. The second topic shows how a pesticide reaches the environment, and how it behaves. In other words, if hits the water, soil, and / or are biodegraded. Finally, the third topic addresses methods of detection of pesticides. Gas chromatography (GC), Liquid Chromatography (LC), and Capillary Electrophoresis (CE) constitute physicochemical methods. Immunoassays and biosensors are methods related to the biological factor.

Currently, there is a pursuit of a sustainable society, generating huge concern for human health just like the environment, this occurs due to action/ persistence of pesticides in the environment, as well as its toxic effects to humans and other living beings. This pursuit for a healthier society tries to combat the toxic effects of pesticides, as they have caused a large reduction in biodiversity (mainly insects pollinators), and affect humans causing genetic mutations, Mutagenicity and carcinogenicity, reproductive damages as well as disturbances behavioral (depression and suicides). Faced with this problem, many governments have sought to measures to limit access to these compounds, aimed at protecting human and environmental health, such as work done by the governments of India, Western Samoa and Finland, which restricted access to pesticides and reduced cases of suicides in their countries [42, 134].

This concern can also be viewed on the growing interest of researchers and regulatory agencies regarding research related to biopesticides and biological control of pests, also seeking the quality of environmental and human health mainly in the near future [135].

REFERENCES

1. USEPA - United States of Environmental Protection Agengy. About pesticides. U.S. EPA. http://www.epa.gov/pesticides/about/index.htm (Accessed 27 August 2014).

2. Unsworth 2010 - History of pesticide use. International Union of pure and applied chemistry (IUPAC). 2010. http://agrochemicals.iupac.org/ index.php?option=com_sobi2&sobi2Task=sobi2Details&catid=3&sobi 2Id=31 (Accessed 2 September 2014).

3. Gupta PK. Toxicity of herbicides. In: GUPTA, R. C. (Ed) Veterinary toxicology: Basic and clinic principles. USA: Elsevier. 2007. p.567-586.

4. Ratcliffe D. Decreases in eggshell weight in certain birds of prey. Nature 1967; 215: 208-210.

5. Levengood JM, Beasley VR. Principles of ecotoxicology. In: Gupta RC (ed.) Veterinary toxicology: basic and clinical principles. Academic press, Amterdan. 2007. p. 689-708.

6. Fenik J, Tankiewicz M, Biziuk M. Properties and determination of pesticides in fruits and vegetables. Trends in Analytical Chemistry 2011; 30 [6]: 814-826.

7. Knutson RD, Taylor CR, Penson JB, Smith EG. Economic impacts of

reduced chemical use. Knutson and Associates, College Station, Texas. 1990.

8.   Delaplane K S. Pesticide Usage in the United States: History, Benefits, Risks, and Trends. The University of Georgia, Athens, Georgia 1996. http://ipm.ncsu.edu/safety/factsheets/pestuse.pdf (Accessed 20 August 2014).

9.   Mostafalou S, Abdollahi M. Pesticides and human chronic diseases: Evidences, mechanisms, and perspectives. Toxicology and Applied Pharmacology 2013; 268: 157-177.

10. Hernández AF, Parrón T, Tsatsakis AM, Requena M, Alarcón R, López-Guarnido O. Toxic effects of pesticide mixtures at a molecular level: Their relevance to human Health. Toxicology 2013; 307: 136-145.

11. Poletta GL, Larriera A, Kleinsorge E, Mudry MD. Genotoxicity of the herbicide formulation Roundup® (glyphosate) in broad-snouted caiman (Caiman latirostris) evidenced by the Comet assay and the Micronucleus test. Mutation Research 2009; 672: 95-102.

12. EPA - Environmental Protection Agency. Overview of the Ecological Risk Assessment Processin the Office of Pesticide Programs,U.S., Environmental Protection Agency. 2004. http://www.epa.gov/espp/consultation/ecorisk-overview.pdf (Acessed 13 August 2014).

13. Matthews G, Bateman R, Miller P. Book Pesticide application methods. John Wiley & Sons. 2008. Available from http://onlinelibrary.wiley.com/book/10.1002/9780470760130 (Accessed 4 September 2014).

14. Barr D B, Needham LL. Analytical methods for biological monitoring of exposure to pesticides: a review. Journal of Chromatography B 2002; 778: 5-29.

15. Ellenhorn MJ, Barceloux DG. Medical Toxicology. Diagnosis and Treatment of Human Poisoning. Ed. Elsevier, New York, 1988. 1512p.

16. Costa LG. Toxic Effects of Pesticides. In: Klaassen CD. (ed). Cassarett and Doull's Toxicology *The Basic Science of Poisons*. 7 Ed, McGraw-Hill, New York, 2008. p.883-930.

17. Davies JE, Barquet A, Freed VH, Haque R, Morgade C, Sonneborn RE, Vaclavek C. Human pesticide poisonings by a fat soluble organophosphate insecticide. Archives of Environmental Health 1975; 30: 608-613.

18. Ho IK, Fernando JC, Sivam SP, Hoskins B. Roles of Dopamine and GABA in neurotoxicity of organophosphorous cholinesterase inhibitors. Proceedings of the Western Pharmacoly Society 1984; 27: 177-180.

19. Paudyal BP. Organophosphorus poisoning. Journal of Nepal Medical

Association 2008; 47(172): 251-258.

20. Thiermann H, Worek F, Kehe K. Limitations and challenges in treatment of acute chemical warfare agent poisoning. Chemico Biological Interactions 2013; 206[3]: 435-43.

21. Ecobichon DJ. Carbamate insecticides. In: Krieger R (ed). Handbook of Pesticide Toxicology. San Diego, CA: Academic Press; 2001. p.1087-1106.

22. Jokanovic M. Medical treatment of acute poisoning with organophosphorus and carbamate pesticides, Toxicology Letters 2009; 190: 107-115.

23. Marrs TC, Maynard RL. Neurotransmission system as targets for toxicants: a review. Cell Biology and Toxicology 2013; 29: 381-396.

24. Rosman Y, Makarovsky I, Bentur Y, Shrot S, Dushnistky T, Krivoy A. Carbamate poisoning: treatment recommendations in the setting of a mass casualties event. American Journal of Emergence Medicine 2009; 27[9]:1117-1124.

25. Lee DH, Lee IK, Song K, Steffes M, Toscano W, Baker BA, Jacobs DR. A strong dose-response relation between serum concentrations of persistent organic pollutants and diabetes. Diabetes Care 2006; 29: 1638-1644.

26. Snedeker S. Pesticides and breast cancer risk: a review of DDT, DDE and dieldrin. Environmental Health Perspectives 2001; 109 (suppl 1): 35-47.

27. Simpson WM Jr, Schuman SH. Recognition and management of acute pesticide poisoning. American Family Physician. 2002; 65: 1599-1604.

28. Reigart JR, Roberts JR. Recognition and Management of Pesticide Poisonings. 5ª Ed. United States Environmental Protection Agency, 1999. http://www.epa.gov/oppfead1/safety/healthcare/handbook/handbook.pdf (Accessed 4 September 2014).

29. Zhang Q, Zhang W, Wang X, Li P. Immunoassay Development for the Class-Specific Assay for Types I and II Pyrethroid Insecticides in Water Samples. Molecules 2010; 15: 164-177.

30. Sogorb MA, Vilanova E. Enzymes involved in the detoxification of organophosphorus, carbamate and pyrethroid insecticides through hydrolysis. Toxicology Letters 2002; 128: 215-228.

31. Ray DE, Fry JR. A reassessment of the neurotoxicity of pyrethroid insecticides. Pharmacology & Therapeutics 2006; 111[1]: 174-193.

32. Bradberry SM, Cage SA, Proudfoot AT, Vale JA. Poisoning due to Pyrethroids. Toxicology Reviews 2005; 24 [2]: 93-106.

33. Sathiakumar N, MacLennan PA, Mandel J, Delzell E. A review of epidemiologic studies of triazine herbicides and cancer. Critical Review

in Toxicology 2011; 41: 1-34.

34. Mayhew DA, Taylor GD, Smith SH, Banas DA. Twenty-four month combined chronic oral toxicity and oncogencity study in rats utilizing atrazine technical grade. Lab Study No.: 410-1102, Accession No. 262714-262727, American Biogenics Corp., Decatur, 1986. p. 2-6.

35. Zeljezic D, Garaj-Vrhovac, V, Perkovic P. Evaluation of DNA damage induced by atrazine and atrazine-based herbicide in human lymphocytes in vitro using a comet and DNA diffusion assay. Toxicology in vitro 2006; 20: 923-935.

36. Bradberry SM, Proudfoot AT, Vale A. Glyphosate Poisoning. Toxicology Reviews 2004; 23: 159-167.

37. Honore P, Hantson P, Fauville JP, Peeters, A. Manieu, P. Paraquat poisoning: State of the art. Acta Clinica Belgica 1994; 49:220-8.

38. Toygar M, Aydin I, Agilli M, Aydin FN, Oztosun M, Gul H, Macit E, Karslioglu Y, Topal T, Uysal B, Honca M. The relation between oxidative stress, inflammation, and neopterin in the paraquat-induced lung toxicity. Human and Experimental Toxicity, 2014. In press. Ver volume ou colocar link.

39. Bateman DN. Pharmacological treatments of paraquat poisoning. Human Toxicology 1987; 6[1]: 57-62.

40. Neves FF, Sousa RB, Pazin-Filho A, Cupo P, Elias Júnior J, Nogueira-Barbosa MH. Severe paraquat poisoning: clinical and radiological findings in a survivor. Jornal Brasileiro de Pneumologia 2010; 36[4]: 513-516.

41. Ronnen M, Klin B, Suster S. Mixed diquat/paraquat-induced burns. International Journal of Dermatology 1995; 34[1]: 23-5.

42. Sarchiapone M, Mandelli L, Iosue M, Andrisano C, Roy A. Controlling access to suicide means. International Journal of Environmental Research and Public Health. 2011; 8: 4550-4562.

43. Kwiatkowska M, Nowacha-Krukowska H, Bukowska B. The effects of glyphosate, its metabolites and impurities on erythrocyte acetylcholinesterase activity. Environmental Toxicology and Pharmacology 2014; 37: 1101-1108.

44. Campbell AW. Glyphosate: its effects on humans. Alternatives Therapies in Health and Medicine 2014; 20: 9-11.

45. de Liz Oliveira Cavalli VL, Cattani D, Heinz Rieg CE, Pierozan P, Zanatta L, Benedetti Parisotto E, Wilhelm Filho D, Mena Barreto Silva FR, Pessoa-Pureur R, Zamoner A. Roundup disrupted male reproductive

functions by triggering calcium-mediated cell death in rat testis and Sertoli cells. Free Radicals Biology and Medicine 2013; 65: 335-346.

46. Williams GM, Kroes R, Munro IC. Safety evaluation and risk assessment of the herbicide Roundup and its active ingredient, glyphosate, for humans. Regulatory Toxicology and Pharmacology 2000; 31: 117-65.

47. Chan PC, Mahler JF. NTP technical report on the toxicity studies of glyphosate (CAS no. 1071-83-6) administered in dosed feed to F344/N rats and B6C3F1 mice. Toxicity Reports Series 1992; 16: 1-D3.

48. Sandrini JZ, Rola RC, Lopes FM, Buffon HF, Freitas MM, Martins Cde M, da Rosa CE. Effects of glyphosate on cholinesterase activity of the mussel Perna perna and the fish Danio rerio and Jenynsia multidentata: in vitro studies. Aquatic Toxicology. 2013; 130-131:171-173.

49. Koller VJ, Furhacker M, Nersesyan A, Mišik M, Eisenbauer M, Knasmueller S. Cytotoxic and DNA-damaging properties of glyphosate and Roundup in humanderived buccal epithelial cells. Archieves of Toxicology 2012; 86[5]:805-813.

50. Alonzo HGA, Corrêa CL. Praguicidas. In: Oga S, Camargo MMA, Batistuzzo JA. (eds) Fundamentos de Toxicologia. 4ª. Ed. Editora Atheneu, São Paulo; 2014. p.323-341.

51. Inoue Y, Onodera M, Fujita T, Fujino Y, Kikuchi S, Endo S. Factors associated with severe effects following acute glufosinate poisoning. Clinical Toxicology 2013; 51: 846-849.

52. Park JS, Kwak SJ, Gil HW, Kim SY, Hong SY. Glufosinate Herbicide Intoxication Causing Unconsciousness, Convulsion, and 6th Cranial Nerve Palsy. Journal of Korean Medical Science 2013; 28: 1687-1689.

53. Hori Y, Tanaka T, Fujisawa M, Shimada K. Toxicokinetics of DL-glufosinate enantiomer in human BASTA poisoning. Biological and Pharmaceutical Bulletin 2003; 26: 540-543.

54. Sampogna RV, Cunard R. Roundup intoxication and a rationale for treatment. Clinical Nephrolpgy 2007; 68[3]: 190-6.

55. Belpoggi F, Soffritti M, Guarino M, Lambertini L, Cevolani D, Maltoni C. Results of Long-Term Experimental Studies on the Carcinogenicity of Ethylene-bis- Dithiocarbamate (Mancozeb) in Rats. Annals of the New York Academy of Sciences 2006; 982: 123-136.

56. Dearfield KL, McCarroll NE, Protzel A, Stack HF, Jackson MA, Waters M.D. A survey of EPA/OPP and open literature on selected pesticide chemicals. II. Mutagenicity and carcinogenicity of selected chloroacetanilides and related compounds. Mutatation Research 1999;

443: 183-221.

57. Ashby J, Kier L, Wilson AG, Green T, Lefevre PA, Tinwell H, Willis GA, Heydens WF, Clapp MJ. Evaluation of the potential carcinogenicity and genetic toxicity to humans of the herbicide acetochlor. Human & Experimental Toxicology 1996; 15: 702-735.

58. Heydens WF, Lamb IC, Wilson AGE. Chloroacetanilides. In: Krieger R (ed). Handbook of Pesticide Toxicology. San Diego, CA: Academic Press; 2001. p. 1543–1558.

59. Ermler S, Scholze M, Kortenkamp A. Seven benzimidazole pesticides combined at sub-threshold levels induce micronuclei in vitro. Mutagenesis 2013; 28: 417-426.

60. Kookana RS, Boxal AB, Reeves PT, Ashauer R, Chaudhry Q, Cornelis G, Fernandes TF, Gan J, Kah M, Lynch I, Ranville J, Sinclair C, Spurgeon D, Tiede K, Van Den Brink PJ. Nanopesticides: guiding principles for regulatory evaluation of environmental risks. Journal of Agriculture and Food Chemistry 2014; 62: 4227-4240.

61. Bergeson LL. Nanosilver: US EPA's pesticide office considers how best to proceed. Environmental Quallity Management 2010; 19 [3]: 79–85. http://onlinelibrary.wiley.com/doi/10.1002/tqem.20255/pdf (Accessed 2 September 2014).

62. Sekhon BS. Nanotechnology in agri-food production: an overview. Nanotechnology, Science and Applications 2014; 7: 31-53.

63. Buzea C, Pacheco II, Robbie K. Nanomaterials and nanoparticles: sources and toxicity. Biointerphases. 2007; 2: 17-71.

64. Mostafalou S, Abdollahi M. Pesticides and human chronic diseases: Evidences, mechanisms and perspectives. Toxicology and Applied Pharmacology. 2013; 268: 157-177.

65. Gallo MA, Lawryk NJ. Organophosphorus pesticides. In: Hayes WJ Jr.; Laws ER Jr. (Eds) Handbook of Pesticide Toxicology. San Diego: Academic Press. 1991. 2: 917-1123.

66. Shen, L.; Wania, F. Compilation, Evaluation, and Selection of Physical-Chemical Property Data for Organochlorine Pesticides. J. Chem. Eng. Data, 50, p. 742 – 768, 2005.

67. Bjorling-Poulse M, Andersen HR, Grandjean P. Potential developmental neurotoxicity of pesticides used in Europe. Environmental Health. 2008; 7-50.

68. MacBean C. The Pesticide Manual. 6a. Ed. British Crop Production Council. 2012. 557p.

69. Jokanovic M. The Impact of Pesticides. 1a Ed. The Academy Publish, Cheyenne. 2012. 417p.

70. Farenhorst A, Saiyed IM, Goh TB, McQueen P. The important characteristics of soil organic matter affecting 2,4-dichlorophenoxyacetic acid sorption along a catenary sequence. Journal of Environmental Science and Health, Part B. 2010; 45:204-13.

71. Mackay D, Shiu WY, Ma KC, Lee SC. Handbook of Physical-Chemical Properties and Environmental Fate for Organic Chemicals. 2. Ed. Taylor & Francis Group, New York. 2006. 3201p.

72. Almeida EA, Bainyb ACD, Dafrec AL, Gomesa OF, Medeirosa MHG, di Mascioa P. Oxidative stress in digestive gland and gill of the brown mussel Perna perna exposed to air and re-submersed. Journal of Experimental Marine Biology and Ecology. 2005; 318: 21–30.

73. Halliwell B, Gutteridge JMC. Free Radicals in Biology and Medicine, 3ª ed. Oxford University Press, Oxford. 1999.

74. Lushchak VI. Environmentally induced oxidative stress in aquatic animals. Aquatic Toxicology 2011; 101: 13-30.

75. Pereira LC, Souza AO, Pazin M. Dorta, DJ. A mitocôndria como alvo para avaliação da toxicidade de xenobióticos, Revista Brasileira de Toxicologia 2012; 25: 1-14.

76. Kelly SA, Havrilla CM, Brady TC, Abramo KH, Levin ED. Oxidative Stress in Toxicology: Established Mammalian and Emerging Piscine Model Systems. Research Reviews 1998; 106: 375-384.

77. Griffiths HR, Dias IHK, Willetts RS, Devitt A. Redox regulation of protein in plasma. Redox Biology 2014; 2: 430-435.

78. Wallace KB. Target organ toxicology series. Taylor & Francis Ltd, Washington 442p. 1997.

79. Valavanidis A, Vlahogianni T, Dassenakis M, Scoullos M. Molecular biomarkers of oxidative stress in aquatic organisms in relation to toxic environmental pollutants. Ecotoxicology and Environmental Safety 2006; 64: 178-189.

80. Abdollahi M, Ranjbar A, Shadnia S, Nikfar S, Rezaie A. Pesticides and oxidative stress: a review. Medical Science Monitor 2004;10:141-147.

81. Wafa T, Nadia K, Amel N, Ikbal C, Insaf T, Asma K, Hedi MA, Mohamed H. Oxidative stress, hematological and biochemical alterations in farmers exposed to pesticides. Journal of Environmental Science and Health, Part B 2013; 48: 1058-1069.

82. Milatovic D, Gupta RC, Aschner M. Anticholinesterase toxicity, oxidative

stress. Science World Journal 2006; 6:295-310.

83.   Sayeed I, Parvez S, Pandey S, Bin-Hafeez B, Haque R, Raisuddin S. Oxidative stress biomarkers of exposure to deltamethrin in freshwater fish, Channa punctatus Bloch. Ecotoxicological and Environmental Safety 2003; 56: 295-301.

84.   Hai DQ, Varga SI, Matkovics B. Effects of diethyl-dithiocarbamate on antioxidant system in carp tissue. Acta Biologica Hungarica 1997; 48: 1-8.

85.   Bergman A, Heindel JJ, Jobling S, Kidd KA, Zoeller T. State of the Science of Endocrine Disrupting Chemicals – 2012. United Nations Environment Programme and the World Health Organization 2013. http://www.who.int/ceh/publications/endocrine/en (Accessed 3 September 2014).

86.   USEPA - United States of Environmental Protection Agengy. Special Report on Environmental Endocrine Disruption: An Effects Assessment and Analisys, U.S. Environmental Protection Agency, Report No. EPA/630/R-96/012, Washington D. C, 1997. http://www.epa.gov/raf/publications/pdfs/ENDOCRINE.PDF (Accessed 27 August 2014).

87.   Fortin MC, Bouchard M, Carrier G, Dumas P. Biological monitoring of exposure to pyrethrins and pyrethroids in a metropolitan population of the Province of Quebec, Canada. Environmental Research 2008; 107: 343-350.

88.   Freire C, Koifman S. Pesticides, depression and suicide: a systematic review of the epidemiological evidence. International Journal of Hygiene and Environmental Health 2013; 216[4]:445-460.

89.   WHO - World Health Organization. Suicide prevention (SUPRE). http://www.who.int/mental_health/prevention/suicide/suicideprevent/en/ (Accessed 16 August 2014).

90.   Kumar A, Kumar A, Murty OP, Gupta VP, Das S. A rare case of homicidal insecticide (organochloro compound) poisoning by intraperitoneal injection. Medicine, Science and the Law. 2012; 52[4]: 231-233

91.   Hour BT, Belen C, Zar T, Lien YH. Herbicide roundup intoxication: successful treatment with continuous renal replacement therapy. American Journal of Medicine 2012; 125[8]: e1-2.

92.   Van Der Berg F, Kubiak R, Benjey WG, Majewski MS, Yates SR, Reeves GL, Smelt JH. Emission of pesticides into the air. Water, Air & Soil Pollution. 1999; 115: 195-210.

93.   Lourencetti C, Marchi MRR, Ribeiro, ML. Determination of sugar cane herbicides in soil and soil treated with sugar cane vinasse by solid-phase

extraction and HPLC-UV. Talanta 2008; 77: 701-709.

94. Agriculture and Natural Resources. Water quality: Controlling Nonpoint Source (NPS) Pollution. Pesticide Management To Protect Water Quality. Understanding Pesticides And How They Affect Water Quality 1999. Available from http://www.aces.edu/pubs/docs/A/ANR-0790/ WQ4.5.1.pdf (accessed 2 September 2014).

95. Bedos C, Cellier P, Calvet R, Barriuso E, Gabrielle B. Mass transfer of pesticides into the atmosphere by volatilization from soils and plants: overview. Agronomie 2002; 22: 21–33.

96. Akcha F, Spagnol C, Rouxel J. Genotoxicity of diuron and glyphosate in oyster spermatozoa and embryos. Aquatic Toxicology 2012; 106-107: 104-113.

97. ÇAVAS TC, KONEN S. Detection of cytogenetic and DNA damage in peripheral erythrocytes of goldfish (Carassius auratus) exposed to a glyphosate formulation using the micronucleus test and the comet assay. Mutagenesis 2007; 22: 263-268.

98. Van Dyk JS, Pletschke B. Review on the use of enzymes for the detection of organochlorine, organophosphate and carbamate pesticides in the environment. Chemosphere 2011; 82: 291-307.

99. Jarrard HE; Delaney KR; Kennedy CJ. Impacts of carbamate pesticides on olfactory neurophysiology and cholinesterase activity in coho salmon (Oncorhynchus kisutch). Aquatic Toxicology 2004; 69: 133-148.

100. Kitamura S, Sugihara K, Fujimoto N. Endocrine Disruption by Organophosphate and Carbamate Pesticides. In: Gupta RC (ed.) Toxicoly of Organophosphonate and Carbamate Compounds. Elsevier Academic Press, New York, 2006. p. 481-194.

101. Sirotkina M, Lyagin I, Efremenko E. Hydrolysis of organophosphorus pesticides in soil: New opportunities with ecocompatible immobilized His6-OPH. International Biodeterioration & Biodegradation 2012; 68: 18-23.

102. Vorkamp K, Rigét FF. A review of new and current-use contaminants in the Arctic environment: Evidence of long-range transport and indications of bioaccumulation. Chemosphere 2014; 111: 379-395.

103. Durmaz H, Sevgiler Y, Üner N. Tissue-speciWc antioxidative and neurotoxic responses to diazinon in Oreochromis niloticus. Pesticide Biochemistry and Physiology 2006; 84: 215–226.

104. Llasera MPG, González MB. Presence of carbamate pesticides in Environmental waters from the northwest of mexico: determination by

liquid chromatography. Water Research 2001; 35 [8]: 1933-1940.

105. Tien CJ, Lin MC, Chiu WH, Chen CS. Biodegradation of carbamate pesticides by natural river biofilms in different seasons and their effects on biofilm community structure. Environmental Pollution 2013; 179: 95-104.

106. Li Y, Niu J, Shen J, Zhang C, Wang Z, He T. Spatial and seasonal distribution of organochlorine pesticides in the sediments of the Yangtze Estuary. Chemosphere 2014; 114: 233-240.

107. Singh B.K, Walker A. Microbial degradation of organophosphorus compounds. FEMS Microbiol Reviews 2006; 30: 428-471.

108. Pérez-Fernández V, Garcia MA, Marina ML. Characteristics and enantiomeric analysis of chiral pyrethroids. Journal of Chromatography A 2010; 1217: 968–989.

109. Anadón A, Martínez-Larrañaga MR, Martínez MA. Use and abuse of pyrethrins and synthetic pyrethroids in veterinary medicine. The Veterinary Journal 2009; 182: 7-20.

110. Albaseer SS, Nageswara Rao R, Swamy YV, Mukkanti K. Analytical artifacts, sample handling and preservation methods of environmental samples of synthetic pyrethroids. Trends in Analytical Chemistry 2011; 30 [11]: 1771-1780.

111. Cai Z, Wang D, Ma WT. Gas chromatography/ion trap mass spectrometry applied for the analysis of triazine herbicides in environmental waters by an isotope dilution technique. Analytica Chimica Acta 2004; 503: 263–270.

112. Fenoll J, Vela N, Garrido I, Pérez-Lucas G, Navarro S. Abatement of spinosad and indoxacarb residues in pure water by photocatalytic treatment using binary and ternary oxides of Zn and Ti. Environmental science and pollution research international. 2014; *In press.*

113. Nwani CD, Nagpure NS, Kumar R, Kushwaha B, Kumar P, Lkra WS. Mutagenic and genotoxic assessment of atrazine-based herbicide to freshwater fish Channa punctatus (Bloch) using micronucleus test and single cell gel electrophoresis. Environmental toxicology and pharmacology 2011; 31 [2]: 314-322.

114. Cserháti T, Forgács E. Phenoxyacetic acids: separation and quantitative determination. Journal of Chromatography B 1998; 717: 157-178.

115. Grabinska-Sotaa E, Wisniowska E, Kalka J. Toxicity of selected synthetic auxines—2,4-D and MCPA derivatives to broad-leaved and cereal plants. Crop Protection 2003; 22: 355-360.

116. Peterson HG, Boutin C, Freemark KE, Martin PA. Toxicity of hexazinone and diquat to green algae, diatoms, cyanobacteria and duckweed. Aquatic Toxicology 1997; 39: 111-134.

117. Roberts T R, Dyson JS, Lane MCG. Deactivation of the Biological Activity of Paraquat in the Soil Environment: a Review of Long-Term Environmental Fate. Journal of Agriculture and Food Chemestry 2002; 50: 3623-3631.

118. Padhye LP, Kim JH, Huang CH. Oxidation of dithiocarbamates to yield N-nitrosamines by water disinfection oxidants. Water research 2013; 47: 725-736.

119. Kubrak OI, Atamaniuk TM, Husak VV, Drohomyretska I Z, Storey JM, Storey KB, Lushchak V. Oxidative stress responses in blood and gills of Carassius auratus exposed to the mancozeb-containing carbamate fungicide Tattoo. Ecotoxicology and Environmental Safety 2012; 85: 37-43

120. Van Wezel AP, Van Vlaardingen P. Environmental risk limits for antifouling substances. Aquatic Toxicology 2004; 66: 427-444.

121. Osano O, Admiraal W, Klamerc HJC, Pastor D, Bleeker EAJ. Comparative toxic and genotoxic effects of chloroacetanilides, formamidines and their degradation products on Vibrio fischeri and Chironomus riparius. Environmental Pollution 2002; 119: 195-202.

122. Abass K, Reponen P, Turpeinen M, Jalonen J, Pelkonen O. Characterization of diuron N-demethylation by mammalian hepatic microsomes and cDNA-expressed human cytochrome P450 enzymes. Drug Metabolism and Disposition. 2007; 35: 1634-1641.

123. Narciso ES, Nakagawa LE. Análise de praguicidas por bioensaio com mosca drosophila melanogaster e cromatografia em camada delgada. Arquivos do Instituto Biológico 2009; 76[2]: 313-316.

124. Niessen WM. Fragmentation of toxicologically relevant drugs in negative-ion liquid chromatography-tandem mass spectrometry. Mass Spectrometry Reviews 2012; 31[6]: 626-65.

125. Wen Y, Fu Z, Xu J, Tang S, Wang Q, Li H, Xie G, Zhu Y, Gu Y, Tan F. Determination of 2, 4-dichlorophenoxyacetic acid in air of workplace by high-performance liquid chromatography. Zhonghua Lao Dong Wei Sheng Zhi Ye Bing Za Zhi 2014; 32[6]: 458-459.

126. Martins JG, Amaya Chávez A, Waliszewski SM, Colín Cruz A, García Fabila MM. Extraction and clean-up methods for organochlorine pesticides determination in milk. Chemosphere 2013; 92[3]: 233-46.

127. Santaladchaiyakit Y, Srijaranai S, Burakham R. Methodological aspects of sample preparation for the determination of carbamate residues: a review. Journal of Separation Science 2012; 35[18]: 2373-2389.

128. Skoog D, Leary J, Principles of Instrumental Analysis, 4th Edition, Saunders College Publishing. 1992.

129. Andreu V, Picó Y. Determination of currently used pesticides in biota. Analytical and Bioanalytical Chemistry 2012; 404[9]:2659-81.

130. Assunção NA, Bechara EJH, Simionato AVC, Tavares MFG, Carrilho E. Capillary electrophoresis coupled to mass spectrometry (CE-MS): twenty years of development. Química Nova. 2008; 31: 2124-2133

131. Poole CF. Derivatization reactions for use with the electron-capture detector. Journal of Chromatografy A 2013; 1296: 15-24.

132. Di Stefano V, Avellone G, Bongiorno D, Cunsolo V, Muccilli V, Sforza S, Dossena A, Drahos L, Vékey K. Applications of liquid chromatography-mass spectrometry for food analysis. Journal of Chromatography A 2012; 1259: 74-85.

133. Oliveira, G.A.R.; Lapuente, J.; Leme, D.M.; Ferraz, E.R.A, Meireles, G.; Oliveira, D.P. New paradigms for the environmental assessment: an ecotoxicological and genetic approach. 2012. In: Advances in Environmental Research. Nova Science Publishers.

134. Hernke MT, Podein RJ. Sustainability, health and precautionary perspectives on lawn pesticides, and alternatives. Ecohealth. 2011; 8:223-32.

135. Czaja K, Góralczyk K, Struciński P, Hernik A, Korcz W, Minorczyk M, Lyczewska M, Ludwicki JK. Biopesticides - towards increased consumer safety in the European Union. Pest Manag Sci. 2014; In press.

Chapter 3

PESTICIDES AND HUMAN HEALTH

B. Alewu and C. Nosiri

Shehu Idris College of Health Science and Technology, Makarfi, Nigeria

INTRODUCTION

Pesticides are those chemicals that are used to destroy unwanted forms of life or organisms. They include insecticides, rodenticides, herbicides, fungicides, fumigants etc. They are expected to have a selective action or toxicity to animals and man. This has made the government of many countries to introduce legislation which prevent the use of the more dangerous and persistent chemicals as pesticides and to demand withdrawal from the market any licensed chemical found to be harmful to animals or man.

In 1977, it was estimated that every year 20,000 fatalities result from the use of pesticides. Most of these occurred in developing countries (Forget, 1991). The 1981 estimate by OXFAM gave a figure of 40,000 fatalities from about two million cases of poison per year (Akubue, 1997).

Pesticides are those substances which, on entering the body by whatever route, e.g. ingestion, inhalation, or absorption through intact skin, produce harmful effects. The effect may be in the form of damage to the tissues or as a disturbance of the functioning of the body.

According to the project of cooperative extension officers of cornel, University, Michigan state university, the science of toxicology is based on the principle that there is a relationship between a toxic reaction (response) and the amount of the poison received (the dose). An important assumption in this relationship is that there is almost always a dose below which no response occurs or can be measured. A second assumption is that once a maximum response is reached any further increases in the dose will not result in any increased effect. In a particular instance, a dose – response relationship does not hold true in regard to true allergic relations. Allergic reactions are special kind of changes in the immune system; they are not really toxic responses. The difference between allergies and toxic reactions is that a toxic effect is directly, the result of the toxic chemical acting on the cells. Allergic responses are the result of a chemical stimulating the body to release natural chemicals which are in turn directly responsible for the effects seen. Thus in allergic reaction, the chemical acts as a trigger, not as bullet.

For all types of toxicity of chemicals to humans, knowing the dose response relationship is a necessary part of understanding the cause and effect relationship between chemical exposure and illness. As Paracelsus in the 16th century once wrote, "The right dose differentiates a poison from a remedy". Note that the toxicity of a chemical is an inherent quality of the chemical and cannot be changed without changing the chemical to another form. The toxic effects on humans are related to the amount of exposure.

Measure of Exposure

Exposure to poisons can be intentional or unintentional. The effects vary with the amount of exposure. Contamination of food or water with varying doses of chemicals can be obtained each time when contaminated food and drink are taken. Some commonly used measures for expressing levels of contaminants are; Parts per million (PPM), which in metric equivalent is in milligrams per kilogram (mg/kg). This is approximately the amount in the water as one teaspoon per 1,000 gallons.

Parts per billion (PPB), which in metric equivalent is in micrograms per kilogram (μg/kg). This is approximately the amount in water as one teaspoon per 1,000,000 gallons.

However, as an example, individual's sensitivity to alcohol varies, as do individual sensitivity to other poisons. In testing the effects of poison to human health, lower animals are used.

In one particular measure of effects e.g. ED_{50} which means effective dose for 50 percent of animal tested. The ED_{50} of any poison varies depending on the effect measured. In general, the less severe the effect measure, the lower the ED_{50} for that particular effect. Obviously, poisons are not tested in humans in such a fashion. Instead animals are used to predict the toxicity that may occur in humans.

One of the most commonly used measures of toxicity is the LD_{50}. The LD_{50} (the lethal dose for 50 percent of the animals tested) of a poison is usually expressed in milligrams of chemical per kilogram of body weight (mg/kg). A chemical with a small LD50 (like 5mg/kg) is highly toxic. A chemical with a large LD_{50} (1,000 to 5,000 mg/kg) is practically non-toxic. The LD_{50} says nothing about non-lethal effects though. A chemical may have a large LD_{50}, but may produce illness at very small exposure levels. It is incorrect to say that chemicals with small LD_{50}'s are more dangerous than chemicals with large LD_{50}'s. They are simply more toxic. The danger or risk of adverse effect of chemicals is mostly determined by how they are used, not by the inherent toxicity of the chemical itself.

The LD_{50} of different poisons may easily be compared; however, it is always necessary to know which species was used for the tests, the age, sex and how the poison was administered (the route of exposure), since the LD_{50} of a poison may vary considerably based on these factors. Some pesticides (poisons) may be extremely toxic if swallowed (oral exposure) and not very toxic at all if splashed on the skin (dermal exposure). If the oral LD_{50} of a poison were 10mg/kg, 50 percent of the animals that swallowed 10mg/kg would be expected to die and 50 percent to live.

The potency of a pesticide is a measure of its strength compared to other poisons. The more potent the poison, the less it takes to kill; the less potent the pesticide, the more it take to kill. The potencies of pesticides are often compared using signal words or categories as 'Danger-poison' (skull and cross bones) – highly toxic.

Moderately toxic – WARNING

Slightly toxic – CAUTION

Practically non – toxic – none required.

Toxicity assessment is quite complex, many factors can affect the result of toxicity tests. Some of these factors include variables like temperature food, light, and stressful environmental conditions. Others factors related to the animals itself include age sex, health, and hormonal status.

The NOEL (No Observable Effect Level) is the highest dose or exposure level of a poison that produces no noticeable toxic effect on animals.

In toxicology, residue tolerance levels of poisons that are permitted in food or in drinking water for instance, are usually set from 100 to 1,000 times less than the NOEL to produce a wide margin of safety for humans.

The TLV (Threshold Limit Value) for a chemical is the airborne concentration of the chemical (expressed in PPM) that produces no adverse effects in workers exposed for eight hours per day, five days per week. The TLV is usually set to prevent minor toxic effects like skin or eye irritation.

Very often, people compared poisons based on their LD_{50}s and use the decisions about the safety of a chemical based on this number. This is over simplified approach to comparing chemical because the LD_{50}'s and base decisions about the safety of a chemical based on this number. This is an over simplified approach to comparing chemicals because the LD_{50} is simply one point on the dose-response curve that reflects the potential of the compound to cause death. What is more important in assessing chemical safety is the threshold dose, and the slope of the dose-response curve, shows how fast the response increases as the dose increases. It is quite possible that a chemical will produce a very undesirable toxic effect (such as reproductive toxicity or birth defects) at doses which cause no death at all.

A true assessment of chemical toxicity involves comparisons of numerous dose-response curves covering many different types of toxic effects. The determination of which pesticides will be restricted use pesticides involves this approach. Some restricted use pesticides have very large LD_{50s} (low acute oral toxicity) however, they may be very strong skin or eye irritants and thus require special handling.

Although there is no direct extrapolation of animal studies to man, the knowledge gained from dose-response studies in animals is used to set standards for human exposure and the amount of chemical residue that is allowed in the environment. As mentioned previously, numerous dose-response relationships must be determined, in many different species. Without this information, it is impossible to accurately predict the health risks associated with chemical exposure. With adequate information, we can make informed decisions about chemical exposure and work to minimize the risk to human health and the environment.

Pesticides differ from any other chemical substances because they are deliberately spread into the environment. As a consequence, a great part of the human population may be exposed either in the general environment or in the working setting. The occupational exposure involving the manufacturing and the use of pesticides, takes place mainly through dermal or respiratory route. While the environmental exposure, involving general population is mainly due to the ingestion of the contaminated foods and water. The environmental and occupational exposure determines the detrimental effect that this exposure could have on reproductive function. In women, if primordial follicles are destroyed extensively, they cannot be regenerated. This can cause premature ovarian failure and early menopause.

Note the information contained in this topic is not a substitute for pesticide labels. The trade names used here are for convenience only, no endorsement of products is intended nor criticism of unnamed products implied.

In California, suspected pesticide-related illness and suspected work related illnesses and injuries are reportable conditions. In 1998, the occupational Health branch of the California department of health services (CDHS) received a report from California department of pesticide regulation (CDPR) of a pesticide exposure incident in fresno county involving 34 farm workers. CDHS investigated this incident by reviewing medical records of the 34 workers and interviewing 29. The finding indicated that the workers became ill after early re-entry into a cotton field that had been sprayed with a cholinesterase inhibiting carbamate pesticide.

Prenatal Exposure

The study of the effects of low level exposure to environmental neurotoxic agents has been one of the most important aspects of environmental research. The developing organism may be more susceptible to many of these chemicals than the adults. This susceptibity varies during different stages of development, but severe behavioural neurochemical, and neurophysiological abnormalities can be measured in adults that were exposed during neuronal development. Apparently, one of the problems of toxicity testing is to correlate these behavioral alterations with the anatomical changes induced by the neonatal exposure.

The large scale pollution of our environment by chemicals released from industrial and other sources has resulted in many well-publicized episodes of mass poisonings of populations in modern world. While these episodes have highlighted the consequences of unchecked industrial growth, they have not shed light on what may be a much more urgent problem facing mankind: the problem of low level exposure to a variety of chemicals throughout the life cycle. This problem is of completely different nature from the problem of acute mass poisoning in that it not only exposes all age groups but also exposes these groups to several chemicals simultaneously whose individual toxicities even today are largely unknown. The realization that biological systems and ecosystem may not be able to cope with the continuously rising tide of chemical pollution has led government agencies to pass legislation before toxicity thresholds had been adequate determined. In fact, it has resulted in a dilemma for the toxicologist in that more is known about the toxicity of many of the chemicals from cases of human poisoning than is known from animal studies. The production of a new untested chemicals to the already existing pool at an ever-increasing rate also seem to the possibility of careful dose-response studies on identified toxic agents because of the pressure to respond to today's emergency rather than yesterday's head line.

In response to this emergency, biologists have tried to develop rapid screening tests that would serve as early indicators of toxicity. Such tests show a great deal of promise in the screening of potential carcinogens and of neurotoxic substances.

The minimata episode in the 1950's in Japan has made it clear that pregnant women who are exposed to low concentrations of pesticides (e.g. methimercury) may experience few symptoms and yet can give birth to severely retarded children. This effect has since been documented in animal studies and has been shown to be due to the increased fetal brain concentration of mercury (null et al; 1973). As was stated earlier, the purpose of many animal studies, therefore, has been to define the threshold level of exposure to mercury below which no neurological abnormalities will be observed in the offspring.

PRINCIPLES OF BIOLOGICAL TESTS FOR TOXICITY

Toxicology has been defined as the study of the effects of chemical agents on biological material with special emphasis on harmful effects. It basically involves an understanding of all effects of essentially all chemicals on all types of living matter. There is ample evidence to indicate that every chemical is capable, under some conditions of producing some type of effect on every biological tissue. Toxicologic tests are therefore the tests that define the conditions that must be present when a biological cell is affected by a given chemical entity, and the nature of the effect which is produced. As far as the conditions that must be present are concerned, they may vary from being practically unattainable under ordinary circumstances to being so readily attained that simple exposure of living tissue to certain chemical produces destruction of the cells. As far as the nature of any effect of a chemical on living tissue is concerned, effects may be of such minor significance that the tissue is able to carry on its ordinary function in a normal manner so that it is only under conditions of stress or critical tests that a chemical induced effect is even detectable. Effects may result from small amounts of some chemicals and required to produce any positive findings. Generally it is a simple matter to separate those relatively few chemicals that in small amounts produce prompt effects that are distinctly harmful to living cells from those that are practically harmless when exposure is over a short period of time, but it becomes difficult to demonstrate that small amounts of some compounds do not produce some types of toxicity when animals are exposed over a long period of time.

Most of the biological methods which have been developed in toxicology are the result of the practical need to obtain as much information as possible about the effects of chemical in so far as they may be pertinent to man's continued physical well-being. The continuing economic progress of the human race has been accompanied by a continuing increase in the numbers of chemical entities to which man is either in intentionally or unintentionally exposed. A person may be exposed through direct industrial of domestic occupational contact, through contact with the clothes or devices he wears, the food, he used and drugs he consumes and the atmosphere he inhales. It is necessary not only to understand the toxicities that can occur but also to obtain assurance that exposure of man to large number of chemical entities will mot lead to obvious direct or insidious indirect detrimental effects. Consequently, it is essential that chemicals which are to be intentionally administered to man, such as food additives, food substitutes or drugs, it is necessary to obtain as much toxicity data as is economical possible.

Because of the moral ethical and legal restrictions regarding the use of humans for experimental purposes in order to acquire toxicological data, only limited amounts of such data are available. Information regarding the effects of chemicals on human is obtained only after a chemical is used by human or from limited types of experimental procedures that may be conducted on humans. Biological methods in toxicology therefore generally involve the use of expendable species of animals on the hypothesis that toxicity studies in suitable species have an extrapolative value for man.

Several of the procedures involved in testing for toxicity involve the use of non-mammalian species and even cell cultures. It would be of great advantage to be able to utilize such species as bacteria, neurospora, daphnia, drosophila, the various echinoderms or fish for evaluation of toxicity because of the economic advantage and abundance of such populations of living cells. Furthermore some of these species lend themselves to accurate and simple procedures such as those that make use of their accurately defined and measurable genetic characteristics, reproductive processes and enzymatic performance. The main drawbacks associated with the use of such species are the dissimilarities in translocation barriers as compared to man and differences in or the lack of biotransformation mechanisms that are present in man. These factors preclude extrapolation of the data obtained on most non-mammalian species to man. Never the less such tests serve the purpose of alerting the investigator to potential toxic hazards which can then be further studied in mammalian species.

However when any chemical is used in massive quantities such as in agriculture and becomes available in the general environment, it is necessary to evaluate the toxicity of that agent in many species which may directly or indirectly influence the overall welfare of man.

It should be recognized that there are many variations in both short and long term chemical-induced toxicity between various mammalian species of animals, however careful complete evaluation of the effects of chemicals on animals have been shown to be the most rational, acceptable and successful means of determining most types of toxicity for purposes of extrapolation to man. The principal exception is the rather unsuccessful evaluation of immunogenic types of toxicity. It is interesting to know that many workers before the 19th century described the actions of poisons and their antidotes, these studies seemed to lack the scientific approach. The first to undertake scientific studies on the harmful effects of chemicals on biological systems was M.J.B. Orfilia (1787-1853) a Spaniard at the University of Paris (USA today, 1989). He is regard as the father of modern toxicology and was the first to introduce quantization in the study of the actions of chemicals on animals and to consider toxicology as a separate discipline from pharmacology. Orfilia was the author of the first book on harmful effects of chemicals (in 1815). He not only studied and reported on the effect of chemical but also on the treatment of poisoning due to such chemicals.

Toxicology as a science has its basis in the science of chemistry and biochemistry and is dependent on the knowledge of physiology. Pathology is often regarded as part of toxicology because the effect of the chemical on the biological system may appear as macro-or microscopical deviations of the normal cell or organ.

Toxicology is an offshoot of, and closely related to pharmacology because a pharmacologist attempts to understand the beneficial effects of the chemical when used therapeutically as well as its harmful or adverse effects. There are three main divisions of toxicology namely, economic, forensic and environmental toxicology, each with its own specialist toxicologist.

Economic Toxicology

This concerns the harmful effect of chemicals, administered to man or animals in order to produce a specific effect. This includes drugs which are administered to modify physiological functions or eliminate some bacteria/parasitic organisms in the body. The study of drug toxicity (including tests for toxicity) is a major area of economic toxicology. It includes studies on the safety of food additives and cosmetics. Some chemicals have selective action on biological organisms and are used by man to eliminate pests and insects (as pesticides and insecticides) which become the uneconomic species. The human which is protected from the effect of the pest and insects becomes the economic species. The effect of the chemical on both economic and uneconomic species also forms a part of economic toxicology.

Forensic Toxicology

This concerns the medical and legal aspects of the harmful effects of chemical on animals, including man. The medical aspects refer to the diagnosis and treatment of the effects of the chemical and the harmful effects produced by it. In forensic toxicology, an attempt is made to identify the chemical in the tissue by chemical analysis and in case of death, to establish the cause or circumstances of death. Both intentional and accidental exposures to the chemical are of interest to forensic toxicologists. They develop methods for the management and treatment of acute and chronic poisoning including the use of antidotes. These are specifically referred to as clinical toxicologists.

Environmental Toxicology

This deals with the harmful effects to man and animals of chemicals that are present as contaminants of the environment. The chemical may be present in the air, water, soil or food. In the urban centers, because of the industrial activities, the environment may be polluted by particles or by gasses. Environmental toxicology is concerned with hazardous substances in the air, water or soil, the disposal of industrial waste and the protection of the environment and peoples either at home or at manufacturing sites from industrial emissions. The environmental toxicologist deals with the evaluation of the effects of and the establishment of the limits of safety of exposure to these chemicals i.e. estimating the health risks of a particular chemical.

FACTORS THAT AFFECT THE TOXICITY OF CHEMICALS

Among the factors that affect the toxicity of a chemical are the chemical, biological, genetic factors and the route of administration or exposure to the chemicals.

Chemical Factors

The chemical structure determines the ability of the chemical to interact with specific receptors responsible for the observed effects. Biotransformation mechanism is dependent on the structure and may produce a metabolite that is more toxic or less toxic than the parent chemical. It may influence the excretion process of the chemical or its metabolite. It is known that some chemicals do induce or inhibit metabolizing enzymes and in this way modify their activities and those of other chemicals metabolized by the same enzymes.

Biological Factors

Here the factors include biotransformation and elimination mechanisms, plasma protein binding, storage of the biological membrane through which it passes. Each of this will influence the toxicity. Age, sex, nutritional status of the individual and species of aminal also play roles too.

Genetic Factors

Genetic differences may have a great effect on toxicity. For example, some individuals are genetically deficient in blood psendocholinesterase enzymes. Thus succinylcholine (a muscle relaxant) normally hydrolyzed by psendocholinesterase, will induce prolonged muscle relaxation in a person genetically deficient in the enzyme contrary to expectation.

The anti-cholinesterase's used in therapeutics' are generally those which reversibly inactivate cholinesterase for a few hours. Insecticides of the carbamate type act by reversible inhibition of cholinesterase but organophosphorous insecticides inhibit the enzyme almost or completely irreversible so that recovery depends on formation of fresh enzymes. This process may take weeks although clinical recovery is usually evident in days. Cases of poisoning are usually, after agricultural, industrial or transport accidents. Substances of this type have also been studied for use in war (nerve gas). The prominence of individual effects varies with different agents, e.g. sweating and salivation are not usual in dyflos poisoning.

A typical case of poisoning by cutaneous absorption, will, perhaps after a delay, develop headache confusion, anorexia and a sense of unreality. The patient is often giddy, apprehensive and restless. Conspicuous salivation, rhinorrhoea and sweating follow, with respiratory wheeze and dyspnoea indicating the onset of broncho constriction and excessive bronchial secretion. Miosis may occur and cause the headache, but it is not invariable nor is it an index of severity, for it may be due to a local effect of the poison entering via the conjunctiva. Vomiting and cramping abdominal pains may lead to diarrhoea and tenesmus, and there may also be urinary incontinence. Muscle twitching typically begins in the eyelids, tongue and face, then extends to the neck and limbs and is accompanied by severe weakness. Progressive respiratory difficulty leads to convulsions and coma. Death is due to a combination of the actions in the central nervous system, to paralysis of the respiratory muscles by peripheral neuromuscular block, and to excessive bronchial secretions causing respiratory failure. At autopsy, ideal intrsusception are commonly found.

ROUTE OF ADMINISTRATION OR ENTRY INTO THE BODY

Chemicals may enter the body through inhalation, by contact with the skin and by oral route. In addition, drugs may be administered by porenteral routes. The toxicity of a chemical may be many times greater through one route than through the other. A typical example is curare which is not absorbed orally and hence induces no toxicity. Its toxicity manifests itself when administered parenterally.

TOXICOLOGICAL EFFECTS

The toxicological actions of a chemical or materials may induce acute or chronic toxic effects or poisoning. Acute effects arise form an exposure to a chemical or an over dosage of drugs and the poisoning may be accidental, suicidal or homicidal. This is a toxicological emergency and demands emergency management, care and treatment.

Chronic poisoning is caused by ingestion of or exposure to the chemical over a period. In certain cases of occupational poisoning, ingestion of polluted water or inhalation of insecticides by farm workers results in chronic poisoning. It takes times to manifest itself and an equally long time to treat.

However chronic poisoning may produce an acute toxic effect which will necessitate toxicological emergency action and a chronic management of the patient until the body load is reduced or poison eliminated form the body.

THE ENVIRONMENT

The human environment contains many chemicals that are toxic to man and animals. It is one of the sources of health hazard and is responsible for various acute toxicities and many chronic illnesses. The environmental chemicals may be present in the atmosphere as air pollutants, in soil or water including the under ground water as contaminants and in food as residue or contaminants.

The sources of environmental hazardous chemicals are two, namely, natural sources and man-made sources. For the sake of this topic, emphasis is laid on man-made source.

Man-Made Sources

These arise from human activities and include chemicals that reach the atmosphere as a result of industrial activities. There are many chemicals that are present or are used in work places and therefore constitute occupational hazard.

Many industrial activities pollute the atmosphere with particulate matters and gasses like carbonmonoxide, sulphurdioxide, hydrocarbon e.t.c. Some factory smoke stacks release particles which are deposited onto vegetables crops that are consumed by man.

Some potential toxic chemicals are normally found in the home as drugs, as pesticides. These are responsible for accidental poisoning among children. Studies in developed countries have shown that poisoning is the second or third most important cause of fatal accidents in the home (Backett, 1965). Pesticides used by farmers for enhanced food production may appear in the food and the same applies to food preservatives. In Nigeria, some farmers and traders used pesticides to preserve grains like beans, maize, rice, etc, in spite of the fact that pesticides are not approved for the preservation of grains. Through this source, pesticides are consumed by the unsuspecting public with possible acute or often chronic consequences. The pesticides used in the farm may be carried by rain, run-off to contaminate the streams and rivers in countries without the necessary controls.

There are many activities that routinely release hazardous chemicals to the environment. The petrol station attendant inhales benzene as petrol is put in the car and benzene is also released to the air. The exhaust fumes form motor vehicles release a number of chemicals including particulate matters in to the atmosphere. Such substance in the air create problems in the some developed countries like the USA to control and set limits of emission from motor vehicles for permit to ply the road.

The Natural Resources Defense Council (NRDC) of the USA estimated in 1989 that industry is pumping more than 361 million pounds of cancer producing chemicals into the air yearly (USA Today, 1989).

Over the years, human beings have tended to depend on some sudden and unexpected episodes to realize the dangers of environmental pollution from man-made sources. There are many instances which point to the fact that pollution can cause serious illness or death. In 1952 in London, a dense fog (SMOG) due to environmental pollution settles over the city for 4 days. This resulted in about 3,500 to 4,000 deaths in greater London alone (Klassen, 1990). Ten years later a similar episode occurred in London and caused many deaths particularly amongst the elderly and children (Akubue, 1997).

In the Meuse valley in Belgium in 1930, a heavy fog associated with very stable air mass caused severe respiratory symptoms and death, the death rate in the community during the period was 10 times more than normal. The air pollution was said to have come from the industrial plants in the neighbor hood,

In June, 1996 in Santiago, Chile, there was part of SMOG as air pollution reached an alarming level. There were many deaths from respiratory diseases. The city authorities ordered 300,000 cars off the roads in an effort to improve the quanlity of air.

There are also many instances of major accidental release of toxic chemicals in to the atmosphere.

In 1976, there was a massive exposure of the city and people of seveso, in Italy to 2, 3, 7, 8-tetra chlorobenzodioxin (TCDD) (Ottobin, 1991). This was as a result of an explosion in a manufacturing plant. Many people particularly children suffered chlodacne and thousands of animals like chicken, birds, dogs and horse died a few days after the explosion. It is believed that dioxins contaminants of herbicide, agents' orange, were responsible for many health problems of Vietnam Veterans (Ottoni, 1991). Dioxins of which TCDDTS, the most toxic and most persistent poison, are by-product contaminants of many chemicals reactions and are formed during the manufacture of chemicals like trichlorophenols and the combustion of waste materials. Studies have shown that if TCDD is not a cancer inducer, it is a promoter. In Nigeria, there is no doubt that TCDD is formed as a by-product of chemical processes being carried out in the industries and as a product of combustion of waste materials. Hence, TCDD is in the air around us. The question that needs to be answered is, what have we done to bring its level to the bear rest minimum?

An explosion at an industrial plant in Bhopal, India in 1985 released into the environment methyl-isocyanate and this was responsible for the death of about 1,500 people (Keritage, 1992). This would not have happened if adequate safety measures were in place in the industry. Other toxic chemicals worthy of note are discussed below.

Dichlorodiphenyltrichloroethane

DDT is an organochlorine insecticides (also referred to as halobenzene derivative) which was widely used in agriculture and in malaria control. It is highly soluble in fats and poorly soluble in water. In the body it is stored in the fat depot. It is only slowly eliminated from the body. DDT is usually used as solution in organic solvents, especially kerosene. It is established that DDT increases the incidence of liver cancer in mice (Innes, 1969) there is yet no evidence for such an effect in humans. However, its use has been limited or withdrawn in many countries because of uncertainly of the effect of prolonged exposure and storage in man, beside, some insects developed resistance to it.

Symptoms

Symptoms of acute poisoning include vomiting tremor, and convulsion. There is anesthesia of the tongue, lips and face with marked apprehension and excitement. Diarrhoea may occur.

A study indicated that carbonate pesticides namely aldicarb, aldicarb sulfoxide, baygon, penthiocarb, carbofuran, 3-hydrocarbofuram, carbaryl, desmedipham, methiocab, methomyl, thiodicarb, oxamyl, and propham was made in ground and surface water from an agricultural zone of the Yaqui valley located in northwest Mexico. From the result of trace determinations made by liquid chromatograph (LC) with post-Column fluorescence detection, it showed that the level of contamination with methiocarb was about 5.4 mg/l in a ground water sample and that for 3-hydroxycarbofuran was 18mg/l in a surface water sample (Garcia-de-Llasera, 2001).

Carbofuran was estimated for an Acceptable Daily intake (ADI) in 1978 and 1979 and a temporary ADI for man was estimated to be 0-0.003mg/kg body weight (FAO/WHO, 1977; FAO.1980). The available data reflected that carbofuran is a highly toxic carbonate ester whose metabolic profile has been well defined. Carbofuran is a potent, reversible cholinesterase inhibitor. Chlolinesterase inhibition and acute toxic signs of poisoning are subject to rapid spontaneous reversal and recovery. The measurement and evaluation of cholinesterase depression induced by carbofuran, because of the rapid reversibility, is difficult and required substantial care.

Similarly, in a group of mice (100 male and 100 females, Charles River CD-1mice/group were fed carbofuran in the diet of dosage levels of 0, 20, 125, or 500 mg/kg for two years. It was reported that a localized hair loss and reddening of the ear(s) frequently followed by scabbing or sloughing of portions of the ears was noted with greater frequency in the treated mice. (Pesticide residues in food: 1950 evaluation) In a similar study using fenamiphos, a carbonate pesticide like carbofuran, exposure of rats to technical grade fenamiphos (purity, 92.2%) diluted with a 1:1 mixture of ethanol and polyethylene glycol 400 for aerosolozation in a dynamic flow inhalation chamber at doses of 0,0.03,0.25, or 3.5 mg/l for 6hrs per day, five days per week for three weeks showed a significant decrease in (48-7.9%) in plasma cholinesterase activity and a slight decrease (9-18%) in erythrocyte acetyl cholinesterase activity in animals of each sex at 3.5µg/l (Thysen, 1979b)

Benzene Hexachloride and Lindane

Benzene hexachloride (BHC) is a mixture of eight isomers with gamma isomer (Lindane) being the most toxic and active but most rapidly excreted. The toxic effects of Lindane resemble those of DDT. It is a CNS stimulant and a potent inducer of hepatic microsomal enzymes. BHC is said to cause aplastic anaemia (Akubue, 1997)

Polycyclic Chlorinated Substance Chlorinated Cyclodienes

The substances are many with varying toxicities. Examples are aldrin, dieldrin, heptachlor, Endosulfan and chlordane. They stimulate the CNS and induce convulsion. Before this, there may be headache with nausea, dizziness, vomiting and mild chronic jerking, and tremor ataxia. The CNS stimulation may be followed by depression which may end in respiratory failure. The insecticides have potentials as carcinogens and cause haematoma in mice. Hence their use in some countries (e.g. USA) is banned.

Organophosphorous Insecticide

Organophosphorous insecticides are extensively used in agriculture. Through they do not persist in the environment; they can cause serious toxic effect in man, being potent and irreversible inhibitors of cholinesterase. The symptoms of poisoning are due to muscarinic, nicotinic and CNS effects.

In acute (mild to moderate) poisoning from ingestion or inhalation, the following symptoms can be as observed (headache, dizziness, tremor of the tongue and eyelids, miosis and impaired vision. These symptoms are followed by nausea, vomiting, salivation, tearing, abdominal cramps, and sweating, slow, pulse and muscle fasciculation. In severe poisoning, diarrhoea, pinpoint pupils not reactive to light, respiratory difficulty, cyanosis, convulsion, coma and heart block may be observed. Chronic poisoning may occur and inhibition of cholinesterase can persist for up to 6 weeks. Delayed neuropathy occasionally develops after poisoning.

RODENTICIDES

Many drugs had been used to kill small animal like rats and mice. Amongst these are fluoroacetate, x-naphthylthiourea(Antu) and pindone (warfarin- like anticoagulant).

Fluoroacetate Sodium

Fluoro acetate sodium is extremely toxic to rodents and to man and other animals. Fluoroacetate is present in the plant known as *Dichapetalum cymosum* which grows in Nigeria and is used as rat poison. Fluoroacetate is too toxic for use in the home. It has been withdrawn from the market in some countries but fluoro actamide, which has the same toxicity, is still on the market.

The most prominent effects of a cute poisoning in man from ingestion or inalation of fluoroacetate are vomiting and convulsions. It induces irregular heart beat, exhaustion and coma. Death is usually due to respiratory failure.

Pindone

This is a coumarin anticoagulant with actions similar to warfarin which is also used as rodenticide. In large doses, they can cause vascular collapse.

HERBICIDES

Herbicides are now used extensively to destroy noxious weeds. The most common ones are the following;

- 2, 4-dichlorophenoxyacetic acid (2, 4-D)
- 2, 4, 5- trichlorophenoxyacetic acid (2, 4, 5-T)
- Dinitrophenols
- Paraquat.

Chlorophenoxy Compounds (2, 4-D and 2, 4, 5,-T)

Chlorophenoxy compounds are used to control broad-leaf weeds. They rarely cause toxicity in man though contact dermatitis is known to occur. The effect is due to contaminant called TCDD(2,3,7,8- tetrachlorodibenzo-p-dioxin).

Dinitrophenols

Dinitrophenols are also used extensively in weed control. The toxic effect is due to the uncoupling of oxidative phosphorylation. Thus the metabolic rate is increased with a subsequent rise in body temperature. Other symptoms of acute poisoning include nausea, restlessness, sweating tachycardia, fever rapid respiration, fever and cyanosis.

FUNGICIDE

Most fungicides that are used to control fungal diseases on seeds and plants are not particularly toxic except the mercurials. However, some do produce toxic effects in man. A good example is seen in carbamates e.g aldicarb, aminocarb, carbofuran, propoxur and bendiocarb. Health hazards to man are mainly as a result of occupational exposure.

Dithiocarbamates

Both dimethyldithiocarbamate and ethylenebisdithiocarbamate have low acute toxicity. However, thay may have teratogenic as well as carcinogenic effect and can produce disulfiram-like effects when alcohol is ingested. It is known that in humans, in the environment and during cooking of contaminated food, the ethylenebisdithiocarbamate is broken down to form ethylenethiourea, which is carcinogenic and teratogenic.

Other chemicals used as fungicides include hexachlorobenzene and penta chlophenol. The toxic effects of pentacholophenol resemble those of nitrophenols which increase the metabolic rate by uncoupling oxidative phosphorylation.

FUMIGANTS

Funigants are used in gaseous form as pesticides to reach areas that are inaccessible. Examples include cynide, crabondisulphide, carbon tetrachloride, thylene oxide, methylbromide and phosphine.

Cyanide

Cyanide as hydrogen gas, is used as a fumigant in ships and buildings. It is also used in industry in chemical synthesis and electroplating and as household pesticides and rodenticide.

Large doses of hydrogen cyanide can cause rapid respiration, convulsion. Loss of consciousness and death within 5minutes: lower doses produce headache, staggering gait, dilated pupils, palpitation, unconsciousness, violent convulsion and death. Combustion of plastic materials releases hydrogen cycanide. This was the cause of death of 303 pilgrims in 1980 in Riyadh, Saudi Arabia (Wager, 1983).

Cyanide in the body complexes with cytochrome oxidase and is responsible for cellular oxygen transport. In this way it interferes with oxygen uptake by the body cells.

Methyl Bromide

Methyl bromide is a colourless gas used as a fumigant for soil, stored dried food stuffs and for disinfection of fresh fruits and vegetables.

Methyl bromide causes headache, blurred vision, weakness, oliguria, anuria, confusion, drowsiness, convulsion, and coma. There may be circulatory collapse or respiratory failure. With low doses, the symptoms may take 12 to 24 hours to appear.

Phosphine (PH$_3$)

This is used to fumigate grains. It is slowly released from aluminum phosphine tablets in the presence of atmospheric moisture. Poisoning by inhalation may cause a fall in blood pressure, pulmonary oedema, collapse, convulsion and coma. The first sign of chronic poisoning is toothache followed by swelling of the jaw and necrosis of the mandible (phossy jaw). There may be anaemia and spontaneous fractures.

CARCINOGENS

Chemical substances have been associated with cancer in man long before it was demonstrated experimentally in animals. For example the high incidence of cancer of the scrotum was reported amongst chimney workers in 1775 (Wolf, 1953). This was thought to be due to exposure to soot. Cancer was also recognized as an industrial hazard among coal tar workers. It is now generally accepted thatmany human cancers are directly and indirectly due to environmental factors which induces exposure to chemicals and ionizing radiations (Tomatis, 1976)

Chemicals known to be responsible for certain human and animal cancer are referred to as carcinogens. Some of these chemicals produce the same type of cancer in animals and in man but with some other chemicals there is species variation for example, 2- methylamine causes bladder cancer in man and dogs (WHO, 1972). Occupational exposure to chemicals is known to cause cancer mostly that of the skin, bladder, lungs and sinuses (WHO, 1972).

Cancer develops slowly and there may be along latent period (15-40 years) before the cancer is detected. It can occur at the site of application or at a site far away from the site of contact. Certain carcinogenic agents are not themselves carcinogenic until metabolized to an active product which is called proximate carcinogen.

It is also well established that chemical like polycyclic hydrocarbons e.g. 3.4- benpyrene present in coal tar) and 3, 4-benzphenanthrene azo dyes e.g. Dimethylnitrosamine are potent carcinogens. Thus a very wide range of chemical structures can induce carcinogenicity. This has made determination of the mechanism of induction rather difficult. It is believed that most organic carcinogens react covalently with macromolecules in the tissue. Others are converted to reactive metabolites which contain an electrophilic atom. It is this reactive (elctrophilic) group in the carcinogen or its metabolite that reacts with any nucleophilis atom of the target macro molecules of the cell. The nucleophilic site is on the DNA, RNA, and/ or protein. Carcinogens can react covalently with DNA in vivo directly or after conversion to the reactive group.

The problem of carcinogens is very great in developing countries where control measures and legislation are not available to protect the public from the industrial hazard of these chemicals and from the use of such chemicals in foodstuffs, cosmetic and agriculture. Many local industries do not even try to protect their workers and the workers, out of ignorance do not protect themselves. It may be realized that even when chemical is declared non-carcinogenic in other developed country, because of our genetic make-up and because of other promoting factors which may be presented in our environment may be carcinogenic in our surroundings. It is therefore imperative that all chemicals in use for whatever purpose should be monitored for possible carcinogenic effects (Akubue, 1997)

CHEMICAL MUTAGEN

A change in the hereditary constitution (otherwise called genotype) of the individual is known as mutation. This may be produced by chemicals and also by radiation and hence called mutagens. Mutation can occur spontaneously by unknown mechanisms.

Mutagens act by altering the genetic make-up of the individual and this can be handed down to the offspring during cell division. This means the new cell will have heritable characteristics. The mutation may occur as a result of alteration of one or more nucleotide or changes in number as structure of chromosomes.

In general, two kinds of mutation can be recognized: Chromosome and gene mutation. A chromosome mutation may be of several types e.g. Deletion, duplication inversion, translocation (Roberts, 1982).

However, on some occasions, two homologous chromosomes, instead of separating during cell division, called meiosis, go off into the same gamete. This phenomenon is known as non-disjunction and it results in half the gametes having two of the chromosomes whilst the other half has none. The fusion of the first kind of gamete with a normal gamete of the other sex will give an individual with three such chromosomes i.e. the normal pair plus an extra one. This condition is called trisomy.

Quite how important non-disjunction has been in generating useful genetic novelty is uncertain, but there is no doubt it can have profound effect on an organism's development. For example, mongolism (better called Down's syndrome after the clinician who first described it) is now known to be caused by the presence o f an extra chromosome in the cells, chromosome number 21 to be precise. This is one of the smallest of all the human chromosomes, and yet its presence plays havoc with the individual's normal development. Sufferers, form Down`s syndrome, if they survive at all, have a characteristically slit-eyed appearance, reduced resistance to infection and are always mentally deficient, all because of one extra chromosome.

The short description above is a tip of an iceberg of the several havoc pesticides can cause to the body.

A gene mutation arises as a result of a chemical change in an individual gene. An alteration in the sequence of nucleotides in that part of a DNA molecule corresponding to a single gene will change the amino acids making up a protein, and this can have far- reaching consequences on the development of an organisms.

Drastic effects can some times be produced by a seemingly trivial change in the nucleotide sequence of a gene. This can be seen, for example in the formation of hemoglobin. In the inherited disease known as sickle cell anaemia, the red blood cells, normally biconcave discs, are sickle-shaped and the victims suffers all the symptoms of extreme oxygen shortage, weakness, emaciation, kidney and heart failure.

The anemia is not caused by a distorted shape of the red blood cells as such, but by the fact that they contain abnormal haemoglobin, haemoglobin S, which inefficient at carrying oxygen. Gene mutation can be by deletion, inversion substitution. Note that although the addition of whole chromosome as seen above may have been disastrous more often than it has been helpful, it is highly likely that the addition of new genes within individual chromosomes (insertion) has been extremely important in promoting evolutionary novelty. In general terms it is important to appreciate that variation is not always disadvantageous (Roberts, 1982)

Where the action of the mutagen is in the human germ cells (i.e. spermatozoa or ova), the offspring will carry the mutant genes in its cells. This may result in death of the zygote) or abnormal offspring. The effect of the mutation appears only in the offspring and may take a few generations before it manifest itself.

Mutation may also affect somatic cells. The effect is not passed on to future generations but may be responsible for carcinogenesis. For example, the alteration in genetic material may cause cell division in cells that normally do not divide during adult life. Such a division will eventually lead to cancer.

Though it is not known how many gene mutations in man can be attributed to chemicals it is acknowledged that many chemicals can cause gene mutation in micro-organisms and in insects. Also many chemicals can cause chromosome aberrations in man and can interact with nucleic acids. This may be a pointer to the fact that there are many chemical mutagens in our environment.

Among the chemicals shown to be mutagenic in mammalian cells are ethyleneoxide, Ziridine, aminobiphenyl, dimethylsulphate, benzidine e.t.c (Fischbein, 1979)

TERATOGENS

Teratogens are agents that cause abnormalities of foetal development. They usually interfere with the development of the foetus at a dose that produce no serious toxic effect on the mother or impairment of placental function. However, in high doses most of them will cause foetal death followed by abortion or resorption of the foetus.

One of the earliest reports of environmental factors being responsible for the birth of malformed babies came form Australia 1940 (Gregg, 1941). Mothers who suffered form the mild virus disease (rubella) during the first trimester of pregnancy gave birth to blind or deaf children. It was found that if the infection occurred after the third mouth of pregnancy, no abnormality developed. This was followed about twenty years later by the thalidomide disaster of the early sixties when a seemingly non-toxic drug produced teratogenic effect. Some environmental pollutions, like chemical defoliants, 2, 4, 5-T (2, 4, 5-trichlorophenoxyacetic acid), are teratogenic in animals. Their effects in man are not yet fully established.

Evidence for the teratogenic effect in man of organophosphorous insecticides is still inconclusive though such an effect in animal is well established. Lead which is an environmental pollutant may be teratogenic (Scalon, 1972).

From the above account it is obvious that a lot needs to be known about chemical teratogens. It is not unlikely that most of the congenital malformations encountered in a community may have been due to environmental chemicals (Akubue, 1997). As teratogenic effect cannot be reversed or treated, the only right approach is to take necessary steps to prevent it. This is best achieved by avoiding occupational exposure to environmental chemical during pregnancy, particularly during the first trimester.

Developed countries do monitor the activities of industrial companies with regard to pollution and waste disposal. In 1992, Dexter corporation in the USA was fined thirteen million dollars (US$ 13 million for activities in its manufacturing plant in Windsor Locks Connecticut, USA, (EPA, 1990) with regard to illegally disposing of carbondisulphide and discharging hazardous waste and waste water into a river.

The above instances of the effect of environmental pollution and the action taken by a government environmental protection agency against a manufacturing company are presented to draw the attention of all sundry to the time bomb we are setting on. Nigeria and all developing countries must learn from the mistakes of the developed countries. We try to industrialize, we must not repeat the mistakes which other countries made in the past and paid heavily for them.

Apparently, in developing countries like Nigeria where increasing use of pesticides is absolutely vital for it purpose as to protection of economically important crops such as tobacco, cotton, rice and so on with little attention paid to its deleterious effects (Carlson et al., 1998 Roex et al., 2001) and for feeding increasing large populations adequately, some recommendation have made by Iyaniwura, (1991b) in order to maximize the benefits at the least risk to human population. They include;

The increasing need for extension workers to embark on the training of farmers in the choice of pesticides, storage, applications, technique and the use and disposal of spent container become very important in this area.

The need for the development of preventive measures, diagnostic tests and treatment facilities in pesticide poisoning will go along way in helping farmers.

That training of physicians and health workers in high risk areas should includes instruction on the diagnosis and treatment of pesticide poisoning.

Trained scientist specialize in the treatment of victims of pesticide poisoning should be made available to health officials and hospital at those locations where pesticides are being used.

It is worthy of note here that regulations or restrictions of use and production of the chemical itself need to be enforced to avoid damage to man, animals, insects or wild life generally. In this regulations, various tests are performed including; chemical characteristics, toxicological characteristics physiological and biochemical, behavioral, environmental, ecological and tolerance assessment.

Pollution cause immediate health effects, in many cases, the effect may be delayed for 5 to 10 years or more. The greatest danger is that the effect may be unnoticed until irreversible damage is done. This is worrying in our circumstances. In developing countries specifically, Nigeria, an average Nigerian does not feel threatened by any chemical or any operation that has no acute toxic effect. It is known that the health effect may appear long after the person has left the area of exposure. By this time, the health effect may not be associated with the exposure.

Some of the effects that are known to be induced by pesticides in humans includes cancer (various types) infertility, liver damage, kidney damage, premature death, hyperactivity in children, bronchitis, defective sight and blindness, birth defects, blood diseases, nervous system damage, heart defects and sudden death.

These illnesses are very devastating with high morbidity and mortality rates but as they are preventable, they are best prevented. The cost of prevention is next to nothing compared with the cost of treating the ailments or managing the in curable ones.

BIOPESTICIDES AND HUMAN HEALTH

People have been using biotechnology for millennia. This technology is based on the use of microorganisms, which e.g. ferment the sugar in barley to alcohol during beer production. Other examples of everyday products that undergo biotechnological processing are cheese, yogurt, vinegar, wine, yeast, and sourdough. Without knowledge of the exact backgrounds, our ancestors used these methods to discover and improve a range of applications that made their life easier. Genetic engineering is a modern subspecialty of biotechnology. It is concerned with the targeted modification of the genetic material of bacteria or plants, for example to stimulate them to biosynthesize desired products. Today genetic engineering is primarily used in the field of medicine, but is also applied in industry and agriculture.

Biotechnology is the science that modifies the genetic composition of plants, animals and microorganisms. It is used to incorporate genetic material from one living organism to another. Biological pesticides are produced through the use of biotechnology by harnessing the pest-fighting abilities of existing plants and microbes. When these products have unique biological properties, they also pose unique regulatory challenges. In addressing these challenges, the Environmental Protection Agency (EPA), the U.S. Department of Agriculture (USDA), and the Food and Drug Administration (FDA) have shared responsibility for regulating agricultural biotechnology in the United States. For instance, EPA regulates pesticides created through biotechnology as a part of its regulatory jurisdiction over all pesticides marketed and used in the United States. As such, EPA has tailored its basic regulatory framework to fit the distinctive characteristics of these genetically engineered biological pesticides (EPA, 2003). In theory, biotechnology could be used to prevent pest problems and thus reduce the need for pest management and pesticide use. Since the beginning of agriculture, plant breeders have developed crop varieties that were resistant to or tolerant of particular pests. For example, cotton varieties with long, twisted bracts (called frego bracts) around the bolls are resistant to boll weevil damage and solid stemmed wheat varieties are not damaged by the wheat stem sawfly (Cox, 1993).

The tools of biotechnology could be used to make plant breeding easier and quicker. Genetically engineered crop plants will be in farmers' hands in the next few years. We are therefore at a critical point where we need to evaluate this new technology and the impact it will have to our agricultural systems and human health. What problems will the new technology bring? However, genetic engineering has the tendency to move agriculture in the opposite direction, towards maintaining or increasing present pesticide use patterns. The total world production of biopesticides is over 3,000 tons/yr, which is increasing at a rapid rate. India has a vast potential for biopesticides as it utilizes more than 100,000 tons (1992-93) of pesticides annually. Most (80%) of the pesticides are used on cotton (45%), rice (30%) and vegetables (5%), the remaining crops receiving only 20% as share.

Benefits of Biopesticides

International organisations and bodies such as the United Nations Food and Agriculture Organisation (FAO), the World Health Organisation (WHO) and the International Council for Science, as well as a number of national food safety authorities and medical associations, have all positively commented on the safety and/or benefits of agricultural biotechnology. There are direct environmental benefits which arise from the different management techniques plant biotechnology makes possible. For example, herbicide-tolerant crops facilitate the use of no-till agriculture, which reduces both soil erosion and energy inputs. At the same time, soil organic matter is maximized, which reduces agriculture's contribution to global emissions of greenhouse gases, linked to climate change. Pest-resistance reduces the need for spraying, with consequent benefits to non-target organisms and overall biodiversity. Also, by making farming more efficient on limited land area, plant biotechnology contributes significantly to preventing habitat destruction – the biggest single threat to biodiversity.

Organic food sales are increasing because the public is willing to pay more for pesticide-free vegetables, fruits and dairy products. Biotech based pesticides therefore are becoming viable alternatives to chemical pesticides. For instance a biocidal product of plant origin that can combat insects as well as fungus and bacteria would find ready acceptance in the hand of farmers for the control of diseases and pests. It could be grown by farmers themselves and the seeds can be turned to organic pesticides. Is it the spraying of yeast formulation on fruits before they are transported to selling centres to keep the fruits fresh? Biotech pesticides or biopesticides tend to harness nature for solving health problems of agricultural crops. These substances used in controlling pests in crops are ground water and environment friendly. They are used in small quantities.

Crop yield are not affected by them. As they are of natural origin, they mutate hence their use result in substantial labor cost savings.

It is important to know the mode of action of biopesticide as compared to traditional pesticide prior to using it. A biopesticide cannot be considered as non toxic because it is natural in origin, but can generally be considered only as less toxic to humans.

REFERENCES

1. P. I. Akubue, (1997). Poisons in our environment and drug overdose. SNAAP press. Nigeria, 44 57

2. Z. Annau, C. Eccles, 1983Prenatal exposure to environmental chemicals as a test system for neurotoxicology. Invitro toxicity testing of environmental agents. Eds. By Alan, R.K.; Thomas, K.W.; Lester D.G.; Robert S.D.I and Thomas, J. H. Premium publishing corporation. 383 400

3. R. W. Carlson, et al. 1998 Neurological effects on startle responses and escape from predation by medaka exposed to organic chemicals. Aquatic Toxicology 51 68

4. C. Cox, 1993Biotechnology and agricultural pesticide use : an interaction between genes and poisons", 13 3Fall 1993, 4 11

5. L. Fishbein, 1979 Potential industrial carcinogens and mutagensElsvier publishing co. Amsterdam. 80

6. EPA, 2003http://www.epa.gov/oppbppd1/biopesticides/reg_of_biotech/eparegofbiotech.htm

7. Food and Agricultural Organization/World Health Organization (FAO/WHO). 1977evaluations of some pesticide residues in food. Rome, Food and Agriculture Organisation of the United Nations, Joint meeting of the FAO Panel of Experts on pesticide Residues in Food and the Environment and the WHO Expert Group on pesticide Residues.

8. Food and Agriculture Organization/World Health Organization (FAO/WHO). 1980Pesticide residues in food. Data and recommendations of the Joint Meeting on Pesticide Residues, Rome, FAO Plant Production and Protection. Rome (1981). Paper 26Suppl.) 6 15October.

9. G. Forget, 1991 Pesticides and the Third world. J. of Toxicol. and Environmental Health, 32:11 EOF 31 EOF

10. P. C. Fraser, 1967Physical and chemical agents. In:Rubin, A. (1967). Ed. Handbook of congenital malformations. W. B. Saunders London. 365

11. liasera. M. P. Garcia de, Gonzalez. M. Bernal, 2001Presence of carbamate

pesticides in environmental waters from the northwest of mexico: Determation by liquid chromatography. Water research, 35 (8): 1933 1940

12. A. Goldstein, et al. (1974). Principles of drug action. 2nd Ed. 207

13. N. N. Gregg, 1941Congenital cataract following German Measles in the mother. Trans Opthalmol. Soc. 3:35. Australia http://pmep.cce.cornell. edu/profiles/extonexnet/TIB/dose-response.html

14. J. R. M. Innes, 1969Bioassay of the pesticides and industrial chemicals for immunogenicity in mice. A preliminary note. J. of Natural Cancer institute, 42:1101. In: Akubue, P. I. (1997). Poisons in our environment and drug overdose. SNAAP press. Nigeria, 44 57

15. W. Kuhlman, H. G. Fronme, G. M. Heege, W. Ostertag, 1968Mutagenic action of caffeine in higher organisms. Cancer Resources. 28:2375

16. T. A. Loomis, 1975 Acute and prolonged toxicity tests. J. Assc. Official Analytical Chem., 58:645 EOF 9 EOF

17. N. Prakash, Baligar, B. K. Basappa, 2002 Reproductive toxicity of carbofuran to the female mice: Effect on estrous cycle and follicles. Industrial health 345 352

18. E. W. M. Roex, et al. 2001Reproductive impairment in the zebrafish Danio rerio, upon chronic exposure to 1, 2, 3-trichlorobenzene. Ecotoxicology and Environmental Safety, 48: 196 201

19. J. Thyssen, 1979bNemacur® active ingredient. Sub-acute inhalational toxicity studies. Institute of toxicology, Bayer AG, Wuppertal Elberfield, Germany. Submitted to WHO by Bayer AG, Leverkusen, Germany. P. SRA 3886

20. N. P. Wolf, (1952). Chemical induction of cancer. Cassel and co. Ltd. London 24

21. World Health Organisation (WHO) 1972 Health hazards of human environment. WHOGeneva. 229

Chapter 4

PESTICIDES: ENVIRONMENTAL IMPACTS AND MANAGEMENT STRATEGIES

Harsimran Kaur Gill[1] and Harsh Garg[2]

[1]University of Florida, Gainesville, FL, USA

[2]The University of Sydney, NSW, Australia

INTRODUCTION

Increase in food production is the prime-most objective of all countries, as world population is expected to grow to nearly 10 billion by 2050. Based on evidence, world population is increasing by an estimated 97 million per year (Saravi and Shokrzadeh, 2011). The Food and Agricultural Organization (FAO) of the United Nations has in-fact issued a sobering forecast that world food production needs to increase by 70%, in order to keep pace with the demand of growing population. However, increase in food production is faced with the ever-growing challenges especially the new area that can be increased for cultivation purposes is very limited (Saravi and Shokrzadeh, 2011). The increasing world population has therefore put a tremendous amount of pressure on the existing agricultural system so that food needs can be met from the same current resources like land, water etc. In the process of increasing crop production, herbicides, insecticides, fungicides, nematicides, fertilizers and soil amendments are now being used in higher quantities than in the past. These chemicals have mainly come into the picture since the introduction of synthetic insecticides in 1940, when organochlorine (OCl) insecticides were first used for pest management. Before this introduction, most weeds, pests, insects and diseases were controlled using sustainable practices such as cultural, mechanical, and physical control strategies.

Pesticides have now become an integral part of our modern life and are used to protect agricultural land, stored grain, flower gardens as well as to eradicate the pests transmitting dangerous infectious diseases. It has been estimated that globally nearly $38 billion are spent on pesticides each year (Pan-Germany,

2012). Manufacturers and researchers are designing new formulations of pesticides to meet the global demand. Ideally, the applied pesticides should only be toxic to the target organisms, should be biodegradable and eco-friendly to some extent (Rosell et al., 2008). Unfortunately, this is rarely the case as most of the pesticides are non-specific and may kill the organisms that are harmless or useful to the ecosystem. In general, it has been estimated that only about 0.1% of the pesticides reach the target organisms and the remaining bulk contaminates the surrounding environment (Carriger et al., 2006). The repeated use of persistent and non-biodegradable pesticides has polluted various components of water, air and soil ecosystem. Pesticides have also entered into the food chain and have bioaccumulated in the higher tropic level. More recently, several human acute and chronic illnesses have been associated with pesticides exposure (Mostafalou and Abdollahi, 2012). Below, we have detailed the effect of pesticides on target and non-target organisms including earthworms, predators, pollinators, humans, fishes, amphibians, and birds. Additionally, impact of pesticides on soil, water and air ecosystems is also discussed. Furthermore, an eco-friendly practice (Integrated Pest Management (IPM) approach) has been detailed as a strategy that could minimize the use of pesticides.

EFFECTS OF PESTICIDES ON TARGET ORGANISMS

Over the past era there has been an increase in the development of pesticides to target a broad spectrum of pests. The increased quantity and frequency of pesticide applications have posed a major challenge to the targeted pests causing them to either disperse to new environment and/or adapt to the novel conditions (Meyers and Bull, 2002; Cothran et al., 2013). The adaptation of the pest to the new environment could be attributed to the several mechanisms such as gene mutation, change in population growth rates, and increase in number of generations etc. This has ultimately resulted in increased incidence of pest resurgence and appearance of pest species that are resistant to pesticides.

Pesticide Resistance

"Resistance may be defined as a heritable change in the sensitivity of a pest population that is reflected in the repeated failure of a product to achieve the expected level of control when used according to the label recommendation for that pest species" (IRAC, 2013). Resistant individuals tend to be rare in a normal population, but indiscriminate use of chemicals can eliminate normal susceptible populations and thereby providing the resistant individuals a selective advantage in the presence of a pesticide. Resistant individuals continue to multiply in the absence of competition and eventually become the dominant portion of the population over generations. As majority of the individuals of a population are resistant, the insecticide is no longer effective thus causing the appearance or development of insecticide resistance.

Resistance is the most serious bottleneck in the successful use of pesticides these days. The intensive use of pesticides has led to the development of resistance in many targeted pest species around the globe (Tabashnik et al., 2009). Number of resistant insects and mite species had risen to 600 by the end of 1990, and increased to over 700 by the end of 2001. This trend is likely to be continued in 21[st]century as well. Resistance has been found in different insecticides groups e.g., 291 species have developed cyclodiene resistance, followed by DDT (263 species), organophosphates (260 species), carbamates (85 species), pyrethroids (48 species), fumigants (12 species), and other (40 species) (Dhaliwal et al., 2006). Important crop pests, parasites of livestock, common urban pests and disease vectors in some cases have developed resistance to such an extent that their control has become exceedingly challenging (Van Leeuwen et al., 2010; Gondhalekar et al., 2011). However, many factors such as genetics, biology/ecology and control operations influence the development of pesticide resistance (Georghiou and Taylor, 1977).

Insecticide bioassays using whole insects continue to be one of the most widely used approaches for detecting resistance (Brown and Brogdon, 1987; Gondhalekar et al., 2013) despite some associated drawbacks. In the past two decades, however, several new methods employing advanced biochemical and molecular techniques, and combination of insecticide bioassays have been developed for detecting insecticide resistance (Symondson and Hemingway, 1997; Scharf et al., 1999; Zhou et al., 2002). Some examples of these techniques are enzyme electrophoresis, enzyme assays, immuno-assays, allele-specific polymerase chain reaction (PCR) etc.

Pest Resurgence

Pest resurgence is defined as the rapid reappearance of a pest population in injurious numbers following pesticide application. Use of persistent and broad spectrum pesticides that kills the beneficial natural enemies is thought to be the leading cause of pest resurgence. However, resurgence is known to occur due to several reasons, for example, increase in feeding and reproductive rates of insect pests, due to application of sub-lethal doses of pesticides, and sometimes elimination of a primary pest provides favorable conditions for the secondary pests to become primary/key pests (Dhaliwal et al., 2006). There are many pesticide-induced pest outbreaks reported in walnut (*Juglans regia*) (Bartlett and Ewart, 1951), hemlock (*Conium maculatum*) (McClure, 1977), soybeans (*Glycine max*) (Shepard et al., 1977), and cotton (*Gossypium hirsutum*) (Bottrell and Rummel, 1978). Among these, brown plant hopper (BPH) (*Nilaparvata lugens* (Stal)) in rice (*Oryza sativa* L.) cultivation has gained a major importance in Asian countries (Chelliah and Heinrichs, 1984). In general, natural BPH populations were kept under check by natural enemies including mirid bugs, ladybird beetles, spiders and various pathogens. However, pesticides have not only destroyed the natural enemies (Fabellar and Heinrichs, 1986), but have influenced the fecundity of BPH females (Wang et al., 2010) further enhancing their resurgence. Additionally, the resurgence of bed bug, *Cimex lectularius* (Davies et al., 2012) and cotton bollworm *Helicoverpa armigera* (Mironidis et al., 2013) have been reported due to insecticide resistance and indiscriminate use of pesticides.

EFFECTS OF PESTICIDES ON NON-TARGET ORGANISMS

The effect of pesticides on non-target organisms has been a source of worldwide attention and concern for decades. Adverse effects of applied pesticides on non-target arthropods have been widely reported (Ware, 1980). Unfortunately, natural insect enemies e.g., parasitoids and predators are most susceptible to insecticides and are severely affected (Aveling, 1977; Vickerman, 1988). The destruction of natural enemies can exacerbate pest problems as they play an important role in regulating pest population levels. Usually, if natural enemies are absent, additional insecticide sprays are required to control the target pest. In some cases, natural enemies that normally keep minor pests under check are also affected and this can result in secondary pest outbreaks. Along with natural enemies, population of soil arthropods is also drastically disturbed because of indiscriminate pesticide application in agricultural systems. Soil invertebrates including nematodes, springtails, mites, micro-arthropods, earthworms, spiders, insects and other small organisms make up the soil food web and enable decomposition of organic compounds such as leaves, manure, plant residues etc. They are essential for the maintenance of soil structure, transformation and mineralization of organic matter. Pesticide effects on above mentioned soil arthropods therefore negatively impact several links in the food web. The following are the examples of non-target organisms that are adversely impacted by pesticides.

Earthworms

Earthworms represent the greatest proportion of terrestrial invertebrates (>80%) (Yasmin and D'Souza, 2010) and play a significant role in improving soil fertility by decomposing the organic matter into humus. Earthworms also play a major role in improving and maintaining soil structure, by creating channels in soil that enable the process of soil aeration and drainage. However, their diversity, density and biomass are strongly influenced by soil management. They are considered as an important indicator of soil quality in agricultural ecosystems (Paoletti, 1999). Earthworms are affected by various agricultural practices and indiscriminate use of pesticides is one of the leading practices affecting them (Pelosi et al., 2013).

Pesticide applications can cause decline in earthworm populations. For example, carbamate insecticides are very toxic to earthworms and some organophosphates have been shown to reduce earthworm populations (Edwards, 1987). Similarly, a field study conducted in South Africa has also reported that earthworms were influenced detrimentally due to chronic and intermittent exposures to chlorpyrifos and azinphos methyl, respectively (Reinecke and Reinecke, 2007). Various scientific studies reported that pesticides influence earthworm growth, reproduction (cocoon production, number of hatchlings per cocoon, and incubation period) in a dose-dependent manner (Yasmin and D'Souza, 2010). Earthworms exposed to different kind of pesticides showed rupturing of cuticle, oozing out of coelomic fluid, swelling, and paling of body that led to softening of body tissues (Solaimalai et al., 2004). Similarly a study carried out in France showed that the combination of insecticides and fungicides at different concentrations caused neurotoxic effects in earthworms (Schreck et al., 2008). Increased exposure period and higher dose of insecticides can also cause physiological damage (cellular dysfunction and protein catabolism) to earthworms (Schreck et al., 2008).

Predators

Predators are organisms that live by preying on other organisms and they play a very crucial role in keeping pest populations under control. Predators (beneficial organisms) are also an important part of the "biological control" approach which is one component of the integrated pest management strategy discussed later. In some of the examples cited below, pesticides were the main cause for decline in predator population:

- In brinjal (*Solanum melongena* L.) ecosystem, spraying with cypermethrin and imidacloprid caused higher mortality of coccinellids, braconid wasps and predatory spiders compared to when sprayed with bio-pesticides and neem (*Azadirachta indica*) based insecticides (Ghananand et al., 2011).
- Species diversity, richness and evenness of collembola, and numbers of

spiders were found to be lower in chlorpyrifos treated plots compared with control, in grassland pastures in UK (Fountain et al., 2007).

- Studies were carried out to investigate the effects of chemicals on soil arthropods in agricultural area near Everglades National Park, USA. It was found that higher number of arthropods (including predators such as coccinelids and spiders) were present in non-sprayed fields compared to fields sprayed with insecticides and herbicides (Amalin et al., 2009).

- In foliar application, all the systemic neonicotinoids such as imidacloprid, clothianidin, admire, thiamethoxam and acetamiprid were found highly toxic to natural enemies in comparison with spirotetramat, buprofezin and fipronil (Kumar et al., 2012).

Additionally, pesticides can also affect predator behavior and their life-history parameters including growth rate, development time and other reproductive functions. For example, in the eastern USA, glyphosate-based herbicides affected behavior and survival of spiders and ground beetles, apart from affecting arthropod community dynamics that can also influence biological control in an agroecosystem (Evans et al., 2010). Similarly, dimethoate was shown to significantly decrease the body size, haemocyte counts and reduction of morphometric parameters on carabid beetle (*Pterostichus melas italicus*), in Calabria, Italy (Giglio et al., 2011).

Pollinators

Pollinators are biotic agents that play a very important role in pollination process. Some of the recognized pollinators are different species of bees, bumble bees (*Bombus* spp.), honey bees (*Apis* spp.), fruit flies, some beetles, and birds (e.g., hummingbirds, honeyeaters, and sunbirds etc.). Pollinators can be used as bioindicators of ecosystemic processes (process by which physical, chemical, biological events help connecting organisms with their environment) in many ways as their activities are affected by environmental stress caused by parasites, competitors, diseases, predators, pesticides and habitat modifications (Kevan, 1999). However, using pesticides causes direct loss of insect pollinators and indirect loss to crops because of the lack of adequate populations of pollinators (Fishel, 2011).

Pesticide application also affects various activities of pollinators including foraging behaviour, colony mortality and pollen collecting efficiency. Most of our current knowledge about effects of pesticides on change in pollinator behaviour has come from various bee studies as they comprise 80% of the insect pollinator population. For instance, many laboratory studies have demonstrated the lethal and sub-lethal effects of neonicotinoid insecticides (imidacloprid, acetamiprid, clothianidin, thiamethoxam, thiacloprid, dinotefuran and nitenpyram) on foraging behavior, learning and memory abilities of bees (Blacquie`re et al., 2012). Worker bee (female bees that lack full reproductive capacity and play many other roles in bee colony) mortality, decreased pollen collecting efficiency and eventually colony collapse occur due to pesticides (neonicotinoid and pyrethroid) application (Gill et al., 2012). In addition to this, non-lethal exposure of honey bees to neonicotinoid insecticide (thiamethoxam) causes high mortality due to homing failure at a level that could put a risk of colony collapse (Henry et al., 2012). Sub-lethal doses of imidacloprid (the most commonly used pesticide worldwide) affected longevity and foraging in honey bees (*A. mellifera*). *Nosema ceranae* (*Nosema* invades the intestinal tracts of adult bees causing colony collapse disorder (CCD) and nosema disease/nosemosis, which consequently lead to decrease in honey production). Microsporidial infections increased significantly in gut of bees from imidacloprid treated hives. It has been anticipated that interactions between pathogens and imidacloprid pesticide could be a main reason for worldwide honey bee colony mortality, including CCD (Pettis et al., 2012; Wu et al., 2012). There are also reports that imidacloprid reduced brood production due to decline in the fecundity of bumble bees (*B. terrestris*) (Laycock et al., 2012; Whitehorn et al., 2012). On the other hand, little work has been done on the impact of pesticides on wild pollinators. For example, a field study carried out in Italy on an agricultural field found lower bumblebee and butterfly species richness associated with pesticide application. They also found that bees (insect pollinators) were at higher risk from pesticide use (Brittain et al., 2010).

Humans

The deleterious effects of pesticides on human health have started to grow due to their toxicity and persistence in environment and ability to enter into the food chain. Pesticides can enter the human body by direct contact with chemicals, through food especially fruits and vegetables, contaminated water or polluted air. Both acute and chronic diseases can result from pesticide exposure and these are summarized below:

Acute Illness

Acute illness generally appears a short time after contact or exposure to the pesticide. Pesticide drift from agricultural fields, exposure to pesticides during application and intentional or unintentional poisoning generally leads to the acute illness in humans (Dawson et al., 2010; Lee et al., 2011b). Several symptoms such as headaches, body aches, skin rashes, poor concentration, nausea, dizziness, impaired vision, cramps, panic attacks and in severe cases coma and death could occur due to pesticide poisoning (Pan-Germany, 2012). The severity of these risks is normally associated with toxicity and quantity of the agents used, mode of action, mode of application, length and frequency of contact with pesticides and person that is exposed during application (Richter, 2002). About 3 million cases are reported worldwide every year that occur due to acute pesticides poisoning. Out of these 3 million pesticide poisoning cases, 2 million are suicide attempts and the rest of these are occupational or accidental poisoning cases (Singh and Mandal, 2013). Suicide attempts due to acute pesticide poisoning are mainly the result of widespread availability of pesticides in rural areas (Richter, 2002; Dawson et al., 2010). Several strategies have been proposed to reduce the incidences that occur due to acute pesticide poisoning such as restricting the availability of pesticides, substituting the pesticide with a less toxic but with an equally effective alternative and by promoting use of personal protection equipment (Murray and Taylor, 2000; Konradsen et al., 2003). Strict laws regulating pesticide sales along with preventive health programs and community development efforts are needed to enforce such strategies.

Chronic Illness

ontinued exposure to sub-lethal quantities of pesticides for a prolonged period of time (years to decades), results in chronic illness in humans (Pan-Germany, 2012). Symptoms are not immediately apparent and manifest at a later stage. Agricultural workers are at a higher risk to get affected, however general population is also affected especially due to contaminated food and water or pesticides drift from the fields (Pan-Germany, 2012). Incidences of chronic diseases have started to grow as pesticides have become an increasing part of our ecosystem. There is mounting evidence that establish a link between pesticides exposure and the incidences of human chronic diseases affecting nervous, reproductive, renal, cardiovascular, and respiratory systems (Mostafalou and Abdollahi, 2012). The list of chronic diseases that are linked to prolonged pesticide exposure by various studies is summarized in Table 1.

Table 1. The List of chronic diseases that are linked to the exposure to pesticides

Diseases	References
Cancer (Childhood and adult brain cancer; Renal cell cancer; lymphocytic leukaemia (CLL); Prostate Cancer)	Lee et al., 2005; Shim et al., 2009; Heck et al., 2010; Xu et al., 2010; Band et al., 2011; Cocco et al., 2013
Neuro degenerative diseases including Parkinson disease, Alzheimer disease	Elbaz et al., 2009; Hayden et al., 2010; Tanner et al., 2011

Cardio-vascular disease including artery disease	Abdullah et al., 2011; Andersen et al., 2012
Diabetes (Type 2 Diabetes)	Son et al., 2010; Lee et al., 2011a
Reproductive disorders	Petrelli and Mantovani, 2002; Greenlee et al., 2003
Birth defects	Winchester et al., 2009; Mesnage et al., 2010
Hormonal imbalances including infertility and breast pain	Xavier et al., 2004
Respiratory diseases (Asthma, Chronic obstructive pulmonary disease (COPD))	Chakraborty et al., 2009; Hoppin et al., 2009

Several mechanisms have been illustrated that link development of chronic diseases with pesticide exposure. Direct interaction of pesticides with genetic material resulting in DNA damages and chromosomal aberration is considered to be one of the primary mechanisms that lead to the chronic diseases such as cancer etc (Mostafalou and Abdollahi, 2012). In this context, several studies report an increase in frequency of chromosomal aberration, sister chromatid exchange, and breakage in DNA strand in pesticide applicators who worked in agricultural fields (Grover et al., 2003; Santovito et al., 2012). Similar to this, pesticides are also known to induce epigenetic changes (heritable changes without any alteration in DNA sequences) through DNA methylation, histone modifications and expression of non-coding RNAs. For example, neurotoxic pesticide paraquat has been implicated to induce the Parkinson's disease (PD) through epigenetic changes by promoting histone acetylation (Song et al., 2010). Pesticides may also induce oxidative stress by increasing reactive oxygen species (ROS) through altering levels of antioxidant enzymes such as superoxide dismutase, glutathione reductase and catalase (Agrawal and Sharma, 2010). Several health problems such as Parkinson disease, disruption of glucose homeostasis have been linked with pesticides induced oxidative stress (Mostafalou and Abdollahi, 2012).

PESTICIDES AND SOIL ENVIRONMENT

A major fraction of the pesticides that are used for agriculture and other purposes accumulates in the soil. The indiscriminate and repeated use of pesticides further aggravates this soil accumulation problem. Several factors such as soil properties and soil micro-flora determine the fate of applied pesticides, owing to which it undergoes a variety of degradation, transport, and adsorption/desorption processes (Weber et al., 2004; Laabs et al., 2007; Hussain et al., 2009). The degraded pesticides interact with the soil and with its indigenous microorganisms, thus altering its microbial diversity, biochemical reactions and enzymatic activity (Hussain et al., 2009; Munoz-Leoz et al., 2011). A summary of the effects of pesticides on its various components are given below:

- Pesticides that reach the soil can alter the soil microbial diversity and microbial biomass. Any alteration in the activities of soil microorganisms due to applied pesticides eventually leads to the disturbance in soil ecosystem and loss of soil fertility (Handa et al., 1999). Numerous studies have been undertaken which highlight these adverse impacts of pesticides on soil microorganisms and soil respiration (Dutta et al., 2010; Sofo et al., 2012). In addition to this, exogenous applications of pesticides could also influence the function of beneficial root-colonizing microbes such as bacteria and arbuscular mycorrhiza (AM), fungi and algae in soil by influencing their growth, colonization and metabolic activities etc (Debenest et al., 2010; Menendez et al., 2010; Tien and Chen, 2012). The pesticides that reach the soil can interact with soil microflora in several ways:

  a. It can adversely affect the growth, microbial diversity or microbial biomass of the soil microflora. For example, sulfonylurea herbicides- metsulfuron methyl, chlorsulfuron and thifensulfuron methyl were reported to reduce the growth of the fluorescent bacteria *Pseudomonas* strains that were isolated from an agricultural soil (Boldt and Jacobson, 1998). The *Pseudomonas* spp. is known to play an important ecological role in the soil habitat (Boldt and Jacobson, 1998), and hence its reduction can adversely affect soil fertility. Similarly, benomyl, captan and chlorothalonil were reported to suppress the peak soil respiration (an indicator of microbial biomass) in an unamended soil by 30–50% (Chen et al., 2001b).

  b. Pesticide application may also inhibit or kill certain group of microorganisms and outnumber other groups by releasing them from the competition (Hussain et al., 2009). For example, increase in bacterial biomass by 76% was reported in response to endosulfan application and that reduced the fungal biomass by 47% (Xie et al., 2011).

  c. Applied pesticide may also act as a source of energy to some of the microbial group which may lead to increase in their growth and disturbances in the soil ecosystem. For example, bacterial isolates collected from wastewater irrigated agricultural soil showed the capability to utilize chlorpyriphos as a carbon source for their growth (Bhagobaty and Malik, 2010).

d. Pesticides can alter and/or reduce the functional structure and functional diversity of microorganisms, but increase the microbial biomass (Lupwayi et al., 2009). In contrast, application of pesticides can also reduce the microbial biomass while increasing the functional diversity of microbial community. For example, methamidophos and urea decreased the microbial biomass and increased the functional diversity of soil as determined by microbial biomass and community level physiological profiles (Wang et al., 2006).

• Pesticides may also adversely affect the soils vital biochemical reactions including nitrogen fixation, nitrification, and ammonification by activating/deactivating specific soil microorganisms and/or enzymes (Hussain et al., 2009; Munoz-Leoz et al., 2011). The synergistic and additive interactions between pesticides, micro-organisms and soil properties ultimately govern increase or decrease in rate of soil biochemical reactions. For example, populations of the *Azospirillum* spp. bacteria and the rate of ammonification was reported to increase at a particular pesticide concentration (i.e 2.5 to 5.0 kg ha^{-1}) in both laterite and vertisol soils planted to groundnut (*Arachis hypogaea* L.). But the tested pesticides exerted antagonistic interactions on the population of *Azospirillum* spp. and ammonification at higher concentrations (7.5 and 10.0 kg ha^{-1}) (Srinivasulu et al., 2012a).

• Pesticides have also been reported to influence mineralization of soil organic matter, which is a key soil property that determines the soil quality and productivity. For example, a significant reduction in soil organic matter was found after the application of four herbicides (atrazine, primeextra, paraquat, and glyphosate) (Sebiomo et al., 2011). However, soil organic matter then increased after continuous application from the second to the sixth week of herbicide treatment.

• Pesticides that reach the soil may also disturb local metabolism or can alter the soil enzymatic activity (Gonod et al., 2006; Floch et al., 2011). Soil in general contains an enzymatic pool which comprises of free enzymes, immobilized extracellular enzymes and enzymes excreted by (or within) microorganisms that are indicator of biological equilibrium including soil fertility and quality (Mayanglambam et al., 2005; Hussain et al., 2009). Degradation of both pesticides and natural substances in soil is catalyzed by this enzymatic pool (Floch et al., 2011; Kizilkaya et al., 2012). Due to this, measuring the change in enzymatic activity has now been classified as a biological indicator to identify the impact of chemical substances including pesticides on soil biological functions (Garcia et al., 1997; Romero et al., 2010). In fact, it has generally been assumed that measuring the change in enzyme activity is an earlier indicator of soil degradation as compared to the chemical or physical parameters (Dick et al., 1994). Several studies have already been undertaken which indicate both increase and decrease in activities of soil enzymes such as hydrolases, oxidoreductases, and dehydrogenase (Ismail et al., 1998; Megharaj et al., 1999). A description of pesticides interactions with soil enzymes has been summarized in Table 2.

Table 2. A summary of the effects of pesticides on different soil enzymes

Enzyme (Function in soil)	Examples of the pesticides applied	Comments
Nitrogenase (An enzyme used by organisms to fix atmospheric nitrogen gas).	Carbendazim, Imazetapir, Thiram, Captan, 2,4-D, Quinalphos, Monocrotophos, Endosulfan, γ-HCH, Butachlors	Pesticide reduced or inhibited the nitrogenase activity in laboratory or field conditions (Chalam et al., 1996; Martinez-Toledo et al., 1998; Niewiadomska, 2004; Niewiadomska and Klama, 2005; Prasad et al., 2011)/Pesticides stimulated the nitrogenase activity (Patnaik et al., 1995)
Phosphatase (hydrolyzes organic P compounds to inorganic P)	2,4-D, Nitrapyrin, Monocrotophos, Chlorpyrifos, Mancozeb and Carbendazim	Inhibited (Tu, 1981); Activity increased, but higher concentration or increasing incubation period has inhibitory effects (Madhuri and Rangaswamy, 2002; Srinivasulu et al., 2012b)
Urease (catalyzes the hydrolysis of urea into CO_2 and NH_3 and is a key component in the nitrogen cycle in soils)	Isoproturon, Benomyl, Captan, Diazinon, Profenofos	Increase in urease activity (Chen et al., 2001a; Nowak et al., 2004), Pesticide reduced/inhibited urease activity (Abdel-Mallek et al., 1994; Ingram et al., 2005)
Dehydrogenase (DHA): (an oxidoreductase enzyme that catalyzes the removal of hydrogen)	Azadirachtin, Acetamiprid, Quinalphos, Glyphosate	Positive/stimulatory influence on the DHA (Singh and Kumar, 2008; Kizilkaya et al., 2012)/Initially inhibited but later on activity was restored (Andrea et al., 2000; Mayanglambam et al., 2005)

Invertase (hydrolyzes sucrose to fructose and glucose)	Atrazine, Carbaryl, Paraquat	Inhibited invertase activity (Gianfreda et al., 1995; Sannino and Gianfreda, 2001)
β-glucosidase (hydrolyzes disaccharides in soil to form β-glucose)	Metalaxyl, Ridomil gold plus copper	Enzyme activity increased and then decreased (Sukul, 2006) or inhibited (Demanou et al., 2004)
Cellulase (hydrolyzes cellulose to D-glucose)	Benlate, Captan, Brominal	Inhibited enzyme activity (Arinze and Yubedee, 2000; Omar and Abdel-Sater, 2001)
Arylsulphatase (an enzyme that hydrolyzes aryl sulfates)	Cinosulfuron, Prosulfuron, Thifensulfuron methyl, Triasulfuron	Decreased enzyme activity (Sofo et al., 2012)

Several environmental factors control the bioavailability, degradation and effect of pesticides on soil microorganisms in addition to the persistence, concentration and toxicity of the applied pesticides. These include soil texture, presence of organic matter, vegetation and cultural practices (Murage et al., 2007). For instance, a mixture of compost and straw was found to have the capability of bio-degrading different mixtures of fungicides that are usually applied in vineyards when tested under laboratory conditions (Coppola et al., 2011). Similarly, persistence of the herbicide imazapyr was reported to be different in three Argentinean soils (Tandil, Anguil, and Cerro Azul sites) and its half-life was negatively associated with soil pH, iron and aluminum content, and positively related with clay content (Gianelli et al., 2013). Additionally, level of soil moisture is also one of the most important factors that regulates pesticide bioavailability and degradation, as water acts as solvent for pesticide movement and diffusion, and is essential for microbial functioning (Pal and Tah, 2012). For example, degradation of herbicide saflufenacil was found to be faster at field capacity for Nada, Crowley and Gilbert soils as compared to the saturated soil conditions (Camargo et al., 2013).

It is important to monitor the response of soil microbial communities and various enzymatic activities to pesticide exposure in order to reduce their deleterious effects. A combination of both cultivation-dependent (e.g., community-level physiological profiling (CLPP), measuring overall rates of microbial activity) and cultivation-independent (e.g., DNA sequence information, proteomics of environmental samples) methods can be applied to measure and interpret the effects of pesticide exposure (Imfeld and Vuilleumier, 2012). With the advent of efficient new sequencing techniques and metagenomics, the scope of deploying cultivation independent methods for measuring bacterial diversity and function in soil ecosystem has been further increased. Metagenomics approach has been applied already to measure microbial diversity for a range of soil systems including contaminated sites (Ono et al., 2007) and land managed with different cultural practices (Souza et al., 2013). Such high-tech approaches hold the key for future methods to measure the mode of adaptation ecosystem to different pesticides and in development of new methods to better manage pesticide applications.

A careful screening of pesticide effects on soil microflora should be done in laboratory before their field applications. This is because pesticides tend to accumulate in soil due to repeated applications over time and can pose adverse effects on soil microflora even though they are applied at recommended doses (Ahemad et al., 2009). For instance, Ahemad and Khan (2011) reported the highest toxicity to plant growth promoting characteristics of the *Bradyrhi zobium* sp. when its strain MRM6 was grown with three times the recommended field rates of glyphosate, imidacloprid and hexaconazole. Similarly,Dunfield et al. (2000) assessed the effects of the fungicides captan and thiram at rates of 0.25-2 g a.i. kg^{-1} on the survival and phenotypic characteristics of bacteria *Rhizobium leguminosarum* bv. *viceae*, strain C1. They found that even though both captan and thiram significantly reduced the numbers of rhizobia recovered from seed and altered the FAME (fatty acid methyl ester) and biological profiles of recovered rhizobia, it was only the highest concentrations of captan that affected nodulation and plant growth. Similarly, herbicide mesotrione affected soil microbial communities, but the effects were only detected at doses far exceeding the recommended field rates (Crouzet et al., 2010). Overall, it is crucial to comprehend the role of pesticides in perturbing soil environment, so that the risk of pesticide contamination and its consequent adverse impacts on soil environment can be evaluated.

PESTICIDES IN WATER AND AIR ECOSYSTEM

Pesticide residues in water are a major concern as they pose a serious threat to biological communities including humans. There are different ways by which pesticides can get into water such as accidental spillage, industrial effluent, surface run off and transport from pesticide treated soils, washing of spray equipments after spray operation, drift into ponds, lakes, streams and river water, aerial spray to control water-inhibiting pests (Carter and Heather, 1995; Singh and Mandal, 2013). Pesticides generally move from fields to various water reservoirs by runoff or in drainage induced by rain or irrigation (Larson et al., 2010). Similarly, the presence of pesticides in air can be caused by number of factors including spray drift, volatilization from the treated surfaces, and aerial application of pesticides. Extent of drift depends on: droplet size and wind speed. The rate of volatilization is dependent on time after pesticide treatment, the surface on which the pesticide settles, the ambient temperature, humidity and wind speed and the vapor pressure of the ingredients (Kips, 1985). The volatility or semi-volatility nature of the pesticide compounds similarly constitutes an important risk of atmospheric pollution of large cities (Trajkovska et al., 2009). For instance, organophosphorus (OP) pesticides were identified from environmental samples of air and surface following agricultural spray applications in California and Washington (USA) (Armstrong et al., 2013). In Italian forests, indiscriminate use of pesticides and its active metabolites has led to the contamination of water bodies and ambient air, possibly affecting the health of aquatic biota fishes, amphibians and birds (Trevisan et al., 1993). The following section describes the effect of pesticides on fishes, amphibians and birds.

Fishes

Fishes are an important part of marine ecosystem as they interact closely with physical, biological and chemical environment. Fishes provide food source for other animals such as sea birds and marine mammals and thus fishes form an integral part of the marine food web. A lot of research has been carried out to examine the impact of pesticides on decline in fish population (Scholz et al., 2012). Pesticides have been directly linked to causing fish mortality worldwide. For example, 27 freshwater fish species are found to be affected by "plant protection products" (PPP) in Europe (Ibrahim et al., 2013). Another pesticide pentachlorophenol (NaPCP) is reported to cause large numbers of fish mortality in the rice fields of Surinam (Vermeer et al., 1970). Pesticides not only impact the fish but also food webs related to them. The persistent pesticides (organochlorine pesticides and polychlorinated biphenyls) have already been found in the major Arctic Ocean food webs (Hargrave et al., 1992). A survey was conducted to examine the influence of pesticides on aquatic community in West Bengal, India. Many body tissues of the fish such as gills, alimentary canal, liver and brain of carp and catfish were found drastically damaged by pesticides. It was reported that such level of pesticides in fish could harm the fish consumers as well (Konar, 2011).

Several examples are available where pesticides impacted the vital fish organs and behavior. Organophosphate pesticide "Abate" has the potential to alter the vitellogenesis (the process whereby yolky eggs are produced) of catfish (*Heteropneustes fossilis* (Bloch.)), which can severely affect catfish farming (Kumari, 2012). Another major effect of toxic contaminants is on olfaction in fishes since it can affect activities such as mating, locating food, avoiding predators, discriminating kin and homing etc (Tierney et al.,

2010). Simultaneous exposure of trematode parasite (*Telogaster opisthorchis*), freshwater fish (*Galaxias anomalus*) and snails to high glyphosate concentrations significantly reduced their survival and development. Within 24 hrs of exposure to higher glyphosate concentrations, 100% mortality of individuals was found (Kelly et al., 2010).

The impact of pesticides within an aquatic environment is influenced by their water solubility and uptake ability within an organism (Pereira et al., 2013). For example, Clomazone, a popular herbicide, is particularly water soluble; a property that increases its likelihood of contaminating surface and groundwater. The hydrophilic (water-loving) or lipophobic (fat-hating) nature of this pesticide makes it less available in the fatty tissues of an organism (Pereira et al., 2013). Further to this, the toxicity of chemical (e.g., endosulfan in this case) in juvenile rainbow trout (*Oncorhynchus mykiss*) was affected by alkalinity, temperature of water and size of the fish (Capkin et al., 2006).

Pesticides in natural water within the acceptable concentration range can still pose harmful effects. Kock-Schulmeyer et al. (2012) found that even if the pesticide levels found in Llobregat River basin of Spain were within the European Union Environmental Quality Standards, they still accounted for a low to high ecotoxicological risk for aquatic organisms, especially algae and macro-invertebrates. Proper measures should be taken while disposing of expired pesticides, so that their discharge into the water bodies does not danger the aquatic life. This is because the alteration in water pH by expired insecticides can lead to acute toxicity of different fish (Satyavani et al., 2011).

Amphibians

Amphibians are ectothermic, tetrapod vertebrates of class Amphibia. They inhabit a wide variety of habitats, with most species living within fossorial, arboreal, terrestrial, and freshwater aquatic ecosystems. The global decline in the amphibian population has become an environmental concern worldwide. Many amphibian species are on the brink of extinction with 7.4% listed as critically endangered, and at least 43.2% experiencing some sort of population decrease (Stuart et al., 2004). There could be multitude of reasons for decline in amphibian species diversity, but pesticides appear to be playing an important role. Global warming and climate change are leading to more variable and warmer temperatures which may have increased the impact of pesticides on amphibian populations (Relyea, 2003; Johnson et al., 2013).

Many studies showed that amphibians are susceptible to environmental contaminants due to their permeable skin, dual aquatic-terrestrial cycle and relatively rudimentary immune system (Kerby et al., 2010). Several studies showing the impact of pesticides on amphibians are being mentioned here. It has been reported that the world's most commonly used herbicide (Roundup (Glyphosate)) may have far reaching effects on non-target amphibians (Relyea, 2012). Roundup, a globally used herbicide caused high mortality of larval tadpoles (3 different species in North America) and juvenile frogs under natural conditions in an outdoor pond mesocosm (Relyea, 2005a). Most of the evidence supported the toxic effects of pesticides on juvenile European common frogs (*Rana temporaria*) in an agricultural field that was over sprayed. Mortality of frogs ranged from 100% after 1 hour to 40% after 7 days at the recommended concentrations of pesticides (Bruhl et al., 2013). It was found that population of the wood frog (*Lithobates sylyaticus*) near an agricultural area was more resistant to common insecticide (chlorpyrifos), but not to the common herbicide (Roundup). However, no evidence was reported that resistance carried a performance cost when facing competition and the fear of predation (Cothran et al., 2013).

Further to this, pesticides indirectly affect amphibian populations by influencing growth of aquatic communities such as fungi, zooplankton, and phytoplankton as they are one of their prime energy resources. Malathion is the most commonly used broad-spectrum insecticide in United States. It is legal to spray malathion over aquatic habitats to control mosquitoes (Family: Culicidae), that vector malaria and West Nile Virus. A study found that even low concentration of malathion caused direct and indirect effects on aquatic communities (Relyea, 2012). For example, indirect effect of malathion led to decrease in zooplankton diversity, that led to increase in phytoplankton, a decrease in periphyton, and finally decrease in growth of frog tadpoles (Relyea and Hoverman, 2008). Moreover, it was found that repeated applications of low doses had largest impacts than single high dose application of malathion on an aquatic system (Relyea and Diecks, 2008). A comprehensive study was conducted to examine the effect of globally used pesticides including insecticides (carbaryl, malathion, and herbicides (glyphosate, 2, 4-D)) on aquatic communities (algae, 25 animal species). Species richness reduced differentially, 15% with carbaryl, 30% with malathion, and 22% with roundup, whereas 0% with 2, 4-D. It was found that Roundup completely eliminated two species of tadpoles and led to 70% decline in tadpole species (Relyea, 2005b). Another study demonstrated that frogs (*Rana pipiens*) living in agricultural area, where they experienced higher exposure to chemicals were smaller in size and weight than frogs living in area exposed to low-levels of chemicals. It suggests that frogs living in agricultural areas might have more vulnerability to infections and diseases due to their smaller size and alternation in their immune system (decrease in number of splenocytes and phagocytic activity) (Christin et al., 2013).

Birds

Birds are a diverse group, and apart from their distinct songs and calls, showy displays and bright colors adding enjoyment to lives of humans, they play a very critical role in food chains and webs in our ecosystems. Birds are also called "aerial acrobats" consuming different kinds of insects such as mosquitoes, European corn borer moth (*Ostrinia nubilalis*), Japanese beetles (*Popillia japonica*), and many other insect species that are considered as some of the most serious agricultural and health pests. Birds are important biotic components of an ecosystem and help in maintaining a natural equilibrium of insect populations by predating on them. In absence of birds, outbreaks of insect pest populations would become more common, ultimately leading to increased pesticide use. Pesticides exposure by different means such as direct ingestion of pesticide granules and treated seeds, treated crops, direct exposure to sprays, contaminated water, or feeding on contaminated prey, and baits cause birds mortality (Fishel, 2011; Guerrero et al., 2012). In USA, almost 50 pesticides are known for killing song birds, game birds, seabirds, shorebirds, and raptors (BLI, 2004).

Pesticides have a potential to alter behavior and reproduction of birds. Some of the examples cited here, using different synthetic chemicals including carbamates, organochlorines, and organophosphates can cause a decline in the populations of raptorial birds by altering their feeding behavior and reproduction (Mitra et al., 2011). A large area in the world is under rice and therefore cultivation and volume of pesticides applied in rice field is quite significant. Many different kinds of organochlorines, cholinesterase-inhibiting insecticides including carbofuran, monocrotophos, phorate, diazinon, fenthion, phosphamidon, methyl parathion and azinphos-methyl along with fungicides, herbicides and molluscicides are being used in rice fields. Some of these chemicals are highly toxic to birds causing mortality and some chemicals even have the potential to affect their reproductive systems (Parsons et al., 2010). Indirect effects of pesticides, through food chain have been proposed as a possible factor in decline of farmland bird species. Insecticides applied in breeding season can affect breeding performance of corn bunting (*Miliaria calandra*) and yellowhammer (*Emberiza citrinella*) (Boatman et al., 2004).

Pesticides, especially insecticides such as carbamates and organophosphates have the potential to cause bird mortality due to their high toxicity (Hunter, 1995). Further to this, insecticides and fungicides pose a most prominent threat to ground-nesting farmland birds as compared to other agricultural practices. The decline of US grassland birds is attributed to acute pesticide toxicity and not agricultural intensification as previously thought (Mineau and Whiteside, 2013). An estimate suggests that 672 million birds are directly exposed to pesticides every year on farmlands, and 10% of these birds die due to acute toxic effects of pesticides (Williams, 1997). A study was conducted in rice fields of Surinam to examine the effects of pesticides, pentachlorophenol (NaPCP) on birds. NaPCP was sprayed for the purpose of killing *Pomacea* snails. Large numbers of dead sick/dead egrets, herons and jacana birds were found during the period of pesticide application. Pentachlorophenol and endrin levels in these birds suggested that ingestion of contaminated food was the probable cause of sickness and mortality (Vermeer et al., 1970).

PESTICIDES AND BIOMAGNIFICATION

The increase in concentration of pesticides due to its persistent and non-biodegradable nature in the tissues of organisms at each successive level of food chain is known as biomagnification. Due to this phenomenon, organisms at the higher levels of food chain experience greater harm as compared to those at lower levels. Several studies have been undertaken that demonstrate enhanced amount of toxic compounds with increase in trophic levels. For example, out of 36 species collected from three lakes of northeastern Louisiana (USA) that were found to contain residues of 13 organochlorines, tertiary consumers such as green-backed heron (*Butorides striatus*), and snakes etc., contained the highest residues as compared to secondary consumers (bluegill (*Lepomis macrochirus*), blacktail shiner (*Notopis venustus*)) (Niethammer et al., 1984). Similarly, significantly higher concentrations of dichlorodiphenyltrichloroethane (4,4'-DDE) were found in the top consumer fish in Lake Ziway, catfish (*Clarias gariepinus*) than in lower consumers, Nile tilapia (*Oreochromis niloticus*), tilapia (*Tilapia zillii*) and goldfish (*Carassius auratus*) (Deribe et al., 2013). Some of the adverse effects of pesticides on non-target organisms such as fish, amphibians and humans discussed in the above section have also occurred as a result of biomagnifications of the toxic compounds. For example, reproductive failure and population decline in the fish-eating birds (e.g., gulls, terns, herons etc.) was observed as a result of DDE induced eggshell thinning (Grasman et al., 1998). The extent of biomagnifications increases with increase in persistence and lipophilic (fat-loving) characteristics of the particular pesticide. As a result of this, organochlorines are known to have higher biomaginification rate and are more persistent in a wider range of organisms as compared to organophsphates (Favari et al., 2002). It is important to do the risk assessments associated with the pesticides on the basis of their bioaccumulation and biomagnifications before considering them for agricultural purposes.

STRATEGIES FOR PESTICIDE MANAGEMENT

There are a relatively few pesticide resistance management tactics that have been proposed risk-free and have a reasonable chance of success under a variety of different circumstances. Headmost among these are: monitoring of pest population in field before any pesticide application, alteration of pesticides with different modes of action, restricting number of applications over time and space, creating or exploiting refugia, avoiding unnecessary persistence, targeting pesticide applications against the most vulnerable stages of pest life cycle, using synergists which can enhance the toxicity of given pesticides by inhibiting the detoxification mechanisms. The most difficult challenge in managing resistance is not the unavailability of appropriate methods but ensuring their adoption by growers and pest control operators (Denholm et al., 1998; Dhaliwal et al., 2006).

Pest resurgence is a dose-dependent process and there are ways to tackle this problem using correct dosage of effective and recommended pesticides. Resurgence problem occurs due to a number of reasons. One of them is due to farmers' tendency to apply low-dose insecticides due to economic constraints that lead to inadequate and ineffective control of pests. Pest resurgence also occurs due to reduced biological control (most common with insects), reduced competition (most common with weeds; monocots vs. dicots), direct stimulation of pest (due to sub-lethal dose), and improved crop growth.

In the current scenario, optimized use of pesticides is important to reduce environmental contamination while increasing their effectiveness against target pest. This way we can reduce pesticide resistance as well as pest resurgence problems. This has led to the consideration of rational use of pesticides, and the physiological and ecological selectivity of pesticides. Physiological selectivity is characterized by differential toxicity between taxa for a given insecticide. However, ecological selectivity refers to the modification of operational procedure in order to reduce unnecessary destruction to non-target organisms (Dent, 2000). Farmers should focus to use insecticides that are more toxic to target species than their natural enemies which could help to reduce resurgence to some extent (Dhaliwal et al., 2006).

One should consider adopting an Integrated Pest Management (IPM) approach for controlling pests, as these practices are designed to have minimal environment disturbance. The aim of IPM is not only to reduce indiscriminate pesticide use but also to substitute hazardous chemicals with safe chemistries. IPM is a process of achieving long-term, environmentally safe pest control using wide variety of technology and other potential pest management practices. According to National Academy of Science, "IPM refers to an ecological approach in pest management in which all available necessary techniques are consolidated in a unified program so that populations can be managed in such a manner that economic damage is avoided and adverse side effects are minimized" (NAS, 1969). In European arable systems, applied multi-disciplinary research and farmer incentives to encourage the adoption of innovative IPM strategies are essential for development of sustainable maize-based cropping systems. These IPM strategies can contribute immensely to address the European strategic commitment to the environmentally sustainable use of pesticides (Vasileiadis et al., 2011). The added cost and time to do an IPM approach is sometimes a difficult task for growers, but government and extension services can help in convincing and encouraging growers to go for IPM strategy for eco-friendly and long term pest control. We have already discussed earlier that continuous use of pesticides leads to pesticide resistance and pest resurgence problem. To avoid these issues we can always go for other potential management options that include cultural and physical control, host plant resistance, biocontrol, and the use of biopesticides etc.

Cultural Control

Historically, cultural control methods were the farmer's most important tool of preventing crop losses. Cultural control for pest management has been adopted by growers throughout the world for a long time due to its environmentally friendly nature and minimal costs (Gill et al., 2013). Cultural control practices are regular farm operations, which are used to destroy the pests or to prevent them from causing plant damage. Several methods of cultural control have been practiced, such as crop rotation, sanitation, soil solarization, timed planting and harvest, use of resistant varieties, certified seeds, allelopathy, intercropping or "companion planting", use of farmyard manure, and living and organic mulches (Altieri et al., 1978; Dent, 2000; Dhaliwal et al., 2006). Soil solarization (McSorley and Gill, 2010; Gill and McSorley, 2011b) and organic mulches (Gill and McSorley, 2011a) alone and their integration (Gill and McSorley, 2010) were reported as economical and eco-friendly technique for controlling soil-surface arthropods (various insects, and nematodes) (Gill et al., 2010; Gill et al., 2011) and weeds (Gill et al., 2009; Gill and McSorley, 2011b). More effective cultural control can be achieved by synchronizing existing practices with life cycles of pests. This way the weakest link in their life cycle is subjected to adverse climatic conditions.

Large insect populations are killed automatically by farmers when they expose them to adverse climatic conditions through agricultural practices like weeding, ploughing, and hoeing. Ploughing of agricultural field allows turnover of the upper layer of soil while burying the weeds and residues from last year. For example, in South Africa, about 70% of overwintering populations of spotted stalk borer (*Chilo partellus*) and maize stalk borer (*Busseola fusca*) in grain sorghum (*Sorghum bicolor* L.) and maize (*Zea mays* L.) fields were destroyed by slashing the plants. Ploughing and discing of plant residues after slashing further destroyed 24% population on grain sorghum and 19% on maize (Kfir, 1990). Planting dates (Goyal and Kanta, 2005a), and barrier crops (teosinte (*Zea* spp.) and pearl millet (*Pennisetum glaucum* (L.)) (Goyal and Kanta, 2005b) were found to be effective against maize stem borer (*Chilo partellus*) in India. The brown seaweeds *Spatoglossum asperum* and *Sargassum swartzii* can be used as manure to protect plants (tomato (*Solanum lycopersicum* L.) in this case) from root rotting fungi, (*Macrophomina phaseolina, Rhizoctonia solani* and *Fusarium solani*) and root-knot nematode (*Meloidogyne javanica*) and for providing necessary nutrients to plants (Sultana et al., 2012). In India, rodents are pests in agriculture, horticulture, forestry, animal husbandry as well as in human dwellings and rural and urban storage facilities. Cultural methods, such as clean cultivation, proper soil tillage and crop scheduling, barriers, repellents and proofing that reduce the rodent harbourage, food sources and immigration may have long lasting effects (Parshad, 1999).

Physical and Mechanical Control

Managing pest populations using devices which affect them physically or alter their physical environment is called physical control. Exposure to sun rays, steaming, moisture management especially for stored grain pests, and light traps for attracting various kinds of moths, beetles and other pests are different methods used in physical control. For example steaming woolen winter clothes help in eliminating population of the woolly bear moth, *Antherenus vorax* (waterhouse) (Dhaliwal et al., 2006). Hot water treatment of plant storage products like corns, and bulbs helps to kill many concealed pests such as eelworms and bulb flies. Superheating of empty grain storage godowns to a temperature of 50°C for 10-12 hours helps killing hibernating stored grain pests. Exposure of cotton seeds to sun's heat in thin layers for 2-3 days during summer helps in killing diapausing larvae of pink bollworm (*Pectinophora gossypiella* Saunders) (Dhaliwal et al., 2006).

Mechanical control refers to suppression of pest population by manual devices. It includes various practices such as hand picking, trapping and suction devices, clipping, pruning and crushing of infested shoots and floral parts, and exclusion by screens and barriers to keep away house flies (*Musca domestica*), mosquitoes and other pests. In south-eastern Australia, the common starling (*Sturnus vulgaris*) is an established invasive avian pest that is now making incursions into Western Australia which is currently free of this species. Trapping with live-lure birds is suggested to be the most cost-effective and widely implemented starling control technique (Campbell et al., 2012). Numerous wildlife species such as coyotes (*Canis latrans* Say), squirrels (Sciuridae family), and birds are known pests of California agriculture in the United States. For these pests, different non-lethal control options including habitat modification, exclusionary devices, and baiting are generally preferred (Baldwin et al., 2013). Mechanical weed control is mainly associated with tillage practices which are performed with special tools such as harrows, hoes, and brushes in growing crops. Increased knowledge about side effects of herbicides has further driven the interest in adoption of mechanical weed control thus increasing the prevalence of organic farming (Rueda-Ayala et al., 2010; Jat et al., 2011). Trapping using yellow colored sticky traps is an effective way for controlling tephritid flies (Dhaliwal et al., 2006).

Host Plant Resistance

Host plant resistance (HPR) is the genetic ability of the plant to improve its survival and reproduction by a range of adaptations as compared to the other cultivars when exposed to the same level of pest infestation. HPR offers the most effective, economical and eco-friendly method of pest control (Sharma and Ortiz, 2002), and is considered to be a key element of the IPM strategy. Due to this, identifying and developing HPR has always been a major thrust area of plant breeding, and a number of breeding programs aiming to develop pest resistant crops have been deployed in almost all the cultivated crop species. For example, identification and/or development of resistant varieties in maize against European corn borer (*Ostrinia nubilalis* (Hubner)) (Dhaliwal et al., 2006) , brassica against cabbage butterfly (*Pieris brassicae* Linn.) (Chahil and Kular, 2013), wheat (*Triticum aestivum* L.) and rye (*Secale cereale*L.) against *Fusarium* diseases (Miedaner, 1997) *Brassica* sp. against Sclerotinia disease (Garg et al., 2008), and in rice against bacterial blight (Khush et al., 1989). Additionally, availability and access to various germplasm collections have increased the scope of widening the gene-pool of cultivated crops and further identifying and developing HPR. Wild species are especially known to possess a rich repository of genes against various defense traits as they have evolved under different geographic locations. Considerable progress has been made where identification and/or transfer of resistance gene from wild to cultivated species against various pest species has been achieved such as in potato (*Solanum tuberosum* L.) against late blight (*Phytophthora infestans*) from ten wild *Solanum sp.* (Colon and Budding, 1988), wheat against powdery mildew (*Erysiphe graminis*) from wild emmer wheat (*Triticum dicoccoides*) (Reader and Miller, 1991) and mustard (*Brassica juncea*) against *Sclerotinia sclerotiorum* from *Erucastrum cardaminoides* (Garg et al., 2010).

Use of Biotechnology and Molecular Approaches for Developing Resistant Genotype

The advent of new biotechnological and molecular approaches has opened the way to develop resistant genotype that could not only reduce the pesticides application, but it also has a potential to be a part of IPM. Development of resistant genotypes in classical breeding is met with several challenges such as it is time consuming, desired traits are linked with the undesirable traits (linkage drag) and most importantly lack of resistant genotypes in the gene pool. On the other hand, use of biotechnology in crop improvement ensures the development of pest-resistant genotypes in a comparatively short period of time and minimizes the effects of linkage drag. One of the classic examples where biotechnology was successfully deployed to develop resistant genotype is by the synthesis of transgenic plants which involves modifying plant traits by inserting foreign DNA from a different species (De la Pena et al., 1987). A number of different crops including cotton, rice, mustard, and maize have been modified up to now to engineer the genotypes against various biotic stresses (Ahmad et al., 2012). One of the most successful examples of synthesis of transgenic genotype against pest resistance is in cotton where the gene coding for Bt toxin from the bacterium *Bacillus thuringiensis* (Bt) was inserted leading cotton genotypes to produce Bt toxin in its tissue (Pray et al., 2002; Wu et al., 2008). The lepidopteran larvae that fed on the transgenic plants were killed due to Bt toxin eventually decreasing the amount of pesticide applied to the field. Examples of transgenic crops that have been developed with a potential to reduce pesticides use are abound and few of them include potato lines against potato tuber moth (*Phthorimaea operculella*) expressing Cry1Ab (Kumar et al., 2010), rice against yellow stem borer (*Scirpophaga incertulas*) expressing potato proteinase inhibitor 2 (Bhutani et al., 2006) and oilseed rape lines resistant to various fungal attack over-expressing tomato chitinase gene (Grison et al., 1996).

Another strategy where biotechnology and molecular approaches have been deployed to combat biotic stresses involves the use of RNA interference (RNAi) technique. This technique primarily uses transgenic plants expressing double stranded RNA (dsRNA) and that reduces the messenger RNA (mRNA) levels (with a high specificity and fidelity) of a crucial gene in the target pest upon feeding (Price and Gatehouse, 2008; Kos et al., 2009). This ultimately interferes with the development and survival of the target pest. RNAi has emerged as a powerful functional genomics approach and it has been used to engineer several crops against number of insect-pests. For example, RNAi technique was used in tobacco genotype that targeted the gene "integrase splicing factor" in root knot nematode,*Meloidogyne incognita* nematode eventually leading to the decrease in the number of nematodes 6-7 weeks post inoculation (Yadav et al., 2006). When such an advanced and effective approach is combined with IPM, it has a great potential to decrease chemical use in agricultural and other ecosystems.

Biological Control

The process of using natural enemies of particular pests to reduce their populations to such a level where economic losses are either eliminated or suppressed is called biological control. Traditionally the most important biocontrol agents are parasitoids, predators and pathogens. Biological control involves three major techniques, *viz.,* introduction, conservation, and augmentation of natural enemies. Biocontrol agents include vertebrates, nemathelminthes (flatworms, and roundworms), arthropods (spiders, mites, and insects), pathogens like viruses, bacteria, protozoa, fungi and rickettsiae all of which play a dynamic role in natural regulation of insect and mite populations (Dhaliwal et al., 2006). In 1762, the Indian Mynah, *Acridotheres tristis* (Linnaeus), was introduced to control red locust in Mauritius. First significant success in controlling a pest was achieved on the suggestion of C. V. Riley of California (USA) in 1888. The Vedalia beetle (*Rodolia cardinalis* (Mulsant)), was introduced from Australia into California (USA) for the control of cottony-cushion scale (*Icerya purchasi* maskell) on citrus plants. This scale insect had been accidentally introduced earlier from Australia (Dhaliwal et al., 2006).

Biological control of weeds has been very successful worldwide. There are about 41 species of weeds which have been successfully controlled using insects and pathogens as biocontrol agents. Also, 3 weed species have been controlled using native fungi as mycoherbicides (Mcfadyen, 2000). A total of 12 insects were released in Australia against prickly pear (*Opuntia stricta*), out of these, *Dactylopius opuntiae* and *Cactoblastis cactorum* were responsible for the successful control of prickly pear weed (Julien and Griffiths, 1998). In the past decade, Australia has released 43 species of arthropods and pathogens in 19 different projects for successful biological control of many exotic weeds. Effective biological control was achieved in several projects and outstanding success was achieved in the control of rubber vine (*Cryptostegia grandiflora*), and bridal creeper (*Asparagus asparagoides*) (Palmer et al., 2010).

Examples of biological control are available for other organisms like helminthes, nematodes, fungi, bacteria etc. A nematophagous fungus (*Monacrosporium thaumasium*) was found to be effective in controlling cyathostomin, one of the most important helminthes in tropical region of southeastern Brazil (Tavela et al., 2011). *Trichoderma* species are free-living fungi that have been used to control a broad range of plant pathogenic fungi, viruses, bacteria and nematodes especially root-knot nematodes (*Meloidogyne javanica* and *M. incognita*) (Sharon et al., 2011).

Biorational Pesticides

Biorational pesticides/ biopesticides are considered as third-generation pesticides that are rapidly gaining popularity. The word biorational is derived from two words, "biological" and "rational", which means pesticides of natural origin that have limited or no adverse effects on the environment or beneficial organisms. Biopesticides encompass a broad array of microbial pesticides, plant pesticides and biochemical pesticides which are derived from micro-organisms and other natural sources, and processes involving the genetic incorporation of DNA into agricultural commodities. The most commonly used biopesticides include biofungicides (e.g., *Trichoderma* spp.), bioherbicides (*Phytopthora* spp.), bioinsecticides (spore forming bacteria, *Bacillus thuringiensis*, and *B. popilliae*, Actinomycetes), naturally occurring fungi (*Beauveria bassiana*), microscopic roundworms (Entomopathogenic nematodes), Spinosad, insect hormones and insect growth regulators (Gupta and Dikshit, 2010; Singh et al., 2013).

Applications of microbial insecticide, *Chromobacterium subtsugae* for suppression of pecan weevil (*Curculio caryae* (Horn)), and combination of eucalyptus extract and microbial insecticide, *Isaria fumosorosea* (Wize) for control of black pecan aphid (*Melanocallis caryaefoliae* (Davis)) were found promising as alternative insecticides (Shapiro-Ilan et al., 2013). Entomopathogenic nematodes (EPNs) belonging to the families Heterorhabditidae and Steinernematidae are potentially used in South Africa as biocontrol agents against vine mealybug (*Planococcus ficus* (Signoret)) (le Vieux and Malan, 2013). Spinosad was found effective in controlling Colorado potato beetle (*Leptinotarsa decemlineata*) in Iran, and is recommended for use in IPM program for Colorado potato beetle (Soltani and Agricultural, 2011). In China, entomopathogenic fungus (*Beauveria bassiana*) has shown great potential for the management of some bark beetle species including red turpentine beetle (RTB) (*Dendroctonus valens*LeConte), a destructive invasive pest (Zhang et al., 2011).

The allelopathic properties of plants can be exploited successfully as a tool for weed and pathogen reduction. In a rice field, application of allelopathic plant material @ 1-2 tonne/ha reduced weed diversity by 70% and increased yield by 20%. Numerous growth inhibitors identified from these allelopathic plants are responsible for their allelopathic properties and may be a useful source for the future development of bio-herbicides and pesticides (Xuan et al., 2005). A combination of coleopteran-active toxin, *Bacillus thuringiensis* Cry3Aa protoxin and protease inhibitors, especially a potato carboxypeptidase inhibitor, have efficiency in preventing damage to stored products and grains by stored grain coleopteran pests (Oppert et al., 2011).

Chemical Control

Sometimes cultural and other agro-technical practices are not sufficient to keep pest population below economic injury level (lowest pest population density that will cause economic crop damage). Therefore, the chemical control agents are resorted to both as preventive and curative measures to minimize the insect pest damage. A good pesticide should be potent against pests, should not endanger the health of humans and non-target organisms, and should ultimately break down into harmless compounds so that it does not persist in environment. Both relative and specific toxicities of the pesticide need to be estimated in order to determine its potency.

It is very important to know spray droplet size and density chemical dosage, application timing, which can provide adequate pest control. There is also a need for research into the development of suitable packaging and disposal procedures, as well as refining of the application equipment. All of these shall rationalize the use of pesticides, so that they can be used in an acceptable way.

Very strict laws should be enacted to protect wildlife and other non-target organisms. Following directions on the pesticide label can prevent injury to non-target organisms. However, when these directions are not followed, benefits from pesticides can be outweighed by the harm and risk associated with pesticides (Fishel, 2011). During pesticide application, things that need to be considered are timing of insecticide application, dosage and persistence, and selective placement of insecticides as discussed below.

Timing of Pesticide Application

The timing of pesticide application is an important factor to consider before doing any pesticide application. Appropriate application time can ensure not only maximum impact on the target organisms but also least impact on beneficial organisms. Pesticide application timing mainly depends on availability of weather window, time at which pests can be best controlled, and when least damage will be caused to non-target organisms and environment. Flowering period in crops and middle of the day are the times when bees are more prone to insecticides. Hence, insecticide application should not happen at those times to avoid decline in bee populations. Time of insecticide application should coincide with the most vulnerable stage of insect life cycle. Monitoring of insects in the field is thus extremely important for knowing the stage of insect pest in the field. Monitoring systems are available for most of the insect pests, but spray regime or experiments need to be carried out to determine the most appropriate time for insecticide application for insects for which monitoring systems are not available (Hull and Starner, 1983; Richter and Fuxa, 1984).

Time of the day and season of the year are also important to consider when making pesticide applications. The early morning and evening hours are often the best times for pesticide application because windy conditions are more likely to occur around midday when the temperature warms near the ground level. This causes hot air to rise quickly and mix rapidly with the cooler air above it, favoring drift. During stable conditions, a layer of warm air can stay overhead and not promote mixing with colder air that stays below and closer to the ground. Inversions tend to dissipate during the middle of the day when wind currents mix the air layers. It is very important that applicators recognize thermal inversions and do not spray under those conditions. A temperature or thermal inversion is a condition that occurs naturally and exists when the air at ground level is cooler than the temperature of the air above it. Wind speed is the most important weather factor influencing drift. High wind speeds will move droplets downwind and deposit them off the target. On the other hand, dead calm conditions are never recommended due to likelihood of temperature inversions (Fishel and Ferrell, 2013). Drifting of pesticides increases the possibility of injury to pollinators, humans, domestic animals and wildlife. It is recommended not to spray in wind speed above 2.5 miles/second which otherwise can cause excessive drift and eventually contamination of adjacent areas (Matthews, 1981). Pesticide application should not be made just before rain because pesticides can be washed off by the rain without any impact on the target pest.

Dosage and Persistence

Pesticide dose should be sufficient but no greater than the level required for best results. The pesticide manufacturer sets the dose to ensure an acceptable level of control, producing acceptable residue levels, and maximizing returns per unit of formulated insecticide. Persistent pesticides have their benefit of longer persistence on the target and therefore requires less frequent spraying compared to non- persistent pesticides. But care should be taken while using persistent pesticides since these might diminish benefits from natural enemies even at lower doses. If an insecticide is persistent in nature, chances of insecticide residues being harmful to natural enemies are greatly increased (Dent, 2000).

Selective Placement

Distribution of pesticides in the field should be such that maximum target cover is achieved. Usually only about 1% of the applied pesticides is able to reach its target, while a large amount of it is wasted. Understanding the pest biology and behavior is critical as it can provide information on pest's habitat, fecundity, feeding etc., which can be important considerations before applying pesticides. Most of the pesticides are applied in liquid form and thus the droplet size is very important in determining their effectiveness. Small droplets provide better coverage and greater likelihood of coming in contact with the target compared to larger droplets that can bounce off the plant surface very easily. The disadvantage with smaller and bigger droplets is the increased chance of drift and therefore a balance has to be considered between smaller droplets to obtain the maximum effectiveness and reduced drift.

In situations where crops are grown on beds covered with plastic mulch, pesticides should be injected into soil at the time the plastic is laid or injected afterward through drip irrigation system to achieve maximum pesticide effectiveness. For termite (Order: Isoptera) treatments, sometimes perimeter application of insecticides is required around structures/buildings. Additionally, liquids that form foams following injections can be injected into small spaces that are or might be inhabited by termites or other small creatures.

CONCLUSION

Although, pesticides were used initially to benefit human life through increase in agricultural productivity and by controlling infectious disease, their adverse effects have overweighed the benefits associated with their use. The above discussion clearly highlights the severe consequences of indiscriminate pesticide use on different environmental components. Some of the adverse effects associated with pesticide application have emerged in the form of increase in resistant pest population, decline in on beneficial organisms such as predators, pollinators and earthworms, change in soil microbial diversity, and contamination of water and air ecosystem. The persistent nature of pesticides has impacted our ecosystem to such an extent that pesticides have entered into various food chains and into the higher trophic levels such as that of humans and other large mammals. Some of the acute and chronic human illnesses have now emerged as a consequence of intake of polluted water, air or food.

This is the time that necessitates the proper use of pesticides to protect our environment and eventually health hazards associated with it. Alternative pest control strategies such as IPM that deploys a combination of different control measures such as cultural control, use of resistant genotype, physical and mechanical control, and rational use of pesticide could reduce the number and amount of pesticide applications. Further, advanced approaches such as biotechnology and nanotechnology could facilitate in developing resistant genotype or pesticides with fewer adverse effects. Community development and various extension programs that could educate and encourage farmers to adopt the innovative IPM strategies hold the key to reduce the deleterious impact of pesticides on our environment.

ACKNOWLEDGEMENTS

We would like to say special thanks to Drs. Gaurav Goyal (Territory Agronomist, Monsanto), Ameya D. Gondhalekar (Research Assistant Professor, Purdue University), Siddharth Tiwari (Entomologist, BASF), and Matthew R. Tarver (Research Entomologist, USDA) for their valuable suggestions and comments for improving this manuscript.

REFERENCES

1. Abdel-Mallek AY, Moharram AM, Abdel-Kader MI, Omar SA. Effect of soil treatment with the organophosphorus insecticide profenofos on the fungal flora and some microbial activities. Microbiological Research 1994;149:167-171.

2. Abdullah NZ, Ishaka A, Samsuddin N, Mohd Rus R, Mohamed AH. Chronic organophosphate pesticide exposure and coronary artery disease: finding a bridge. IIUM Research, Invention and Innovation Exhibition (IRIIE) 2011;223.

3. Agrawal A, Sharma B. Pesticides induced oxidative stress in mammalian systems: a review. International Journal of Biological and Medical Research (IJBMR) 2010;1:90-104.

4. Ahemad M, Khan MS, Zaidi A, Wani PA. Remediation of herbicides contaminated soil using microbes. In: Khan MS, Zaidi A, Musarrat J (eds.) Microbes in sustainable agriculture. Nova Science Publishers Inc., New York;2009. p261-284.

5. Ahemad M, Khan MS. Effect of pesticides on plant growth promoting traits of greengram-symbiont, *Bradyrhizobium* sp. strain MRM6. Bulletin of Environmental Contamination and Toxicology 2011;86:384-388.

6. Ahmad P, Ashraf M, Younis M, Hu X, Kumar A, Akram NA, Al-Qurainy F. Role of transgenic plants in agriculture and biopharming. Biotechnology Advances 2012;30:524-540.

7. Altieri MA, Francis CA, Schoonhoven AV, Doll JD. A review of insect

prevalence in maize (*Zea mays* L.) and bean (*Phaseolus vulgaris* L.) polycultural systems. Field Crops Research 1978;1:33-49.

8. Amalin DM, Peña JE, Duncan R, Leavengood J, Koptur S. Effects of pesticides on the arthropod community in the agricultural areas near the Everglades National Park. Proceedings of Florida State Horticultural Society 2009;122:429-437.

9. Andersen HR, Wohlfahrt-Veje C, Dalgård C, Christiansen L, Main KM, Nellemann C, Murata K, Jensen TK, Skakkebaek NE, Grandjean P. Paraoxonase 1 polymorphism and prenatal pesticide exposure associated with adverse cardiovascular risk profiles at school age. PloS One 2012;7:e36830.

10. Andrea MM, Peres TB, Luchini LC, Pettinelli Junior A. Impact of long-term pesticide application on some soil biological parameters. Journal of Environmental Science and Health B 2000;35:297-307.

11. Arinze AE, Yubedee AG. Effect of fungicides on *Fusarium* grain rot and enzyme production in maize (*Zea mays* L.). Global Journal of Pure and Applied Sciences 2000;6:629-634.

12. Armstrong JL, Fenske RA, Yost MG, Galvin K, Tchong-French M, Yu J. Presence of organophosphorus pesticide oxygen analogs in air samples. Atmospheric Environment 2013;66:145-150.

13. Aveling C. The biology of Anthocorids (Heterophera: Anthocoridae) and their role in the integrated control of the damson-hop aphid (*Phorodon humili* Schrank). PhD Thesis, University of London;1977.

14. Baldwin RA, Salmon TP, Schmidt RH, Timm RM. Wildlife pests of California agriculture: regional variability and subsequent impacts on management. Crop Protection 2013;46:29-37.

15. Band PR, Abanto Z, Bert J, Lang B, Fang R, Gallagher RP, Le ND. Prostate cancer risk and exposure to pesticides in British Columbia farmers. Prostate 2011;71:168-183.

16. Bartlett B, Ewart WH. Effect of parathion on parasites of *Coccus hesparidum.* Journal of Economic Entomology 1951;44:344-347.

17. Bhagobaty RK, Malik A. Utilization of chlorpyrifos as a sole source of carbon by bacteria isolated from wastewater irrigated agricultural soils in an industrial area of western Uttar Pradesh, India. Research Journal of Microbiology 2010;3:293-307.

18. Bhutani S, Kumar R, Chauhan R, Singh R, Choudhury VK, Choudhury JB. Development of transgenic indica rice plants containing potato proteinase inhibitor 2 gene with improved defense against yellow stem

borer. Physiological and Molecular Biology of Plants 2006;12:43-52.

19. Blacquiere T, Smagghe G, Cornelis AM, Gestel van, Mommaerts V. Neonicotinoids in bees: a review on concentrations, side-effects and risk assessment. Ecotoxicology 212;21:973-992.

20. BLI. State of the world's birds. Birdlife International 2004. (http://www/ biodiversityinfo.org/default.php.).

21. Boatman ND, Brickle NW, Hart JD, Milsom TP, Morris AJ, Murray AWA, Murray KA, Robertson PA. Evidence for the indirect effects of pesticides on farmland birds. Ibis 2004;146:131-143.

22. Boldt TS, Jacobsen CS. Different toxic effects of the sulphonylurea herbicides metsulfuron methyl, chlorsulfuron and thifensulfuron methyl on fluorescent pseudomonads isolated from an agricultural soil. FEMS Microbiology Letters 1998;161:29-35.

23. Bottrell DG, Rummel DR. Response of *Heliothis* populations to insecticides applied in an area-wide reproduction diapause boll weevil suppression program. Journal of Economic Entomology 1978;71:87-92.

24. Brittain CA, Vighi M, Bommarco R, Settele J, Potts SG. Impacts of a pesticide on pollinator species richness at different spatial scales. Basic and Applied Ecology 2010;11:106-115.

25. Brown TM, Brogdon WG. Improved detection of insecticides resistance through conventional and molecular techniques. Annual Review of Entomology 1987;32:145-162.

26. Bruhl CA, Schmidt T, Pieper S, Alscher A. Terrestrial pesticide exposure of amphibians: an underestimated cause of global decline? Scientific Reports 2013;3. (Article number 1135). DOI:10.1038/srep01135.

27. Camargo ER, Senseman SA, Haney RL, Guice JB, McCauley GN. Soil residue analysis and degradation of saflufenacil as affected by moisture content and soil characteristics. Pest Management Science 2013;69:1291-1297. DOI:10.1002/ps.3494.

28. Campbell S, Cook S, Mortimer L, Palmer G, Sinclair R, Woolnough AP. To catch a starling: testing the effectiveness of different trap and lure types. Wildlife Research 2012;39:183-191.

29. Capkin E, Altinok I, Karahan S. Water quality and fish size affect toxicity of endosulfan, an organochlorine pesticide, to rainbow trout. Chemosphere 2006;64:1793-1800.

30. Carriger JF, Rand GM, Gardinali PR, Perry WB, Tompkins MS, Fernandez AM. Pesticides of potential ecological concern in sediment from South Florida canals: an ecological risk prioritization for aquatic arthropods.

Soil and Sediment Contamination 2006;15:21-45.

31. Carter AD, Heather AIJ. Pesticides in groundwater: In: Best GA, Ruthven AD. (eds.) Pesticides- developments, impacts, and controls. The Royals Society of Chemistry, London, UK;1995. p113-123.

32. Chahil GS, Kular JS. Biology of *Pieris brassicae* (Linn.) on different *brassica* species in the plains of Punjab. Journal of Plant Protection Research 2013;53:53-59.

33. Chakraborty S, Mukherjee S, Roychoudhury S, Siddique S, Lahiri T, Ray MR. Chronic exposures to cholinesterase-inhibiting pesticides adversely affect respiratory health of agricultural workers in India. Journal of Occupational Health 2009;51:488-497.

34. Chalam AV, Sasikala C, Ramana CV, Rao PR. Effect of pesticides on hydrogen metabolism of *Rhodobacter sphaeroides* and *Rhodopseudomonas palustris*. FEMS Microbiology Ecology 1996;19:1-4.

35. Chelliah S, Heinrichs EA. Factors contributing to brown planthopper resurgence. Resurgence of the brown planthopper, *Nilaparvata lugens* (Stål) following insecticide application. Paper presented at the 9th Annual Conference of the Pest Control Council of the Philippines, Philippine International Convention Center, Manila, May 3-6;1984. p36 (mimeo).

36. Chen SK, Edwards CA, Subler S. A microcosm approach for evaluating the effects of the fungicides benomyl and captan on soil ecological processes and plant growth. Applied Soil Ecology 2001a;18:69-82.

37. Chen SK, Edwards CA, Subler S. Effect of fungicides benomyl, captan and chlorothalonil on soil microbial activity and nitrogen dynamics in laboratory incubations. Soil Biology and Biochemistry 2001b;33:1971-1980.

38. Christin MS, Menard L, Giroux I, Marcogliese DJ, Ruby S, Cyr D, Fournier M, Brousseau P. Effects of agricultural pesticides on the health of *Rana pipiens* frogs sampled from the field. Environmental Science and Pollution Research 2013;20:601-611.

39. Cocco P, Satta G, Dubois S, Pili C, Pilleri M, Zucca M, Mannetje AM, Becker N, Benavente Y, Sanjose SDe, Foretova L, Staines A, Maynadie M, Nieters A, Brennan P, Miligi L, Ennas MG, Boffetta P. Lymphoma risk and occupational exposure to pesticides: results of the Epilymph study. Occupational and Environmental Medicine 2013;70:91-98.

40. Colon LT, Budding DJ. Resistance to late blight (*Phytophthora infestans*) in ten wild *Solanum* species. Euphytica 1988;39:77-86.

41. Coppola L, Comitini F, Casucci C, Milanovic V, Monaci E, Marinozzi M, Taccari, M, Ciani M, Vischetti, C. Fungicides degradation in an organic biomixture: impact on microbial diversity. New Biotechnology 2011;29:99-106.

42. Cothran RD, Brown JM, Relyea RA. Proximity to agriculture is correlated with pesticide tolerance: evidence for the evolution of amphibian resistance to modern pesticides. Evolutionary Applications 2013;6:832-841.

43. Crouzet O, Batisson I, Besse-Hoggan P, Bonnemoy F, Bardot C, Poly F, Bohatier J, Mallet C. Response of soil microbial communities to the herbicide mesotrione: a dose-effect microcosm approach. Soil Biology and Biochemistry 2010;42:193-202.

44. Davies TGE, Field LM, Williamson MS. The re-emergence of the bed bug as a nuisance pest: implications of resistance to the pyrethroid insecticides. Medical and Veterinary Entomology 2012;26:241-254.

45. Dawson AH, Eddleston M, Senarathna L, Mohamed F, Gawarammana I, Bowe SJ, Manuweera G, Buckley NA. Acute human lethal toxicity of agricultural pesticides: a prospective cohort study. PLoS Medicine 2010;7:e1000357.

46. De la Pena A, Lörz H, Schell J. Transgenic rye plants obtained by injecting DNA into young floral tillers. Nature 1987;325:274-276.

47. Debenest T, Silvestre J, Coste M, Pinelli E. Effects of pesticides on freshwater diatoms. In: Reviews of environmental contamination and toxicology, Springer, New York 2010;203:87-103.

48. Demanou J, Monkiedje A, Njine T, Foto SM, Nola M, Serges H, Togouet Z, Kemka N. Changes in soil chemical properties and microbial activities in response to the fungicide Ridomil gold plus copper. International Journal of Environmental Research and Public Health 2004;1:26-34.

49. Denholm I, Birnie LC, Kennedy PJ, Shaw KE, Perry JN, Powell W. The complementary roles of laboratory and field testing in ecotoxicological risk assessment. The 1998 Brighton Conferences: Pest and Diseases Volume 2. British Crop Protection Council, Farham, UK;1998. p583-590.

50. Dent D. Cultural and interference methods. In: Insect pest management, 2nd edition. CABI publishing, Cambridge, MA, USA;2000. p235-266.

51. Deribe E, Rosseland BO, Borgstrom R, Salbu B, Gebremariam Z, Dadebo E, Skipperud L, Eklo OM. Biomagnification of DDT and its metabolites in four fish Research 2010;117:51-58species of a tropical lake. Ecotoxicology and Environmental Safety 2013;95:10-18.

52. Dhaliwal GS, Singh R, Chhillar BS. In: Essentials of agricultural entomology. Kalyani Publishers, New Delhi, India;2006.

53. Dick RP. Soil enzyme activities as indicators of soil quality. In: Doran JW, Coleman DC, Bezdicek DF, Stewart BA. (eds.) Defining soil quality for a sustainable environment, Soil Science Society of America, Madison, WI, USA. 1994;p107-124.

54. Dunfield KE, Siciliano SD, Germida JJ. The fungicides thiram and captan affect the phenotypic characteristics of *Rhizobium leguminosarum* strain C1 as determined by FAME and Biolog analyses. Biology and Fertility of Soils 2000;31:303-309.

55. Dutta M, Sardar D, Pal R, Kole RK. Effect of chlorpyrifos on microbial biomass and activities in tropical clay loam soil. Environmental Monitoring and Assessment 2010;160:385-391.

56. Edwards CA. The environmental impact of insecticides. In: Delucchi V (ed.) Integrated pest management, Protection Integrée Quo vadis? An International Perspective. Parasitis 86, Geneva, Switzerland;1987. p309-329.

57. Elbaz A, Clavel J, Rathouz PJ, Moisan F, Galanaud JP, Delemotte B, Alperovitch A, Tzourio C. Professional exposure to pesticides and Parkinson disease. Annals of Neurology 2009;66:494-504.

58. Evans SC, Shaw EM, Rypstra AL. Exposure to a glyphosate-based herbicide affects agrobiont predatory arthropod behavior and long-term survival. Ecotoxicology 2010;19:1249-1257.

59. Fabellar LT, Heinrichs EA. Relative toxicity of insecticides to rice planthopper and leafhoppers and their predators. Crop Protection 1986;5:254-258.

60. Favari L, Lopez E, Martinez-Tabche L, Diaz-Pardo E. Effect of insecticides on plankton and fish of Ignacio Ramirez reservoir (Mexico): a biochemical and biomagnification study. Ecotoxicology and Environmental Safety 2002;51:177-186.

61. Fishel FM, Ferrell JA. Managing pesticide drift. Agronomy department. PI232 University of Florida, Gainesville, FL, USA;2013 (http://edis.ifas.ufl.edu/pi232) (accessed 14 October 2013).

62. Fishel FM. Pesticides effects on nontarget organisms. PI-85. Pesticide information office, Florida Cooperative Extension Service, IFAS, University of Florida, Gainesville, FL, USA; 2011. (http://edis.ifas.ufl.edu/pi122) (accessed 14 October 2013).

63. Floch C, Chevremont AC, Joanico K, Capowiez Y, Criquet S. Indicators

of pesticide contamination: Soil enzyme compared to functional diversity of bacterial communities via Biolog® Ecoplates. European Journal of Soil Biology 2011;47:256-263.

64. Fountain MT, Brown VK, Gang AC, Symondson WOC, Murray PJ. The effect of the insecticide chlorpyrifos on spider and collembolan communities. Pedobiologia 2007;51:147-158.

65. Garcia C, Hernandez T, Costa F. Potential use of dehydrogenase activity as an index of microbial activity in degraded soils. Communications in Soil Science and Plant Analysis 1997;28:123-134.

66. Garg, H, Atri C, Sandhu PS, Kaur B, Renton M, Banga SK, Singh H, Singh C, Barbetti, MJ, Banga, SS. High level of resistance to *Sclerotinia sclerotiorum* in introgression lines derived from wild crucifers and the crop *Brassica* species *B. napus* and *B. juncea*. Field Crops Research 2010;117:51-58.

67. Garg H, Sivasithamparam K, Banga SS, Barbetti, MJ. Cotyledon assay as a rapid and reliable method of screening for resistance against *Sclerotinia sclerotiorum* in *Brassica napus* genotypes. Australasian Plant Pathology 2008;37:106-111.

68. Georghiou GP, Taylor CE. Genetic and biological influences in evolution of insecticide resistance. Journal of Economic Entomology 1977;70:319-323.

69. Ghananand T, Prasad CS, Lok N. Effect of insecticides, bio-pesticides and botanicals on the population of natural enemies in brinjal ecosystem. Vegetos- An International Journal of Plant Research 2011;24:40-44.

70. Gianelli VR, Bedmar F, Luis J. Persistence and sorption of imazapyr in three Argentinean soils. Environmental Toxicology and Chemistry 2013. DOI:10.1002/etc.2400.

71. Gianfreda L, Sannino F, Violante A. Pesticide effects on the activity of free, immobilized and soil invertase. Soil Biology and Biochemistry 1995;27:1201-1208.

72. Giglio A, Giulianini PG, Zetto T, Talarico F. Effects of the pesticide dimethoate on a non-target generalist carabid, *Pterostichus melas italicus*(Dejean, 1828) (Coleoptera:Carabidae). Italian Journal of Zoology 2011;78:471-477.

73. Gill HK, McSorley R, Branham M. Effect of organic mulches on soil surface insects and arthropods. Florida Entomologist 2011;94:226-232.

74. Gill HK, McSorley R, Goyal G, Webb SE. Mulch as a potential management strategy for lesser cornstalk borer, *Elasmopalpus lignosellus*

(Insecta: Lepidoptera: Pyralidae), in bush bean (*Phaseolus vulgaris*). Florida Entomologist 2010;93:183-190.

75. Gill HK, McSorley R, Goyal G. Soil Solarization and organic mulches: tools for pest management. In: Dhawan AK, Singh B, Brar-Bhullar M, Arora R (eds.). Integrated pest management. Scientific Publishers, Jodhpur, India;2013. p301-324.

76. Gill HK, McSorley R, Treadwell DD. Comparative performance of different plastic films for soil solarization and weed suppression. HortTechnology 2009;19:769-774.

77. Gill HK, McSorley R. Effect of integrating soil solarization and organic mulching on the soil surface insect community. Florida Entomologist 2010;93:308-309.

78. Gill HK, McSorley R. Impact of different organic mulches on the soil surface arthropod community and weeds. International Journal of Pest Management 2011(a);58:33-40.

79. Gill HK, McSorley R. Effect of different inorganic/synthetic mulches on weed suppression during soil solarization. Proceeding of Florida State Horticultural Society 2011(b);124:310-313.

80. Gill RJ, Ramos-Rodriguez O, Raine NE. Combined pesticide exposure severely affects individual- and colony-level traits in bees. Nature 2012;491:105-108.

81. Gondhalekar AD, Scherer CW, Saran RK, Scharf ME. Implementation of an indoxacarb susceptibility monitoring program using field-collected German cockroach isolates from the United States. Journal of Economic Entomology 2013;106:945–953.

82. Gondhalekar AD, Song C, Scharf, ME. Development of strategies for monitoring indoxacarb and gel bait susceptibility in the German cockroach (Blattodea: Blattellidae). Pest Management Science 2011;67:262-270.

83. Goyal G, Kanta U. Influence of time and method of sowing of kharif maize on the incidence of *Chilo partellus* (Swinhoe). Journal of Insect Science 2005a;19:24-29.

84. Goyal G, Kanta U. Influence of barrier crops on the incidence of *Chilo partellus* (Swinhoe) in kharif maize. Journal of Insect Science 2005b;19:54-58.

85. Gonod LV, Martin-Laurent F, Chenu C. 2, 4-D impact on bacterial communities, and the activity and genetic potential of 2, 4-D degrading communities in soil. FEMS Microbiology Ecology 2006;58:529-537.

86. Grasman KA, Scanlon PF, Fox GA. Reproductive and physiological effects

of environmental contaminants in fish-eating birds of the Great Lakes: a review of historical trends. In: Trends in levels and effects of persistent toxic substances in the Great Lakes, Springer, Netherlands;1998. p117-145.

87. Greenlee AR, Arbuckle TE, Chyou PH. Risk factors for female infertility in an agricultural region. Epidemiology 2003;14:429-436.

88. Grison R, Grezes-Besset B, Schneider M, Lucante N, Olsen L, Leguay JJ, Toppan A. Field tolerance to fungal pathogens of *Brassica napus*constitutively expressing a chimeric chitinase gene. Nature Biotechnology 1996;14:643-646.

89. Grover P, Danadevi K, Mahboob M, Rozati R, Banu BS, Rahman MF. Evaluation of genetic damage in workers employed in pesticide production utilizing the Comet assay. Mutagenesis 2003;18:201-205.

90. Guerrero I, Morales MB, Oñate JJ, Geige, F, Berendse F, Snoo GD, Tscharntke T. Response of ground-nesting farmland birds to agricultural intensification across Europe: Landscape and field level management factors. Biological Conservation 2012;152:74-80.

91. Gupta S, Dikshit AK. Biopesticides: An eco-friendly approach for pest control. Journal of Biopesticides 2010;3:186-188.

92. Handa SK, Agnihotri NP, Kulshreshtha G. Effect of pesticide on soil fertility. In: Pesticide residues; significance, management and analysis 1999;184-198.

93. Hargrave BT, Harding GC, Vass WP, Erickson PE, Fowler BR, Scott V. Organochlorine pesticides and polychlorinated biphenyls in the Arctic Ocean food web. Archives of Environmental Contamination and Toxicology 1992;22:41-54.

94. Hayden KM, Norton MC, Darcey D, Østbye T, Zandi PP, Breitner JCS, Welsh-Bohmer KA. Occupational exposure to pesticides increases the risk of incident AD The Cache County Study. Neurology 2010;74:1524-1530.

95. Heck JE, Charbotel B, Moore LE, Karami S, Zaridze DG, Matveev V, Janout V, Kollarova H, Foretova L, Bencko V, Szeszenia-Dabrowska N, Lissowska J, Mates D, Ferro G, Chow W-H, Rothman N, Stewart P, Brennan P, Boffetta P. Occupation and renal cell cancer in Central and Eastern Europe. Occupational and Environmental Medicine 2010;67:47-53.

96. Henry M, Bèguin M, Requier F, Rollin O, Odoux J-F, Aupinel P, Aptel J, Tchamitchian S, Decourtye A. A common pesticide decreases foraging success and survival in honey bees. Science 2012;336:348-350.

97.   Hoppin JA, Umbach DM, London SJ, Henneberger PK, Kullman GJ, Coble J, Alavanja MCR, Freeman LEB, Sandler DP. Pesticide use and adult-onset asthma among male farmers in the agricultural health study. European Respiratory Journal 2009;34:1296-1303.

98.   Hull LA, Starner R van. Effectiveness of insecticide applications timed to correspond with the development of rosy apple aphid (Homoptera: Aphididae) on apple. Journal of Economic Entomology 1983;76:594-59.

99.   Hunter K. The poisoning of non-target animals. In: Best GA, Ruthven AD (eds.) Pesticides- developments, impacts, and controls. The Royals Society of Chemistry, London, UK;1995. p74-86.

100. Hussain S, Siddique T, Saleem M, Arshad M, Khalid A. Impact of pesticides on soil microbial diversity, enzymes, and biochemical reactions. Advances in Agronomy 2009;102:159-200.

101. Ibrahim L, Preuss TG, Ratte HT, Hommen U. A list of fish species that are potentially exposed to pesticides in edge-of-field water bodies in the European Union-a first step towards identifying vulnerable representatives for risk assessment. Environmental Science and Pollution Research 2013;20:2679-2687.

102. Imfeld G, Vuilleumier S. Measuring the effects of pesticides on bacterial communities in soil: a critical review. European Journal of Soil Biology 2012;49:22-30.

103. Ingram CW, Coyne MS, Williams DW. Effects of commercial diazinon and imidacloprid on microbial urease activity in soil. Journal of Environmental Quality 2005;34:1573-1580.

104. IRAC. Resistance management for sustainable agriculture and improved public health;2013. (http://www.irac-online.org/).

105. Ismail BS, Yapp KF, Omar O. Effects of metsulfuron-methyl on amylase, urease, and protease activities in two soils. Australian Journal of Soil Research 1998;36:449–456.

106. Jat RS, Meena HN, Singh AL, Surya JN, Misra JB. Weed management in groundnut (*Arachis hypogaea* L.) in India-a review. Agricultural Reviews 2011;32:155-171.

107. Johnson LA, Welch B, Whitfield SM. Interactive effects of pesticide mixtures, predators, and environmental regimes on the toxicity of two pesticides to red-eyed tree frog larvae. Environmental Toxicology 2013;32:2379-2386.

108. Julien MH, Griffiths MW. Biological Control of Weeds: a world catalogue of agents and their target weeds. Wallingford, Oxford: CAB International,

4th ed.;1998. p223.

109. Kelly DW, Poulin R, Tompkins DM, Townsend C. Synergistic effects of glyphosate formulations and parasite infection on fish malformations and survival. Journal of Applied Ecology 2010;47:498-504.

110. Kerby JL, Richards-Hrdlicka KL, Storfer A, Skelly DK. An examination of amphibian sensitivity to environmental contaminants: are amphibians poor canaries? Ecology Letters 2010;13:60-67.

111. Kevan PG. Pollinators as bioindicators of the state of the environments: species, activity and diversity. Agriculture, Ecosystems and Environment 1999;74:373-393.

112. Kfir R. Prospects for cultural control of the stalk borers, *Chilo partellus* (Swinhoe) and *Busseola fusca* (Fuller), in summer grain crops in South Africa. Journal of Entomological Society of Southern Africa 1990;53:41-47.

113. Khush GS, Mackill DJ, Sidhu GS. Breeding rice for resistance to bacterial blight. In: Bacterial Blight of Rice, Manila, Philippines, International Rice Research Institute1989. p207-217.

114. Kips RH. Environmental aspects. In: Haskell PT (ed.) Pesticide application: principles and practice. Clarendon Press, Oxford;1985. p190-201.

115. Kizilkaya R, Akça İ, Aşkın T, Yılmaz R, Olekhov V, Samofalovaf I, Mudrykh N. Effect of soil contamination with azadirachtin on dehydrogenase and catalase activity of soil. Eurasian Journal of Soil Science 2012;1:98-103.

116. Köck-Schulmeyer M, Ginebreda A, González S, Cortina JL, de Alda ML, Barceló D. Analysis of the occurrence and risk assessment of polar pesticides in the Llobregat River Basin (NE Spain). Chemosphere 2012;86:8-16.

117. Konar SK. Pesticides and aquatic environment. Indian Journal of Fisheries 2011;80-85.

118. Konradsen F, Van der HW, Cole DC, Hutchinson G, Daisley H, Singh S, Eddleston M. Reducing acute poisoning in developing countries-options for restricting the availability of pesticides. Toxicology 2003;192:249-261.

119. Kos M, van Loon JJ, Dicke M, Vet LE. Transgenic plants as vital components of integrated pest management. Trends in Biotechnology 2009;27:621-627.

120. Kumar M, Chimote V, Singh R, Mishra GP, Naik PS, Pandey Sk,

Chakrabarti SK. Development of Bt transgenic potatoes for effective control of potato tuber moth by using cry1Ab gene regulated by GBSS promoter. Crop Protection 2010;29:121-127.

121. Kumar R, Kranthi S, Nitharwal M, Jat SL, Monga D. Influence of pesticides and application methods on pest and predatory arthropods associated with cotton. Phytoparasitica 2012;40:417-424.

122. Kumari M. Effects of organophosphate pesticide Abate on the ovary of the cat fish, *Heteropneustes fossilis* (Bloch.). Bangladesh Journal of Zoology 2012;40:207-212.

123. Laabs V, Wehrhan A, Pinto A, Dores E, Amelung W. Pesticide fate in tropical wetlands of Brazil: An aquatic microcosm study under semi-field conditions. Chemosphere 2007;67:975-989.

124. Larson SJ, Capel PD, Majewski M. Pesticides in surface waters: Distribution, trends, and governing factors (No. 3). CRC Press;2010.

125. Laycock I, Lenthall KM, Barratt AT, Cresswell JE. Effects of imidacloprid, a neonicotinoid pesticide, on reproduction in worker bumble bees (*Bombus terrestris*). Ecotoxicology 2012;21:1937-1945.

126. le Vieux PD, Malan AP. An overview of the vine mealybug (*Planococcus ficus*) in South African vineyards and the use of entomopathogenic nematodes as potential biocontrol agent. South African Journal of Enology and Viticulture 2013;34:108-118.

127. Lee DH, Lind PM, Jacobs DR, Salihovic S, van Bavel B, Lind L. Polychlorinated biphenyls and organochlorine pesticides in plasma predict development of type 2 diabetes in the elderly. The Prospective Investigation of the Vasculature in Uppsala Seniors (PIVUS) study. Diabetes Care 2011a;34:1778-1784.

128. Lee SJ, Mehler L, Beckman J, Diebolt-Brown B, Prado J, Lackovic M, Waltz J, Mulay P, Schwartz A, Mitchell Y, Moraga-McHaley S, Gergely R, Calvert GM. Acute pesticide illnesses associated with off-target pesticide drift from agricultural applications: 11 States, 1998-2006. Environmental Health Perspectives 2011b;119:1162.

129. Lee WJ, Colt JS, Heineman EF, McComb R, Weisenburger DD, Lijinsky W, Ward MH. Agricultural pesticide use and risk of glioma in Nebraska, United States. Occupational and Environmental Medicine 2005;62:786-792.

130. Lupwayi NZ, Harker KN, Clayton GW, O'Donovan JT, Blackshaw RE. Soil microbial response to herbicides applied to glyphosate-resistant canola. Agriculture Ecosystem and Environment 2009;129:171-176.

131. Madhuri RJ, Rangaswamy V. Influence of selected insecticides on phosphatase activity in groundnut (*Arachis hypogeae* L.) soil. Journal of Environmental Biology 2002;23:393-397.

132. Martinez-Toledo MV, Salermon V, Rodelas B, Pozo C, Gonzalez-Lopez J. Effects of the fungicide captan on some functional groups of soil microflora. Applied Soil Ecology 1998;7:245-255.

133. Matthews GA. Improved systems of pesticide application. Philosophical Transactions of the Royal Society B 1981;295:163-173.

134. Mayanglambam T, Vig K, Singh DK. Quinalphos persistence and leaching under field conditions and effects of residues on dehydrogenase and alkaline phosphomonoesterases activities in soil. Bulletin of Environmental Contamination and Toxicology 2005;75:1067-1076.

135. McClure MS. Resurgence of the scale, *Fiorinia externa* (Homoptera: Diaspididae) on hemlock following insecticide application. Environmental Entomology 1977;6:480-484.

136. Mcfadyen REC. Successes in biological control of weeds. Proceedings of the X International Symposium on Biological Control of Weeds 3. Montana State University, Bozeman, Montana, USA. Neal RS (ed.);2000. p3-14.

137. McSorley R, Gill HK. Introduction to soil solarization. ENY-062, Department of Entomology and Nematology, Florida Cooperative Extension Service, Institute of Food and Agricultural Sciences, University of Florida, Gainesville, FL, USA;2010. (http://edis.ifas.ufl.edu/in856) (accessed 14 October 2013).

138. Megharaj M, Boul HL, Thiele JH. Effects of DDT and its metabolites on soil algae and enzymatic activity. Biology and Fertility of Soils 1999;29:130-134.

139. Menendez A, Martínez A, Chiocchio V, Venedikian N, Ocampo J A, Godeas A. Influence of the insecticide dimethoate on arbuscular mycorrhizal colonization and growth in soybean plants. International Microbiology 2010;2:43-45.

140. Mesnage R, Clair E, de Vendômois JS, Seralini GE. Two cases of birth defects overlapping Stratton-Parker syndrome after multiple pesticide exposure. Occupational and Environmental Medicine 2010;67:359-359.

141. Meyers L, Bull J. Fighting change with change: adaptive variation in an uncertain world. Trends in Ecology and Evolution 2002;17:551-557.

142. Miedaner T. Breeding wheat and rye for resistance to *Fusarium* diseases. Plant Breeding 1997;116: 201-220.

143. Mineau P, Whiteside M. Pesticide acute toxicity is a better correlate of US grassland bird declines than agricultural intensification. PloS One 2013;8:e57457.

144. Mironidis GK, Kapantaidaki D, Bentila M, Morou E, Savopoulou-Soultani M, Vontas J. Resurgence of the cotton bollworm *Helicoverpa armigera* in northern Greece associated with insecticide resistance. Insect Science 2013;20:505-512.

145. Mitra A, Chatterjee C, Mandal FB. Synthetic chemical pesticides and their effects on birds. Research Journal of Environmental Toxicology 2011;5:81-96.

146. Mostafalou S, Abdollahi M. Concerns of environmental persistence of pesticides and human chronic diseases. Clinical and Experimental Pharmacology 2012;S5:e002. (http://dx.doi.org/10.4172/2161-1459.S5-e002).

147. Munoz-Leoz B, Ruiz-Romera E, Antiguedad I, Garbisu C. Tebuconazole application decreases soil microbial biomass and activity. Soil Biology and Biochemistry 2011;43:2176-2183.

148. Murage EW, Voroney PR, Kay BD, Deen B, Beyaert RP. Dynamics and turnover of soil organic matter as affected by tillage. Soil Science Society of American Journal 2007;71:1363-1370.

149. Murray DL, Taylor PL. Claim no easy victories: evaluating the pesticide industry's Global Safe Use campaign. World Development 2000;28:1735-1749.

150. NAS. Principles of plant and animal pest control, Vol. 3. Insect management and control. National Academy of Sciences, Washington, DC, USA;1969.

151. Niethammer KR, White DH, Baskett TS, Sayre MW. Presence and biomagnification of organochlorine chemical residues in oxbow lakes of northeastern Louisiana. Archives of Environmental Contamination and Toxicology 1984;13:63-74.

152. Niewiadomska A, Klama J. Pesticide side effect on the symbiotic efficiency and nitrogenase activity of Rhizobiaceae bacteria family. Polish Journal of Microbiology 2005;54:43-48.

153. Niewiadomska A. Effect of carbendazim, imazetapir and thiramon nitrogenase activity, the number of microorganisms in soil and yield of red clover (*Trifolium pretense* L). Polish Journal of Environmental Studies 2004;13:403-410.

154. Nowak A, Nowak J, Klodka D, Pryzbulewska K, Telesinski A, Szopa

E. Changes in the microflora and biological activity of the soil during the degradation of isoproturon. Journal of Plant Diseases and Protection 2004;19:1003-1016.

155. Omar SA, Abdel-Sater MA. Microbial populations and enzyme activities in soil, treated with pesticides. Water, Air, and Soil Pollution 2001;127:49-63.

156. Ono A, Miyazaki R, Sota M, Ohtsubo Y, Nagata Y, Tsuda M. Isolation and characterization of naphthalene-catabolic genes and plasmids from oil contaminated soil by using two cultivation-independent approaches. Applied Microbiology and Biotechnology 2007;74:501-510.

157. Oppert B, Morgan TD, Kramer KJ. Efficacy of *Bacillus thuringiensis* Cry3Aa protoxin and protease inhibitors against coleopteran storage pests. Pest Management Science 2011;67:568-573.

158. Pal R, Tah J. Biodegradation and bioremediation of pesticides in Soil: Its objectives, classification of pesticides, factors and recent developments. World Journal of Science and Technology 2012;2:36-41.

159. Palmer WA, Heard TA, Sheppard AW. A review of Australian classical biological control of weeds programs and research activities over the past 12 years. Biological Control 2010;52:271-287.

160. Pan-Germany. Pesticide and health hazards. Facts and figures. 2012;1-16 (www.pan germany.org/download/Vergift_EN-201112-web.pdf) (accessed on 14 October 2013).

161. Paoletti MG. Using bioindicators based on biodiversity to assess landscape sustainability. Agriculture, Ecosystems and Environment 1999;74:1-18.

162. Parshad VR. Rodent control in India. Integrated Pest Management Reviews 1999;4:97-126.

163. Parsons KC, Mineau P, Renfrew RB. Effects of pesticide use in rice fields on birds. Waterbirds 2010;33:193-218.

164. Patnaik GK, Kanungo PK, Moorthy BTS, Mahana PK, Adhya TK, Rao VR. Effect of herbicides on nitrogen-fixation (C2H2 reduction) associated with rice rhizosphere. Chemosphere 1995;30:339-343.

165. Pelosi C, Barot S, Capowiez Y, Hedde M, Vandenbulcke F. Pesticides and earthworms. Agronomy for Sustainable Development 2013;1-30. DOI:10.1007/s13593-013-0151-z.

166. Pereira L, Fernandes MN, Martinez CB. Hematological and biochemical alterations in the fish *Prochilodus lineatus* caused by the herbicide clomazone. Environmental Toxicology and Pharmacology 2013;36:1-8.

167. Petrelli G, Mantovani A. Environmental risk factors and male fertility

and reproduction. Contraception 2002;65:297-300.

168. Pettis JS, van Engelsdorp D, Johnson J, Dively G. Pesticide exposure in honey bees results in increased levels of the gut pathogen *Nosema*. Naturwissenschaften 2012;99:153-158.

169. Prasad SM, Zeeshan M, Kumar D. Toxicity of endosulfan on growth, photosynthesis, and nitrogenase activity in two species of Nostoc (*Nostoc muscorum* and *Nostoc calcicola*). Toxicological and Environmental Chemistry 2011;93:513-525.

170. Pray CE, Huang J, Hu R, Rozelle S. Five years of Bt cotton in China-the benefits continue. The Plant Journal 2002;31:423-430.

171. Price DRG, Gatehouse JA. RNAi-mediated crop protection against insects. Trends in Biotechnology 2008;26:393-400.

172. Reader SM, Miller TE. The introduction into bread wheat of a major gene for resistance to powdery mildew from wild emmer wheat. Euphytica 1991;53:57-60.

173. Reinecke SA, Reinecke AJ. The impact of organophosphate pesticides in orchards on earthworms in the Western Cape, South Africa. Ecotoxicity and Environmental Safety 2007;66:244-251.

174. Relyea RA, Diecks N. An unforeseen chain of events: lethal effects of pesticides on frogs at sublethal concentrations. Ecological Applications 2008;18:1728-1742.

175. Relyea RA, Hoverman JT. Interactive effects of predators and a pesticide on aquatic communities. Oikos 2008;117:1647-1658.

176. Relyea RA. New effects of Roundup on amphibians: predators reduce herbicide mortality; herbicides induce antipredator morphology. Ecological Applications 2012;22:634-647.

177. Relyea RA. Predator cues and pesticides: a double dose of danger for amphibians. Ecological Applications 2003;13:1515-1521.

178. Relyea RA. The lethal impact of Roundup on aquatic and terrestrial amhibians. Ecological Applications 2005(a);15:1118-1124.

179. Relyea RA. The impact of insecticides and herbicides on the biodiversity and productivity of aquatic communities. Ecological Applications 2005(b);15:618-627.

180. Richter AR, Fuxa JR. Timing, formulations and persistence of a nuclear polyhedrosis virus and microsporidium for control of the velvetbean caterpillar (Lepidoptera: Noctuidae) in soybeans. Journal of Economic Entomology 1984;77:1299-1306.

181. Richter ED. Acute human pesticide poisonings. Encyclopedia of Pest

Management 2002;3-6.

182. Romero E, Fernández-Bayo J, Díaz JMC, Nogale, R. Enzyme activities and diuron persistence in soil amended with vermicompost derived from spent grape marc and treated with urea. Applied Soil Ecology 2010;44:198-204.

183. Rosell G, Quero C, Coll J, Guerrero A. Biorational insecticides in pest management. Journal of Pesticide Science 2008;33:103-121.

184. Rueda-Ayala V, Rasmussen J, Gerhards R. Mechanical weed control. In: Precision crop protection-the challenge and use of heterogeneity. Springer Netherlands;2010. p279-294.

185. Singh G, Chahil GS, Jyot G, Battu RS, Singh B. Degradation dynamics of emamectin benzoate on cabbage under subtropical conditions of Punjab, India. Bulletin of Environmental Contamination and Toxicology 2013. DOI 10.1007/s00128-013-1013-8

186. Sannino F, Gianfreda L. Pesticide influence on soil enzymatic activities. Chemosphere 2001;45:417-425.

187. Santovito A, Cervella P, Delpero M. Chromosomal aberrations in cultured human lymphocytes treated with the fungicide, Thiram. Drug and Chemical Toxicology 2012;35:347-351.

188. Saravi SSS, Shokrzadeh M. Role of pesticides in human life in the modern age: a review. In: Stoytcheva M (ed.) Pesticides in the modern world-risks and benefits. InTech;2011. p4-11.

189. Satyavani G, Gopi RA, Ayyappan S, Balakrishnamurthy P, Reddy PN. Toxicity effect of expired pesticides to freshwater fish, *Labeo rohita*. The Journal of Agriculture and Environment 2011;12:1-9.

190. Scharf ME, Meinke LM, Wright RJ, Chandler LD, Siegfried BD. Metabolism of carbaryl by insecticide-resistant and susceptible western corn rootworm populations (Coleoptera: Chrysomelidae). Pesticide Biochemistry and Physiology 1999;63:85-96.

191. Scholz NL, Fleishman E, Brown L, Werner I, Johnson ML, Brooks ML, Michaelmore CL, Schlenk D. A Perspective on modern pesticides, pelagic fish declines, and unknown ecological resilience in highly managed ecosystems. BioScience 2012;62:428-434.

192. Schreck E, Geret F, Gontier L, Treihou M. Neurotoxic effect and metabolic responses induced by a mixture of six pesticides on the earthworm*Aporrectodea caliginosa* noctuma. Chemosphere 2008;71:1832-1839.

193. Sebiomo A, Ogundero VW, Bankole SA. Effect of four herbicides on

microbial population, soil organic matter and dehydrogenase activity. African Journal of Biotechnology 2011;10:770-778.

194. Shapiro-Ilan DI, Cottrell TE, Jackson MA, Wood BW. Control of key pecan insect pests using biorational pesticides. Journal of Economic Entomology 2013;106:257-266.

195. Sharma HC, Ortiz R. Host plant resistance to insects: an eco-friendly approach for pest management and environment conservation. Journal of Environmental Biology 2002;23:111-135.

196. Sharon E, Chet I, Spiegel Y. Trichoderma as a biological control agent. In: Biological control of plant-parasitic nematodes. Springer Netherlands;2011. p183-201.

197. Shepard M, Carner GR, Turnipseed TG. Colonization and resurgence of insects pests of soybean in response to insecticides and field isolation. Environmental Entomology 1977;6:501-506.

198. Shim YK, Mlynarek SP, van Wijngaarden, E. Parental exposure to pesticides and childhood brain cancer: US Atlantic coast childhood brain cancer study. Environmental Health Perspectives 2009;117:1002-1006.

199. Singh B, Mandal K. Environmental impact of pesticides belonging to newer chemistry. In: Dhawan AK, Singh B, Brar-Bhullar M, Arora R (eds.). Integrated pest management. Scientific Publishers, Jodhpur, India;2013. p152-190.

200. Singh DK, Kumar S. Nitrate reductase, arginine deaminase, urease and dehydrogenase activities in natural soil (ridges with forest) and in cotton soil after acetamiprid treatments. Chemosphere 2008;71:412-418.

201. Sofo A, Scopa A, Dumontet S, Mazzatura A, Pasquale V. Toxic effects of four sulphonylureas herbicides on soil microbial biomass. Journal of Environmental Science and Health, Part B 2012;47:653-659.

202. Solaimalai A, Ramesh RT, Baskar M. Pesticides and environment. In: Environmental contamination and bioreclamation; 2004. p345-382.

203. Soltani H, Agricultural H. Efficacy of spinosad against potato Colorado beetle *Leptinotarsa*. Hamedan Agricultural and Natural Resources Research, Hamedan, Iran;2011.

204. Son HK, Kim SA, Kang JH, Chang YS, Park SK, Lee SK, Jacobs Jr. DR, Lee DH. Strong associations between low-dose organochlorine pesticides and type 2 diabetes in Korea. Environment International 2010;36:410-414.

205. Song C, Kanthasamy A, Anantharam V, Sun F, Kanthasamy AG. Environmental neurotoxic pesticide increases histone acetylation

to promote apoptosis in dopaminergic neuronal cells: relevance to epigenetic mechanisms of neurodegeneration. Molecular Pharmacology 2010;77:621–632.

206. Souza RC, Cantão ME, Vasconcelos ATR, Nogueira MA, Hungria M. Soil metagenomics reveals differences under conventional and no-tillage with crop rotation or succession. Applied Soil Ecology 2013;72:49-61.

207. Srinivasulu M, Mohiddin GJ, Madakka M, Rangaswamy V. Effect of pesticides on the population of *Azospirillum* sp. and on ammonification rate in two soils planted to groundnut (*Arachis hypogaea* L.).Tropical Ecology 2012a;53:93-104.

208. Srinivasulu M, Mohiddin GJ, Subramanyam K, Rangaswamy V. Effect of insecticides alone and in combination with fungicides on nitrification and phosphatase activity in two groundnut (*Arachis hypogeae* L.) soils. Environmental Geochemistry and Health 2012b;34:365-374.

209. Stuart SN, Chanson JS, Cox NA, Young BE, Rodrigues ASL, Fischman DL, Waller RW. Status and trends of amphibian declines and extinctions worldwide. Science 2004;306:1783-1786.

210. Sukul P. Enzymes activities and microbial biomass in soil as influenced by metalaxyl residues. Soil Biology Biochemistry 2006;38:320-326.

211. Sultana V, Ehteshamul-Haque S, Ara J, Athar M. Effect of brown seaweeds and pesticides on root rotting fungi and root-knot nematode infecting tomato roots. Journal of Applied Botany and Food Quality 2012;83:50-53.

212. Symondson WOC, Hemingway J. Biochemical and molecular techniques. In: Dent DR, Walton MP (eds.) Methods in ecological and agricultural entomology. CAB International, Wallingford, UK;1997. p293-350.

213. Tabashnik BE, Van Rensburg JBJ, Carrière Y. Field-evolved insect resistance to Bt crops: definition, theory, and data. Journal of Economic Entomology 2009;102:2011-2025.

214. Tanner CM, Kamel F, Ross GW, Hoppin JA, Goldman SM, Korell M, Marras C, Bhudhikanok GS, Kasten M, Chade AR, Comyns K, Richards MB, Meng C, Priestley B, Fernandez HH, Cambi F, Umbach DM, Blair A, Sandler DP, Langston JW. Rotenone, paraquat, and Parkinson's disease. Environmental Health Perspectives 2011;119:866-872.

215. Tavela ADO, Araújo JV, Braga FR, Silva AR, Carvalho RO, Araujo JM, Carvalho G. Biological control of cyathostomin (Nematoda: Cyathostominae) with nematophagous fungus, *Monacrosporium thaumasium* in tropical southeastern Brazil. Veterinary Parasitology 2011;175:92-96.

216. Tien CJ, Chen CS. Assessing the toxicity of organophosphorous pesticides to indigenous algae with implication for their ecotoxicological impact to aquatic ecosystems. Journal of Environmental Science and Health, Part B 2012;47:901-912.

217. Tierney KB, Baldwin DH, Hara TJ, Ross PS, Scholz NL, Kennedy CJ. Olfactory toxicity in fishes. Aquatic Toxicology 2010;96:2-26.

218. Trajkovska S, Mbaye M, Seye MG, Aaron JJ, Chevreuil M, Blanchoud H. Toxicological study of pesticides in air and precipitations of Paris by means of a bioluminescence method. Analytical and Bioanalytical Chemistry 2009;394:1099-1106.

219. Trevisan M, Montepiani C, Ragozza L, Bartoletti C, Loannilli E, Del Re AAM. Pesticides in rainfall and air in Italy. Environmental Pollution 1993;80:31-39.

220. Tu CM. Effects of some pesticides on enzyme activities in an organic soil. Bulletin of Environmental Contamination and Toxicology 1981;27:109-114.

221. Van Leeuwen T, Vontas J, Tsagkarakou A, Dermauw W, Tirry L. Acaricide resistance mechanisms in the two-spotted spider mite *Tetranychus urticae* and other important Acari: A review. Insect Biochemistry and Molecular Biology 2010;40:563-572.

222. Vasileiadis VP, Sattin M, Otto S, Veres A, Palinkas Z, Ban R, Pons X, Kudsk P, Weide van der R, Czembor E, Moonen AC, Kiss J. Crop protection in European maize-based cropping systems: Current practices and recommendations for innovative Integrated Pest Management. Agricultural Systems 2011;104:533-540.

223. Vermeer K, Risebrough RW, Spaans AL, Reynolds LM. Pesticide effects on fishes and birds in rice fields of Surinam, South America. Environmental Pollution 1970;7:217-236.

224. Vickerman GP. Farm scale evaluation of the long-term effects of different pesticide regimes on the arthropod fauna of winter wheat. In: Greeves MP, Grieg-Smith PW, Smith BD (eds.) Field methods for the environmental study of the effects of pesticides. BCPC Monograph No. 40 British Crop Protection Council, Farnham, UK;1988. p127-135.

225. Wang LP, Shen J, Ge LQ, Wu JC, Yang GQ, Jahn GC. Insecticide-induced increase in the protein content of male accessory glands and its effect on the fecundity of females in the brown planthopper *Nilaparvata lugens* Stål (Hemiptera: Delphacidae). Crop Protection 2010;29:1280-1285.

226. Wang MC, Gong M, Zang HB, Hua XM, Yao J, Pang YJ, Yang YH. Effect of methamidophos and urea application on microbial communities

in soils as determined by microbial biomass and community level physiological profiles. Journal of Environmental Science and Health Part B 2006;41:399-413.

227. Ware GW. Effects of pesticides on nontarget organisms. Residue Reviews 1980;76:173-201.

228. Weber JB, Wilkerson GG, Reinhardt CF. Calculating pesticide sorption coefficients (K sub(d)) using selected soil properties. Chemosphere 2004;55:157-166.

229. Whitehorn PR, O'Connor S, Wackers FL, Goulson D. Neonicotinoid pesticide reduces bumble bee colony growth and queen production. Science 2012;336:351-352.

230. Williams T. Silent scourge. Audubon 1997;99:28-35.

231. Winchester PD, Huskins J, Ying J. Agrichemicals in surface water and birth defects in the United States. Acta Paediatrica 2009;98:664-669. DOI:0.1111/j.1651-2227.2008.01207.x

232. Wu JY, Smart MD, Anelli CM, Sheppard WS. Honey bees (*Apis mellifera*) reared in brood combs containing high levels of pesticide residues exhibit increased susceptibility to *Nosema* (Microsporidia). Journal of Invertebrate Pathology 2012;109:326-329.

233. Wu KM, Lu YH, Feng HQ, Jiang YY, Zhao JZ. Suppression of cotton bollworm in multiple crops in China in areas with Bt toxin–containing cotton. Science 2008;321:1676-1678.

234. Xavier R, Rekha K, Bairy KL. Health perspective of pesticide exposure and dietary management. Malaysian Journal of Nutrition 2004;10:39-51.

235. Xie H, Gao F, Tan W, Wang SG. A short-term study on the interaction of bacteria, fungi and endosulfan in soil microcosm. Science of the Total Environment 2011;412:375-379.

236. Xu X, Dailey AB, Talbott EO, Ilacqua VA, Kearney G, Asal NR. Associations of serum concentrations of organochlorine pesticides with breast cancer and prostate cancer in US adults. Environmental Health Perspectives 2010;118:60-66.

237. Xuan TD, Shinkichi T, Khanh TD, Chung IM. Biological control of weeds and plant pathogens in paddy rice by exploiting plant allelopathy: an overview. Crop Protection 2005;24:197-206.

238. Yadav BC, Veluthambi K, Subramaniam K. Host-generated double stranded RNA induces RNAi in plant-parasitic nematodes and protects the host from infection. Molecular and Biochemical Parasitology. 2006;148:219-222.

239. Yasmin S, D'Souza D. Effects of pesticides on the growth and development of earthworm: a review. Applied and Environmental Soil Sciences 2010;1-9. (Article ID 678360).

240. Zhang LW, Liu YJ, Yao J, Wang B, Huang B, Li ZZ, Sun JH. Evaluation of *Beauveria bassiana* (Hyphomycetes) isolates as potential agents for control of *Dendroctonus valens*. Insect Science 2011;18:209-216.

241. Zhou X, Scharf ME, Parimi S, Meinke LJ, Wright RJ, Chandler LD, Siegfried BD. Diagnostic assays based on esterase-mediated resistance mechanisms in western corn rootworms (Coleoptera: Chrysomelidae). Journal of Economic Entomology 2002;95:1261-1265.

Chapter 5

HEALTH PROBLEM CAUSED BY LONG-TERM ORGANOPHOSPHORUS PESTICIDES EXPOSURE - STUDY IN CHINA

Zhi-Jun Zhou

Fudan University, China

INTRODUCTION

Organophosphorus pesticides (OPs), one of the most popular classes of pesticides, are widely used all over the world especially in developing countries, such as China. There are many OPs, with thousands of trade names such as dimethoate, parathion and omethoate, most of which have been used for insect control in residential and agriculture settings. The acute toxicity of OPs are believed to be due primarily to the inhibition of acetylcholinesterase (AChE) resulting in an accumulation of acetylcholine (Ach) with a sustained overstimulation of Ach receptors in the clefts of central and peripheral neuron synapses. They can cause a progression of toxic signs, including hypersecretions, convulsions, respiratory distress, coma and death. However, the heavy usage of OPs has given rise to wide public concern on their chronic toxicity. Generally, long-tem exposure to OPs can be divided into occupational exposure and non-occupational exposure. The former often involves farming population and workers employed in pesticide-related industries. And the latter is more for general population potentially exposed to OPs via a number of different routes including dietary, lifestyle or medicinal.

China is a large country with large demand of pesticides. This means that there are much more Chinese people, both occupational and non-occupational population, whose health are under the threat of OPs exposure. The presence of common and specific metabolites of OPs in urine samples taken from the general population has demonstrated the widespread exposure to OPs in China. Moreover, workers engaged in OPs production are at high risk from OPs exposure, as confirmed by higher levels of OPs metabolites in biological samples compared to those present in individuals from non-agricultural communities. Therefore, a great deal of research has been conducted by Chinese scientists to understand the adverse effects of long-term, low-level exposure to OPs in both general and occupational population.

OPS EXPOSURE ASSESSMENT - BIOLOCIAL MONITORING

OPs exposure in both occupational and general population can be assessed by measurement of esterase activity and by direct measurement of urinary OPs metabolites.

Esterase Activity

The activity of esterases including butyrylcholinesterase (BChE), erythrocyte acetyl cholinesterase (AChE), carboxylesterase (CarbE) and paraoxonase (PonE) can be inhibited by OPs. However, the sensitivity of these four kinds of esterases to inhibition differs. We previously conducted a cross-sectional study among 241 workers from a pesticide plant as directly exposed group, 161 service persons in the same pesticide plant as indirectly exposed group and 150 workers without any records of pesticide exposure in another plant as control group. We measured the esterase activity of all these subjects. The results showed that the CarbE, BChE and PonE activity of subjects in exposed group was significantly lower than subjects in control group (Table 1). The inhibition of AChE activity was related to the type of workshop and work process whereas the inhibition of AChE and BChE activity does not necessarily correlate closely with exposure time and level (Table 2~4). Besides, there was a dose-response relationship between the external exposure dose and CarbE activity (Table 5).

Table 1. Esterase activity (nmol ml^{-1} min^{-1}) of subjects in different groups[*]: the esterase activity of subjects in exposed group are significantly lower than those in Control ($p<0.01$).[#]: the esterase activity of subjects in Indirectly exposed group are significantly lower than those in C ($p<0.01$).

Goup	Number	CarbE	BchE	PON
Direct Exposure	241	513.44±184.5^9*	39.52±17.8^4*	142.75±70.4^9*
Indirect Exposure	161	480.75±115.^{8}#	38.67±15.3^4#	147.96±93.21
Control	150	615.90±149.55	44.05±12.28	167.97±112.04
p value		0.000	0.004	0.021

Table 2. Esterase activity (nmol ml^{-1} min^{-1}) of subjects in different workshops

Esterase	Type of Workshops			pvalue
	Methamidophos (n=87)	Dimethoate (n=83)	Other OPs (n=71)	
CarbE	508.36±194.62	39.21±22.52	488.14±186.19	0.205
BChE	38.65±13.55	137.11±69.62	40.96±16.40	0.710
PonE	150.72±75.91	126.33±9.83	139.57±64.43	0.411
AChE	127.21±8.13	126.33±9.83	139.57±64.43	**0.003**

Table 3. Esterase activity (nmol ml^{-1} min^{-1}) of workers with different jobs in directly exposed group

Esterase	Type of Processes			pvalue
	Packers (n=70)	Operators (n=136)	Inspectors (n=35)	
CarbE	475.23±183.92	526.89±189.88	537.68±156.23	0.115
BChE	39.15±13.61	39.01±14.82	42.26±31.48	0.620
PonE	144.21±68.67	142.84±73.84	139.48±61.95	0.949
AChE	123.31±9.80	126.01±9.23	127.91±7.35	**0.034**

Table 4. Esterase activity (nmol ml^{-1} min^{-1}) of workers with different working time in directly exposed group

Esterase	Working time (years)				Fvalue	pvalue
	1~5 (n=9)	5~10 (n=48)	10~20 (n=97)	"/>20 (n=87)		
CarbE	531.18±283.70	448.54±154.55	509.43±195.27	551.91±167.61	3.377	0.019
BChE	43.52±19.44	41.3±26.24	40.42±15.69	37.12±13.86	0.924	0.430
PonE	146.31±79.60	135.56±67.81	137.09±66.68	152.66±75.06	0.955	0.415
AChE	127.11±7.89	124.21±9.03	125.06±9.47	126.54±9.27	0.840	0.473

Table 5. Relationship between the external exposure level and esterase activity

External Exposure level (mg/m3)	Number	CarbE (nmol·ml^{-1} min^{-1})	BchE (nmol·ml^{-1}min^{-1})	PonE (nmol·ml^{-1}min^{-1})	AChE (U)
0~3	124	485.08±188.90	42.36±20.62	136.75±67.54	136.75±67.54
3~6	63	556.43±175.35	37.33±14.67	152.59±71.18	125.76±9.52
"/>6	54	528.44±176.70	35.56±12.78	145.04±76.04	126.72±8.77
F value		3.417	3.450	1.093	0.812
p value		**0.034**	**0.033**	0.337	0.446

Similar research was done by other Chinese colleagues, for example, they (Lin et al., 2007) investigated 56 parathion exposed workers (as exposed group) and 120 non-exposed persons (as control group) and reported that there were significant differences ($p < 0.001$) of the activity of BChE, AChE, CarbE, and PonE compared with control group, but no difference ($p > 0.05$) in plasma β-glucuronidase (β-GD) activity. And the rates of abnormity (below the lower limit of activity reference range) were 37.5% and 48.2% for CarbE and BChE respectively, which were all significantly higher than that of AChE ($p < 0.001$). But there was no significant difference between PonE activity (5.4%) and AChE activity ($p > 0.05$).

Dialkylphosphate (DAP) Metabolites in Urine

On the other hand, there are clear evidences from biological monitoring studies that dialkylphosphate (DAP) metabolites of OPs can be detected in urine after OPs exposure. Six common DAP metabolites, e.g, dimethylphosphate (DMP), dimethylthiophosphate (DMTP), diethylphosphate (DEP), diethylthiophosphate (DETP), diethyldithiophosphate (DEDTP), and dimethyldiithiophosphate (DMDTP) have been determined. These metabolites are non-specific to a particular organophosphate metabolism of different OPs can give rise to similar urinary metabolites. Urinary DAP metabolites reported in a number of studies are summarized in Table 6.

These metabolites in urine are useful to estimate exposure to several OPs. In the cross-sectional study mentioned above, we found that DMP and DETP concentration of workers in the directly exposed group was significantly higher than that of indirectly exposed group (Table 7). Workers in different workshops have different urinary metabolites whereas the type of job influenced the concentration of urinary metabolites (Table 8 and 9). However, we didn't find that the total exposure time will affect the urine level of DAP metabolites (Table 10).

Table 6. Urinary DAP metabolites of different OPs.

Name	DAP metabolites	Name	DAP metabolites
Dichlorvos	DMP	Malathion	DMP, DMTP, DMDTP
Chlopyrifos	DEP, DETP	Methidathion	DMP, DMTP
Mercaptophos	DEP, DETP	Mevinphos	DMP
Diazinon	DEP, DETP	Paraoxon	DEP
Dichlofenthion	DEP	Parathion	DEP, DETP
Azinphos-methyl	DMP, DMTP, DMDTP	Methyl parathion	DMP
Dimethoate	DMP, DMTP, DMDTP	Phorate	DEDTP

Fenitrothion	DMTP	Diethquinphione	DEP, DETP
Malaoxon	DMP	Metriphonate	DMP

Table 7. Urinary DAP metabolites concentration of subjects in different exposed groups

Group	Number	Median of Urinary DAP metabolites concentration (μg/gCr)				
		DMP	DEP	DETP	DMDTP	DEDTP
Directly Eexposed	161	0.01	1.06×102	9.41	2.18×102	97.48
Indirectly Exposed	122	0.00	8.24×102	8.02	2.21×102	95.10
z value		-4.839	-0.981	-2.733	-0.682	-1.165
p value		**0.000**	0.326	**0.006**	0.495	0.244

Table 8. Urinary DAP metabolites concentration (median) of directly exposed workers in different workshops

DAP metabolites	Type of Workshops			p value
	Methamidophos (n=51)	Dimethoate (n=65)	Other OPs (n=45)	
DMP	0.00	0.00	0.00	0.137
DEP	1292	725	1471	**0.045**
DETP	8	8	8	0.394
DMDTP	342	50	480	**0.004**
DEDTP	90	88	98	**0.037**

Table 9. Urinary DAP metabolites concentration (median) of workers with different job title in directly exposed group

DAP metabolites	Type of Processes			p value
	Packers (n=26)	Operators (n=109)	Inspectors (n=26)	
DMP	0.00	0.00	0.00	0.623
DEP	1307	1180	737	**0.016**
DETP	8	8	15	0.534
DMDTP	222	275	523	0.140
DEDTP	143	88	99	**0.008**

Table 10. Urinary DAP metabolites level (median) of workers with different exposure time in directly exposed group

| DAP metabolites | Working Age Groups (years) | | z value | pvalue |
	≤20 (n=84)	"/>20 (n=77)		
DMP	0.00	0.00	-0.104	0.917
DEP	109	871	-0.338	0.698
DETP	9.41	10.9	-1.080	0.280
DMDTP	232	159	-0.688	0.491
DEDTP	110	95	-0.264	0.792

Another study, done by our research group, investigated in detail 30 workers packaging dimethoate from a pesticide plant. Urine samples of each participant pre- and post- workshift were collected. The results showed that 100% of the workers had at least one DAP metabolite present in both pre-shift and post-shift urine samples. DMP and DMTP were the most frequent metabolites (100%) found, followed by DMDTP, DEP, DETP and finally DEDTP (Table 11). DAP metabolites with dimethyl moieties (DMP, DMTP, and DMDTP) were detected at higher concentrations than those with ethyl moieties (DEP, DETP, and DEDTP) in both time points (pre- and post- workshift). Moreover, DMP, DMTP and DMDTP concentration in the post-shift urine samples were significantly higher than that in the pre-shift urine samples (Table 12).

Table 11. The detection percentage of urinary DAP metabolites of subjects in exposed groups

| Groups | Detection percentage (%) of urinary DAP metabolites | | | | | |
	DMP	DEP	DMTP	DMDTP	DETP	DEDTP
Pre-shift	100.0	40.0	100.0	90.0	20.0	0.0
Post-shift	100.0	53.3	100.0	96.7	26.7	6.7

Table 12. Urinary DAP metabolites concentration (geometric mean) of subjects in exposed groups nd: not detected.*: the urinary DAP metabolites concentration of pre-shift samples are significantly lower than those of pro-shift samples ($p<0.05$).**: the urinary DAP metabolites concentration of pre-shift samples are significantly lower than those of pro-shift samples ($p<0.01$).

| Groups | Urinary DAP metabolites concentration (µg/gCr) | | | | | |
	DMP	DEP	DMTP	DMDTP	DETP	DEDTP
Pre-shift	371±1.^{9*}	102±2.1	891±2.^{4*}	302±2.^{3*}	78±2.7	nd
Post-shift	741±2.1	104±1.5	1479±2.1	832±2.3	74±2.2	47±1.4

Indeed, certain levels of DAP metabolites are also detected in non-occupationally exposed populations. We tested the urine samples of 60 college students and found that more than 86% of them had at least one type of DAP metabolites in the urine. DMDTP was the most frequent metabolite (86.7%) found, followed by DMP, DMTP, DEP, and finally DETP. And the results showed no detectable DEDTP (Table 13). DMTP were detected at much higher concentrations than other metabolites: the geometric mean of DMTP was high as 661 μg/gCr (Table 14).

Table 13. The detection percentage of urinary DAP metabolites of general population

Detection	Detection percentage (%) of urinary DAP metabolites					
	DMP	**DEP**	**DMTP**	**DMDTP**	**DETP**	**DEDTP**
Number	51	30	48	52	18	0
Percentage (%)	85.0	50.0	80.0	86.7	30.0	0.0

Table 14. Urinary DAP metabolites concentration (μg/gCr) of general population

DAP me-tabolites	Range of concentra-tion	25% percen-tile	median	75% percen-tile	geomet-ric mean	geometric standard deviation
DMP	22~1026	100	170	254	166	2.3
DEP	27~383	67	114	197	110	1.9
DMTP	109~3187	404	693	1104	661	2.1
DMDTP	24~784	68	135	219	126	2.3
DETP	24~186	37	51	91	60	2.4
DEDTP	nd	nd	Nd	nd	nd	nd

ADVERSE EFFECTS CAUSED BY LONG-TERM OPS EXPOSURE

Common Illness Caused by Long-Term Ops Exposure

Available evidence suggests that there is a possibility of adverse effects occurring after long-term OPs exposure although these effects may be not clearly related to the inhibition of cholinesterase. Studies on health hazards to farmers who handle, store and use OPs have documented a range of non-specific self-reported symptoms that have been attributed to chronic OPs exposure. These include burning or prickling of the skin; tingling or numbness of hands and face; muscular twitching or cramps in the face, neck, arms and legs; respiratory symptoms such as chest pain, chest stuffiness, cough, runny nose, wheezing, shortness of breath, sore throat; excessive sweating; nausea, vomiting, diarrhoea; excessive salivation; abdominal pain; lacrimation and inflammation of the eyes; difficulty in seeing; restlessness; difficulty in falling asleep; trembling of hands; and irritability.

Zhao and his colleagues use Pittsburgh Sleep Quality Index (PSQI) and Epworth Sleeping Scale (ESS) to investigate and analyze the sleeping status of 482 agricultural workers over 50 years old from 5 counties in Jiangxi province (Zhao et al., 2010). The PSQI scores of these farmers were 5.80±2.81, lower than those of general population. And the ESS scores of these farmers were 7.15±4.99, higher than those of general population. Moreover, the ESS scores of farmers who have been exposed to OPs more than 1000 days were significantly higher than other farmers ($p<0.01$). Zhang observed 284 occupational OPs exposed persons by dynamic ultrasonographic imaging and found a higher prevalence of fatty liver than non-exposed persons (W.P. Zhang et al., 2010).

ECG changes in workers who have been exposed to OPs were also reported. An investigation of 706 exposed workers and 707 non-exposed persons and reported that about 19.69% of the workers had abnormal ECG changes against 12.31% of the non-exposed persons (Tang et al., 2004). The abnormal ECG changes of exposed workers include sinus bradycardia, arrhythmia, incomplete right bundle branch block, and ST-T segment elevation.

Our group once analyzed a series of data of medical examination (particular ECG examination) of 87 workers exposed to three kinds of OPs and found significant differences in the prevalence of ECG abnormalities between exposed and non-exposed groups. Although the prevalence of ECG changes for exposed workers was much higher than that of prior to exposure, it did not increase with the prolongation of the exposure period. And the inhibition of AChE was not correlated to ECG disorders, which indicated that cardiac effects of OPs are not clearly related to the inhibition of AChE (Tables 15and 16).

Table 15. ECG abnormalities of subjects in different groups

Groups	Number	Abnormal ECG rate	Odds ratio of prevalence	pvalue
Control	25	4.0		
Dimethoate	35	20.0	6.0	0.07
Methamidophos	30	13.3	3.7	0.23
Kitazin P	19	21.1	6.4	0.09
Totle	84	17.9	5.2	0.07

Table 16. Types of ECG abnormalities of subjects in different groups[*]: the number of ECG abnormalities in the exposed groups are signicantly different from those of control group ($p<0.01$).

Types of ECG abnormalities		Dimetho-ate	Methami-dophos	Kitazin P	Con-trol
Sinus arrhythmia	Sinus tachycardia	3	4	1	0
	Sinus bradycardia	17	3	8	0
	Sinus irregularity	5	0	0	0
Ectopic arrhyth-mias	Premature beat	0	3	0	0
Conduction ab-normalities	Right bundle branch block	4	0	0	0
Others	Low QRS wave	1	9	10	2
	Left ventricular sypervoltage	5	5	2	0
Left/right axis deviation		10	2	9	0
Total number of abnormalities		42*	26*	21*	2
Total number of subjects		410	360	145	302

Once we collected the information on OPs exposure history and signs and symptoms of the subjects through questionnaires and medical examinations among another exposed population. Then the weighting and total score of the signs and symptoms of neuromuscular system, respiratory system, circulatory system and digestive system was calculated. The results showed that the weighting and total symptom score in directly and indirectly exposed group was higher than that in control group, and there was a dose-response relationship between the internal exposure dose and digestive system score (Table 17~19). A higher percentage of abnormal hemoglobin was found in the workers in directly exposed group, in correlation with exposure time. The workers (working time 5~10 years) in directly exposed group showed a higher percentage of abnormal hemoglobin level, and there was dose-response relationship between the percentage of abnormal hemoglobin and accumulating external exposure dose (liner-liner association analysis ($p<0.05$) (Table 20 and 21). Besides this, some system scores and the percentage of abnormal hemoglobin were related to AChE activity regarded as an exposure dose (Table 22). There was negative correlation between the activity of AChE and signs scores according to correlation analysis. It showed a increasing trend of signs scores and percentage of abnormal hemoglobin with the decrease of AChE activity (Table 23).

Table 17. Total symptom scores of subjects in different groups

Groups	Number	Symptom scores				
		neuro-muscular system	respiratory system	circulatory system	digestive system	Total system scores
Directly Exposed	241	0.66±1.49	0.27±0.84	0.44±0.74	0.21±0.57	1.57±2.44
Indirectly Exposed	161	0.29±0.88	0.11±0.32	0.30±0.64	0.07±0.30	0.63±1.08
Control	150	0.03±0.16	0.05±0.22	0.06±0.27	0.03±0.22	0.16±0.54
H value		49.37	10.87	37.13	23.55	89.01
p value		0.000	0.004	0.000	0.000	0.000

Table 18. Ratio of abnormal symptoms of subjects in different groups

Groups	Total number	Number of person with abnormal symptoms	Number of person without abnormal symptoms	Ratio of abnormal symptoms (%)	X²value	pvalue
Directly Exposed	241	132	109	54.8	91.05	**0.000**
Indirectly Exposed	161	43	108	28.5		
Control	150	15	145	9.4		

Table 19. The symptom scores are affected by internal exposure dose (urinary DETP levels)

DETP (μg/gCr)	Number	Symptom scores				
		neuromuscular system	respiratory system	circulatory system	digestive system	Total system scores
0~7.5	53	0.83	0.15	0.25	0.11	1.34
7.5~15	49	0.80	0.22	0.41	0.18	1.61
"/>15	59	0.80	0.32	0.22	0.39	1.73
H value		1.063	2.642	2.603	6.900	3.674
p value		0.588	0.267	0.272	**0.032**	0.159

Table 20. Medical examinations data of subjects in different groups WBC: white blood cell; Hb: hemoglobin; SBP: systolic pressure; DBP: diastolic pressure

Groups	Number	Abnormalities (%) of medical examinations					
		WBC	Hb	ECG	B ultra-sonic	SBP	DBP
Directly Exposed	241	2.9	33.6	13.7	17.8	12.4	24.1
Indirectly Exposed	161	3.3	5.3	17.9	24.5	20.5	33.1
Control	150	3.1	15.6	17.5	15.1	6.3	19.4
X2		0.053	48.88	1.623	2.536	14.19	8.06
p		0.974	**0.0000**	0.444	0.111	**0.001**	**0.018**

Table 21. Medical examinations data of workers with varied exposure time in directly exposed groups

Rate of ab-normalities (%)	Exposure time (years)				X²value	pval-ue
	1~5 (n=9)	5~10 (n=48)	10~20 (n=97)	"/>20 (n=87)		
WBC	0	6.3	3.1	1.1	3.212	0.360
Hb	11.1	**47.9**	39.2	21.8	13.193	**0.004**
ECG	0	12.5	15.5	13.8	1.744	0.627
B ultrasonic	0	16.7	23.7	13.8	5.252	0.154
SBP	11.1	14.6	11.3	12.6	0.328	0.955
DBP	44.4	27.1	19.6	25.3	3.420	0.331

Table 22. The symptom scores were realted to the AChE activity

AChE activity (U)	Number	Symptom scores				
		neuromus-cular system	respi-ratory system	circula-tory system	diges-tive system	Total system scores
0~120	67	1.07	0.46	0.54	0.34	2.42
120~127	54	0.59	0.15	0.39	0.20	1.33
127~134	74	0.49	0.26	0.49	0.14	1.36
"/>134	46	0.41	0.13	0.28	0.13	0.96
H value		10.018	16.278	3.723	11.564	8.490
p value		0.018	**0.001**	0.293	**0.009**	**0.037**

Table 23. The raiao of medical examination abnormalities were related to the AChE activity

AChE activity (U)	Number	Abnormalities (%) of medical examinations					
		WBC	Hb	ECG	B ultra-sonic	SBP	DBP
0~120	67	1.5	56.7	7.5	16.4	4.5	11.9
120~127	54	5.6	35.2	20.4	14.8	14.8	25.9
127~134	74	1.4	25.7	13.5	17.6	13.5	29.7
"/>134	46	4.3	10.9	15.2	23.9	19.6	30.4
X^2 value		2.724	28.840	4.330	1.591	6.398	7.813
p value		0.436	**0.000**	0.228	0.662	0.094	**0.049**
Trend X^2 value		0.157	28.051	0.878	0.959	5.164	6.330
p value		0.692	**0.000**	0.349	0.327	0.023	**0.012**

We also compared the 686 health surveillance records in 1979 and 1995 in Shanghai Pesticide Factory to understand changes of health status among employees and evaluate the effectiveness of occupational health measures herein. We noted that less symptoms and signs score in 1995 than 1979. Higher percentage of abnormal blood pressure was found among the first year new workers. With the pass of time, the percentage of such change also increased. There were no differences of hemoglobin levels among workers who engaged in different sectors and with different working ages. ANOVA test revealed that the activity of cholinesterase in 1995 was significant higher than 1979. The job code (which dominants the magnitude of OPs exposure) was a main affecting factor to the enzyme activity. Better health status in 1995 than in 1979 was also found based upon the data of 139 workers who had received two-times examinations in 1979 and in 1995. These results confirmed that the general health status of workers exposed to pesticides was better in 1995 than in 1979 in this pesticide factory. It indicated that the occupational health measures taken during this period of time were effective.

In Shanghai Pesticide Factory, we also observed the typical tolerance phenomenon to OPs. The trend of change of ChE and clinical score among the contractor workers exposed to different levels of OPs were carefully studied. The trend of changes in blood ChE and score since starting exposure to 3 or 4 months were expressly present. We found that the ChE and score of packing workers sharply declined since the starting of exposure; there were significant exposure-effect correlations. After withdrawing of those who were poisoned (ca. 2%) in 40-60 days, the ChE and score dropped less steep and then turn to flat. It indicated that body developed tolerance to low-level exposure to OPs in 40-60 days. High level (or higher toxicity) exposure caused poisoning in portion of the workers, but the remainders tolerated the exposure, and kept ChE and score in a steady horizon, though fluctuated and less than normal.

Neurobehavioural Effects Caused by Long-Term Ops Exposure

Some, but not all, epidemiological studies demonstrated that long-term exposure to OPs may be associated with impaired neurobehavioural performance. Clinical features that have been reported include anxiety disorder, depression, psychotic symptoms, dysthymic disorder (DSM-III-R); short-term memory problems, learning disorders, attention-deficit disorders, information processing problems, eye-hand coordination problems and delayed reaction time, and autonomic dysfunction.

Zhang and his colleagues conducted a survey on a representative sample of 9811 rural residents in Zhejiang province (J.M. Zhang et al., 2009). These residents were asked about the storage of pesticides at home and about whether or not they had considered suicide within the 2 years before the interview. The Chinese version of the 12-item General Health Questionnaire (GHQ) was administered to screen for mental disorder. They found that the unadjusted odds ratio (OR) for the association between pesticide storage at home and suicidal ideation over the prior 2 years was 2.12 (95% confidence interval, CI: 1.54–2.93). After adjusting for gender, age, education, socioeconomic status, marital status, physical health, family history of suicidal behaviour, GHQ caseness and study design effects, the OR was 1.63 (95% CI: 1.13–2.35). These results indicated an association between OPs exposure and suicide ideation in rural areas of China.

Effects of Long-Term Ops Exposure on the Human Reproduction

Another important feature of OPs is their endocrine disrupting effects and potential adverse impact on both male and female reproductive function. Studies carried out employing chronic exposure of animals to low doses of the OPs showed a reduction in reproductive function, both female and male. And a number of epidemiology data also demonstrated the deleterious reproductive effects of chronic exposure to OPs in occupational and/or environmental settings.

Lv and her colleagues investigated the cross-sectional association between OPs use and menstrual function among 298 women working at a OPs factory (Lv, 2004). Women were aged 21-45 years, premenopausal, not pregnant or breastfeeding, and not taking oral contraceptives. Menstrual cycle characteristics of interest included symptoms before the menstruation begins; cycle length (short cycles, long cycles, irregular cycles); missed periods (not experiencing a period for more than 6 weeks in the last 12 months); menstruation amount (large, small); and dysmenorrhea. After controlling for age, working time, and education level, the author found that women who used pesticides experienced more pre-menstruation symptoms and increased odds of irregular menstrual cycles compared with women who never used pesticides.

Zhang and her colleagues observed 601 female workers in the first production line of the pesticide factory and 873 unexposed female workers according to the reproduction occupational epidemiological method (S.H. Zhang et al., 2004). Then they reported a significantly higher incidence of premature delivery (8.20%), post-mature delivery (7.64%), spontaneous abortion (2.83%), and pregnancy induced hypertension syndrome (6.41%) in the exposed group than the unexposed group (p=0.000, 0.003, 0.004, 0.035).

Li's investigation also showed an increased incidence of irregular menstruation, spontaneous abortion, and infertility in the OPs exposed group when compared with the control group (G.R. Li et al., 2000).

Li and Zou surveyed 161 male farmers exposed to OPs and 161 unexposed men via epidemiological questionnaires. Then these subjects received genital examinations, and their semen samples were collected for analysis. The authors found a decrease in sperm viability and percentage of sperm with forward progression, and normal sperm morphology. The semen density of farmers in the exposed group was $76.0\pm84.8\times10^6$/mL, significantly lower than those in the unexposed group ($100.0\pm56.4\times10^6$/mL). Logistic regression analysis showed that chronic exposure to OPs would influence the sperm quality (W.Y. Li et al., 2004; Zou et al., 2005).

Effects of Long-Term Ops Exposure on Fetal and Childhood Health

Large amount of evidence have shown that fetuses can be exposed to pesticides. OPs pass through the blood–brain barrier and placenta and have also been found in amniotic fluid. In addition, the young may receive greater exposure than adults, because they eat, drink, and breathe more per unit of body weight. They are closer to the floor and surfaces where pesticides may settle, and have extensive hand-to-mouth contact. Recent studies have shown that fetuses and young children have lower than adult levels of detoxifying enzymes and their brains are developing rapidly. This suggests that the nervous system of the fetus and young children is several-fold more susceptible to potential neurotoxic effects of such low-dose OPs exposure.

Wang and his colleagues investigated the association between neurodevelopment and behavior of 301 children. Child neurodevelopment was assessed by the Gesell Development Schedule at 2 years of age. Developmental quotients (DQs) were obtained in motor, adaptive, language and social areas. They reported that geometric mean (GM) for children DAP metabolites (μg/g) were DMP: 10.38; DMTP: 6.56; DEP: 7.27; DETP: 14.26; DEDTP: 4.46 (Table 24). They found a significant correlation between DAP levels and children neurodevelopment (Table 25 and 26. They also found the DQs were higher in high dose exposure group than in the low dose exposure group. There was highly significant difference between these two groups (p=0.03) (Table 27). In addition, DAP levels were positively associated with 8-OHdG in urine (r=0.594, p=0.000) (Wang, 2009).

Table 24. Creatinine-adjusted OPs urinary DAP metabolites levels among children (μg/g) (n=301)

DAP metabolites	Detection percentage (%)	GM	Range	P25	P50	P75	P95
DMP	41.9	10.38	1.17~724.43	3.95	8.93	23.70	125.60
DMTP	36.5	6.56	0.07~478.63	2.87	5.90	13.12	58.64
DEP	71.8	7.27	0.06~169.82	3.51	7.16	14.79	54.61
DETP	69.1	14.26	1.1~977.24	5.30	12.91	37.15	128.82
DEDTP	2.7	4.46	1.07~72.44	2.46	4.45	7.69	18.36

Table 25. Distribution of GSD DQ score (n=301)

DQ score	Mean±SD	Normal development percentage (%)	Delayed development percentage (%)
Behavioral ability	103.07±7.59	99.67	0.3
Adaptability to environment	107.03±11.87	98.67	1.3
Verbal ability	104.27±16.22	93.7	6.3
Adaptability to people	96.11±7.34	97.3	2.7

Table 26. Adjusted coefficient (â) (95%CI) in points on the Gesell scores of children neurodevelopment for log10 unit increase in pesticide urinary metabolites (n=301)

DAP metabo-lites	Behavioral ability		Adaptability to environment		Verbal ability		Adaptability to people	
	â (95%CI)	p	â (95%CI)	p	â (95%CI)	p	â (95%CI)	p
DMP	-0.20 (-6.88~6.35)	0.94	0.05 (-9.03~11.03)	0.85	0.02 (-13.32~13.45)	0.99	-0.25 (-9.61~3.24)	0.76
DMTP	0.12 (-2.399~6.06)	0.39	0.49 (-5.28~7.53)	0.73	-0.07 (-10.90~6.20)	0.59	-0.15 (-6.20~2.00)	0.31
DEP	-0.19 (-5.13~4.53)	0.90	-0.10 (-9.68~4.98)	0.53	-0.04 (-11.16~8.39)	0.78	-0.18 (-7.40~1.98)	0.26
DETP	-0.47 (-13.16~0.90)	0.09	-0.44 (-19.6~1.71)	0.10	-0.11 (-17.35~11.09)	0.67	-0.16 (-8.90~4.75)	0.55
DEDTP	0.13 (-1.58~7.54)	0.20	0.07 (-4.41 9.42)	0.48	0.06 (-6.12~12.34)	0.51	0.05 (-3.41~5.44)	0.65

Table 27. Gesell scores in two dose groups (n=301)

	DQ scores	High dose group (n=212)	Low dose group (n=89)
Behavioral ability	Mean ± SD (range)	103.36±7.33 (83~125)	102.36±8.17 (90~124)
	Normal (%)	99.53%	100.00%
Adaptability to environment	Mean ± SD (range)	107.34±11.85 (83~136)	106.28±11.94 (79~135)
	Normal (%)	99.06%	97.75%
Verbal ability	Mean ± SD (range)	105.02±15.93 (66~146)	102.5±16.96 (66~138)
	Normal (%)	94.34%	92.13%
Adaptability to people	Mean ± SD (range)	96.99±7.3 (82~133)	94.02±7.02 (71~121)
	Normal (%)	98.11%	95.51%

Wang also collected and analyzed urine samples of 187 pregnant women to evaluate the relationship of maternal prenatal DAP levels with birth outcomes. The results showed that GM of DAP metabolite levels (µg/g) of pregnant women were DMP: 25.75; DMTP: 11.99; DEP: 9.03; DETP: 9.45; DEDTP: 0.75. They did not found the evidence that OP pesticides at current levels adversely affect fetal development.

Luo analyzed the birth outcome data of 5571 prenatal infants in a rural area of Guangdong Province and reported that 1.13% of them were born with deformity including hydrops fetalis syndrome, neural tube defects, hydrocephalus, and congenital equinovarus. Further logistic analysis found a relationship between maternal exposure to OPs and birth defects (Luo, 2004).

Other Health Problems Caused by Long-Term Ops Exposure

By analyzing the death cause data of a cohort including 2270 workers employed for at least 1 year before Jan 1, 1983 and a sub-cohort of 1018 of them worked at OPs exposed workshop in a pesticide factory, we investigated the cause of death and mortality of cancer among OPs exposed workers and evaluated the relationship between long-term occupational OPs exposure and cancer occurrence. This study was followed up from Jan 1, 1983 to Dec 31, 2004. The death cause spectrum of OPs exposed workers was similar to that in reference population locally, but higher mortality of malignant tumor was found in OPs exposed workers. The SMR for all cancer, and malignant cancer were 120.2 and 119.6 respectively. SMR for malignant tumor of bladder, lung and stomach cancer were 303.7, 141.2, and 137.5 respectively ($P<0.01$). Chi-square test showed tumor mortality of exposed workers was higher than that of non-exposed workers ($P<0.01$), indicating the risk of malignant tumor death increased with exposure to OPs (Table 28 and 29).

Hong tested DNA damage in peripheral lymphocytes of workers exposed to OPs via single cell microgel electrophoresis (SCGE) and found that the cometic rate of peripheral lymphocyte among OPs exposed workers was (2.8 ± 1.9)%, significantly higher than that in control group ($p<0.01$). The amount of T lymphocyte α-ANAE in peripheral blood among OPs exposed workers was also significantly higher than that in control group ($p<0.01$). These results suggested that chronic exposure to OPs may lead to genetic damage (Hong et al., 2002).

We studied the M_3 gene expression in peripheral blood lymphocyte of workers exposed to diamethoate and explore its role in the toxic effects of OPs. The lymphocytes in peripheral blood from 33 workers exposed to diamethoate and 15 control people were isolated and treated with saline and diamethoate in vitro, respectively. RT-PCR technique was used in determine M_3 gene expression. Basal and inducible gene expression levels were measured. The result was presented in ratio of optical density of sample mRNA and that of the reference (β-actin) as: $(M_3 \ O.D.\times353)/(248\times\beta\text{-actinO.D.})$. There (OD) no significant difference of basal gene expression level between the exposed group and control group, (1.49 ± 0.20) versus (1.49 ± 0.45); while the inducible gene expression level was significantly higher in exposure group to the control group, (1.92 ± 1.07) versus (1.22 ± 0.19). No difference was found between male and female people in both exposed and control group. The inducible gene expression level was higher in the operators than in the packers, which maybe attribute to the difference of exposure time. The inducible M3 gene expression level showed a gradient increment with the elongation of the working age: <5yr(1.69 ± 0.95), 5~25yr (1.91 ± 1.03), >25yr (2.09 ± 1.25). These indicated that after long-term exposure to OPs, the basal M3 receptor gene expression level in the exposed workers did not show any difference with the control group, but the inducible gene expression level (treated with OPs in vitro) would increase and the level was related to the degree of OPs exposure.

Table 28. The cause of death and mortality of both OPs exposed workers and reference population *: P<0.05.**: P<0.01.

population	Reference population		Cohort of exposed group		Expected deaths	SMR
	Death toll	Mortality	Death toll	Mortality		
All death cause	149511	819.60	263	719.19	300	87.7
All cancer	41484	227.41	100	273.46	83	120.2
Malignant tumors	41306	226.43	99	270.72	83	119.6
Nasopharyngeal cancer	519	2.85	0	0.00	1	0.0
Esophageal Cancer	2285	12.53	5	13.67	5	109.2
Gastric cancer	7258	39.79	20	54.69	15	137.5*
Intestinal cancer	3499	19.18	5	13.67	7	71.3
Liver cancer	5333	29.23	14	38.28	11	131.0
Lung cancer	10248	56.18	29	79.30	21	141.2*
Brest cancer	1229	6.74	2	5.47	2	81.2
Cervical cancer	216	1.18	0	0.00	0	0.0
Bladder cancer	657	3.60	4	10.94	1	303.7**
Leukemia	825	4.52	1	2.73	2	60.5
Benign tumors	81	0.44	1	2.73	0	615.8**
Other tumors	9334	51.17	19	51.96	19	101.5
Other diseases	108027	592.19	164	448.47	217	0.76

Table 29. Constituent ratio of death in OPs exposed population and reference population

Groups	Male			Female		
	Death from tumors	Death from others	To-tal	Death from tumors	Death from others	Total
OPs exposed population	46	54	100	12	8	20
Reference population	36	91	127	3	13	16
Total	82	146	227	15	21	36
	$X^2=7.556$, p=0.006			$X^2=6.223$, p=0.013		

INTERACTION OF GENETIC POLYMORPHISMS AND LONG-TERM OPS EXPOSURE

While this review has focused on health problems caused by long-term OPs exposure via a number of different ways including occupational, dietary, lifestyle or medicinal, it should be recognized that it is likely that polymorphisms within a variety of genes may affect susceptibility to OPs induced toxicity. Much of the work in this field has focused on OPs metabolism and detoxification pathways.

One of our studies examined whether BChE and PonE polymorphisms influenced susceptibility in OPs exposed population. We determined BChE-K, PonE-192 and PonE-55 genotypes of 75 OPs exposed workers using PCR-PFLP. And then their accumulative symptom scores and the whole blood AChE activity (mmol h^{-1} ml^{-1}) were measured as health index. We analyzed their health condition related to single gene site of the three gene loci to determine which kinds of genotype were susceptible. Then, we used the multiple variance analysis to see if there existed interactions among these three gene loci. Finally, we established the multi-factor linear regression equation, considering some other factors that might affect the health status such as age, gender and exposure time. The results showed that the mean AChE activities of the exposed workers with BChE-K genotype UU (61 cases), genotype UK(12 cases)and genotype KK (2 cases) were respectively 105.0±23.0, 84.4±16.4, 79.0±9.9. The accumulative symptom scores were respectively 3.7±3.8, 9.2±3.0, 12.5±0.7. The AChE activities of the exposed workers with PonE-192 genotype BB (37 cases), genotype AB (27 cases) and genotype AA (11 cases)were respectively 116.8±15.1, 91.2±15.6, 72.3±21.4. The accumulative symptom scores were respectively 2.0±3.2, 6.7±3.3, 9.7±1.8. Similarly, the AChE activities of the exposed workers with PonE-55 genotype LL (70 cases) and genotype LM (5 cases) were 102.4±23.0, 82.8±22.0. The accumulative symptom scores were 4.5±4.2, 9.2±3.6. Single variance analysis showed that the accumulative symptom scores of the individuals with abnormal homozygote of these three gene loci were the highest, which indicated that they were most susceptible to OPs exposure. Multiple variance analysis showed there were no interactions among the three gene loci. Age, gender and exposure time had no statistical significance while genotypes of the three gene loci had significant relationship to health status. In conclusion, we found that the genotypes of BChE-K, PonE-192 and PonE-55 are associated with susceptibility to OPs exposure.

Another work of our research group detected the genotypes of enzymes (PonE-192, PonE-55, BChE, P450 and NAT2) and the polymorphic distribution via 7900 genotype detecting system and CMOS Chip technique. We found that the abnormal allele frequency of PonE-192, PonE-55 and BChE was respectively 37.8%, 1.9% and 13.7% whereas the abnormal homozygote frequency of PonE-192 and BChE was 15.0% and 1.6% with no abnormal homozygote of PonE-55 (Table 30). The genotypes of all enzymes reached Hardy-Weinberg balance.

We further analyzed the effects of the genetic polymorphism of enzymes on urinary DAP metabolites, esterase activity, signs and symptoms. The results showed that the polymorphism of P450 metabolic enzymes (CYP1A2, CYP2E1) influenced the concentration of urinary DAP metabolites (DEP, DEDTP) (Table 31). The genotypes of PonE-192 and PonE-55 influenced the activity of PonE. The genotype of PonE-192*AA as well as PonE-55*ML appeared with low activity (Table 32). Lower activity of the same genotype of PonE-192 and PonE-55 (working duration less than 20 years) was found, while the BChE activity of workers more than 20 working years had the higher inhibition. We also found a relationship between PonE, BChE and exposure dose by controling the influence of genetic polymorphism (Table 33). But there was no significant relationship between genetic polymorphism and examination abnormalities of exposed workers (Table 34). The activity of PonE was lowest in the workers with genotype of PonE192*AA + PonE55*ML + BChE*KK, and the AChE activtity was lower while signs scores was higher. The genotype of PONE192*AA + PonE55*ML + BChE*KK was the most sensitive. The liner regression analysis showed the polymorphism of PonE and BChE affected the activity of AChE, indicating that the gene polymorphism influence the health effects caused by OPs exposure (Table 36).

Table 30. The genotypes of enzymes and the polymorphic distribution

Gene loci	Genotypes	Cases	Allele	Allele cases	Allele frequency
	Gln/Gln(AA)	32	Gln	161	0.378
PonE-192	Arg/Gln(BA)	97			
	Arg/Arg(BB)	84	Arg	265	0.622
	Met/Met(LL)	205	Met	418	0.981
PonE-55	Leu/Met(ML)	8			
	Leu/Leu(MM)	0	Leu	8	0.019
	Ala/Ala (UU)	179	Ala	416	0.863
BChE*K	Thr/ Thr (KK)	58			
	Ala/Thr (UK)	4	Thr	66	0.137

CYP1A1	AA	114	A	145	0.797
	A/G	62			
	GG	6	G	37	0.203
CYP1A2	GG	55	G	103	0.575
	G/A	95			
	AA	29	A	76	0.425
CYP2E1	AA	8	A	44	0.243
	A/T	72			
	TT	101	T	137	0.757
NAT2	GG	104	G	125	0.839
	G/A	42			
	AA	3	A	24	0.161

Table 31. The influence of polymorphism of P450 metabolic enzymes on urinary DAP metabolites level

Gene loci	Geno-types	Number of people	Urinary DAP metabolites (μg/gCr)				
			DMP	DEP	DETP	DMDTP	DEDTP
CYP1A1	AA	114	0.00	928	9.95	252	101
	A/G	62	0.00	187	8.36	151	109
	GG	6	0.00	512	103.7	355	60
p value			0.142	**0.015**	0.446	0.606	0.262
CYP1A2	GG	55	0.00	177	9.24	355	104
	G/A	95	0.00	145	9.96	164	101
	AA	29	0.00	402	7.39	149	111
p value			0.988	**0.027**	0.486	0.432	0.931
CYP2E1	AA	8	0.00	844	9.96	245	88.7
	A/T	72	0.00	104	12.9	222	125
	TT	101	0.00	150	7.53	222	94.2
p value			0.189	0.527	0.195	0.795	**0.032**
NAT2	GG	104	0.00	111	9.41	169	109
	G/A	42	0.00	996	9.39	181	86
	AA	3	21.8	191	6.84	655	79.8
p value			0.079	0.920	0.414	0.419	0.164

Table 32. The influence of esterase genetic polymorphism on esterase activity

Gene loci	Geno-types	Number of people	Esterase activity			
			BChE	CarbE	PonE	AChE
BChE	UU	179	33.26±9.13	512.91±186.09	150.81±98.64	122.00±6.68
	UK	58	40.52±17.00	552.31±116.9	148.67±70.05	126.19±9.40
	KK	4	39.34±18.28	500.87±189.18	140.65±70.33	123.60±8.71
pvalue			0.709	0.183	0.735	0.134
PonE-192	AA	32	43.99±31.17	518.04±183.97	94.32±44.18	123.66±10.68
	AB	97	39.43±14.91	503.79±195.26	154.32±71.54	125.69±8.09
	BB	84	39.89±16.25	518.47±193.92	146.04±68.57	125.53±9.56
pvalue			0.475	0.924	**0.000**	0.541
PonE-55	LL	205	40.37±18.98	511.87±190.3	144.25±69.53	125.36±9.25
	LM	8	38.49±7.79	623.61±97.37	85.45±50.75	123.88±7.45
pvalue			0.781	0.101	**0.019**	0.653

Table 33. The influence of esterase genetic polymorphism on esterase activity (nmol/ml min) of workers in different working age groups

Gene loci	Geno-types	Number of people	working age		t value	pvalue
			≤20 years	≥20 years		
PonE-192	AA	32	90.53±33.21	98.11±53.86	-0.479	0.635
	AB	97	137.36±63.34	175.62±76.17	-2.701	**0.008**
	BB	84	141.09±71.92	152.32±64.49	-0.743	0.459
PonE-55	LL	205	133.27±66.14	157.73±71.55	-2.538	**0.012**
	LM	8	109.27±50.97	61.62±43.56	1.421	0.205
BChE	UU	179	42.32±21.92	35.74±11.71	2.562	**0.011**
	UK	58	39.31±15.89	41.74±18.24	-0.542	0.590
	KK	4	33.26±9.13			

Table 34. The examination abnormalities of exposed workers in different genetic polymorphism

Gene loci	Genotypes	Number of people	Abnormalities (%) of medical examinations					
			WBC	Hb	ECG	B ultra-sonic	SBP	DBP
BChE*K	UU	179	2.2	32.4	14.0	19.0	12.3	24.6
	UK	58	5.2	37.9	12.1	15.5	12.1	22.4
	KK	4	0	25.0	25.0	0	25.0	25.0
p value			0.389	0.693	0.988	0.333	0.751	0.783

PonE-192	AA	32	0	28.1	15.6	18.8	12.5	28.1
	AB	97	4.1	32.0	10.3	17.5	14.4	18.6
	BB	84	3.6	35.7	17.9	15.5	9.5	26.2
p value			0.308	0.715	0.334	0.893	0.602	0.361

Table 35. The relationship between multi-genetic polymorphism and esterase activity and symptom scores

BChE*K	PonE-192	PonE-55	Number	PonE (nmol/ml·min)	AChE (U)	Symptom scores
UU	AA	LL	1	83.39	123.00	2.00
UU	BB	LL	1	144.04	131.00	0.00
KK	AA	LL	22	97.91	124.5	1.64
KK	BB	LL	59	143.67	125.85	1.27
KK	BA	LL	70	157.47	126.29	1.70
UU	BA	LL	2	187.90	117.00	1.00
KU	AA	LL	5	114.10	118.20	1.80
KU	BB	LL	23	151.87	125.00	1.39
KU	BA	LL	22	147.88	124.64	1.18
KK	**AA**	**ML**	4	**52.59**	126.00	**5.00**
KK	BA	ML	2	134.83	121.50	0.50
KU	BA	ML	2	101.80	122.00	2.00

CONCLUSION

We present the research results conducted in China by Chinese scientists, mostly our research group. From these, we believe that the health problem caused by OPs exposure can't be ignored, though the exposure-response was not clearly elucidated. It is good that with the economic development towards better, the working condition has been improved and workers have less exposure to OPs. The traditional types of organophosphorus pesticides with high acute toxicity, such as methamidophos, parathion; methyl parathion and phosphamidon were prohibited in China, However, long-term and low level exposure to OPs is still a serious health problem and we should pay more attention to these public problems.

ACKNOWLEDGEMENTS

Many colleagues and graduate students have been involved in these researches, among them I should particularly thank Prof. Shouzheng Xue, who was my tutor. I began my research related to organophosphorus pesticides and health problem when I pursued my master degree in 1985.

REFERENCES

1.   Hong C.J. et al.2002Study on The DNA Damage in Peripheral Lymphocytes of Workers Exposed to Organophosphorus Pesticides Via Single Cell Microgel Electrophoresis. Industrial Health and Occupational Diseases, 285October 2002), 3033041000-7164

2.   Li G.R. et al.2000Study on The Effects of Occupational Exposure to Organophosphorus Pesticides in Female Workers. Chinese Journal of Industrial Medicine, 136December 2000), 3553580100-2221X

3.   Li W.Y. et al.2004Impact of Organophosphorus Pesticides on Semen Quality. Reproduction and Contraception, 245May 2004), 2862300025-3357X

4.   Z. Lin, J. Huang, Q. H. Zhu, 2007Study on New Surveillance Indicator For Workers Exposed to An Organophosphorus Insecticide Parathion. Chinese Journal of Industrial Medicine, 204August 2007), 2142170100-2221X

5.   Luo X.E.2004Study on The Factors of Perinatal Infant Birth Defect in Yingde City. Chinese primary health care, 184April 2004), 490100-1568X

6.   Lv L.P.2004Menstrual Cycle Characteristics among Female Workers Exposed to Organophosphorus Insecticides. Occupation and Health, 2012June 2004), 40411004-1257

7.   Tang Y.Q. & Liu J.F.2004Study on ECG of Workers Exposed to Organophosphorus Insecticides. Industrial Health and Occupational Diseases, 313June 2004), 65661000-7164

8.   P. Wang, 2009Children and Pregnant Women Pesticide Exposure and Effects on Children's Development. (April 2009), Dissertation of Shanghai Jiaotong University

9.   Zhang J.M. et al.2009Pesticides Exposure and Suicidal Ideation in Rural Communities in Zhejiang Province, China. Bull World Health Organ, 87July 2009), 745753Doi

10.   Zhang S.H. et al.2004Study on The Influence of Occupational Pesticides Exposure On Reproductive Function in Female Workers. Modern preventive medicine, 315March 2004), 6646651003-8507

11.   Zhang WP. et al.2010Abdominal Ultrasonic Manifestations of 284 Occupational Organophosphorus Pesticide Contacts, Practical Preventive medicine, 174April 2010), 1006-3110

12. Y. Zhao, et al.2010Survey of Correlation Between Long-term Exposure to Organophosphorus Pesticides and Sleep Quality of Peasants. Occupation and Health, 2618September 2010), 1004-1257

13. Zou X.P. et al.2005Influence of Organophosphorus Pesticides on Semen Quality of Men in Rural Areas of Changshu City. Chinese journal of family planning, 138August 2005), 4764781004-8189

Chapter 6

PLANT BASED PESTICIDES: GREEN ENVIRONMENT WITH SPECIAL REFERENCE TO SILK WORMS

Dipsikha Bora[1], Hiren Gogoi[1], and Bulbuli Khanikor[2]

[1]Department of Life Sciences, Dibrugarh University, Dibrugarh, Assam, India

[2]Department of Zoology, Gauhati University, Guwahati, Assam, India

INTRODUCTION

Pesticides once having entry to an environment either get into the complex web of life through food chain or different components of the environment through physical passages like drifting by air and aquatic runways. Such facts were meticulously described by Rachel Carson [1] in her book 'Silent spring' where she advocated for choosing either the chemical control or biological control to avoid creation of endless problems to mankind owing to pesticide use. Looking back at the history of tremendous potentiality of synthetic chemicals to manage insect pests followed by subsequent cases of failure of chemical control due to development of insect resistance and pest resurgence, here we intend to cite examples of selective toxicity of insecticides and reiterate its importance in management of insect pests. Pesticidal pollution is a global problem. Use of synthetic insecticides to control pest around the world has resulted in disturbances of the environment, secondary pest resurgence, pest resistance to pesticides, lethal effects to non-target organisms as well as direct toxicity to users. It has been reported that about 2.5 million tons of pesticides are used on crops each year and the worldwide damage caused by pesticides reaches $100 billion annually. The reason behind this amount of cost is the high toxicity and non biodegradable properties of pesticides and the residues in soil, water resources and crops that affect public health [2]. Hence search for the environment friendly, highly selective, newer biodegradable pesticides for pest management program has been advocated to be essential for last several decades.

South East Asian countries are the hubs for production of raw silks produced by sericigenous insects namely *Bombyx mori* (mulberry silk worm), *Antheraea assama* (Muga silk worm), *Antheraea mylitta* (Tasar silk worm) and *Philosamia ricini* (Eri silk worm). India holds second position in world's raw silk production and contributes to 13.45 % of the total production [3].Yet the production is not sufficient even to meet the domestic demand of the raw silk. One of the major constraint in silk production is the susceptibility of the silkworms to attack of different pests, parasitoids, predators and pathogens. Evenmulberry silkworms which can be cultured in indoor condition are not free fromsuch constraints. In addition, their host plants are also susceptible to the attack by herbivorous pests(insects, mites) and various pathogens(Nematodes, bacteria, fungus and virus) [4,5]. Growth and development of silkworm to a great extent depend on quality and quantity of food consumed and utilized [6]. In suchsituations, usually farmers take shelter of spurious chemicals for controlling the pest population unless they are well versed with efficacy of pesticides and their hazardous effects. Application of insecticide in sericulture fieldisnot at all advisable as the leaves of host plants are directly consumed by the silk worms and silk worms become affected either through consumption of contaminated food or contact toxicity of the insecticides. *Bombyx mori*are highly susceptible to insecticides and in China its production is reported to be decreasing by almost 30% per annum because of insecticidal poisoning [7]. Fenvalerate-20EC (Sumicidine-20EC), one of the commonly used pyrethroid, reduced the rate of feeding, assimilation and efficiencies of conversion of ingested and digested food into body substance in late instar larvae of *Bombyx mori* [8]. Hexachlorocyclohexane, an organochloride insecticide was reported to cause decrease in fibroin content,pupal and shell weight, adult emergence percentage, fecundityas well as deterioration in quality and quantity of silk thread in *B. mori*[9]. About 50% of normal water intake of eri silkworm(*Philosamia ricini*)was reported to decrease after feeding with leaves of *Ricinus communis* treated with permethrin, a pyrethroid insecticide which might have resulted due to repellency or disruption of feeding physiology[10].The chemicals used if are phytotoxicreduce the nutritive quality of the leaves. It is globally accepted that complete elimination of pesticide drift is impossible. More often sericulture fields are contaminated by insecticides sprayed into other crop systemsin the neighboring areas. For instance many of the nonmulberry- sericulture fields in Brahmaputra Valley of Assamare situated mostly in close vicinity of stretches of crop fields like paddy and tea. Thus those sericulture fields cannot be expected to be spared of indirect contamination caused by widespread use of pesticides in paddy fields and tea gardens that warns precautionary measures (Plates 1).

| Paddy field | Sericulture field | Tea garden |

Plate 1. Components of Seri-ecosystem in Assam.

PESTICIDES IN ENVIRONMENT

Introduction of pesticides into a crop system subjects it into a variety of physico-chemical and biological processes which determine the persistence, fate and the ultimate degradation product. Many workers have shown that only a portion of pesticides sprayed onto crops reach their targets, the rest enter the atmosphere by spray drift, volatilization from soil or water, surface runoff, biotransformation by microorganisms, plants, animals, biomagnification through food chain and photodecomposition [11]. One of the major environmental aspect is the effect of sunlight that may lead to various photoprocesses and to photoproducts which are mostly different from parent pesticides in the environmental properties and toxicological significance. The quantum of light energy emitted from the radiation is absorbed by pesticides in environment and this raises the energy state of the molecule, causes excitation of electrons leading to formation or disruption of chemical bonds. Photolysis of pesticides have been studied in water, soil and plant surface. Various sensitizers present in environment such as riboflavin, humic substance etc. absorb light energy and serve as donor of energy to pesticide acceptor and bring about photodecomposition of pesticides. Most organophosphates whose photochemistry have been studied are phosphorothioate and phosphorodithiotate compounds. Although not highly susceptible to photodegradation by UV light, malathion degrades to different photoproducts such as malaoxon, malathion diacid, o,o-dimethyl phosphorodithioic acid,, o,o-dimethyl phosphorothioic acid and phosphoric acid [12, 13]. Some of the compounds viz. malaoxon are more toxic than the parent compound. Insecticides aimed against pest population may enter non-target arena through spillage at sublethal level but even these sublethal dosages may exert considerable damage on behaviour and activity of non-target population [14]. Pesticidal effect of insecticides at sublethal dosages may have long term effect and they may be expressed at a later part of the insect's life [15,6]. Continuous exposure to sublethal dosages on the environment may on the otherhand help a pest to develop resistance mechanism against the toxic compound. Troitskaya and Chichigina [16] showed that combined use of bacterial and chemical insecticides in silk-producing areas possess a real danger to *Bombyx mori*.

A study was carried out to evaluate the susceptibility of the silkworm, *Antheraea assama* to sublethal dosages of organophosphorous chemicals, malathion and phosphamidon in terms of certain developmental and biochemical parameters. The insecticides were sprayed to leaves at sublethal concentrations based on LD50 values determined earlier and allowed the fifth instar larvae to feed on them in 12 D: 12L photoperiodic condition. The parameters considered were mean larval growth rate, food consumption and utilization computed by the method of Waldbauer (17). In case of malathion treatment although at the lowest dose applied the mean daily consumption, utilization and mean larval weight did not vary significantly they decreased at higher dosages. The correlation coefficient between the mean daily consumption and utilization of food and concentrations of insecticide was -0.826 and -0.812 respectively. This represented a feeding deterrent effect of malathion and its ability to interfere with digestive physiology. Accordingly the growth rate and in later part of developmental period, the percentage of pupation and adult emergence decreased even at lower concentration. In case of treatment with phosphamidon, the daily food consumption and utilization decreased but they did not vary at the highest dose. The correlation coefficient between the mean daily consumption and concentrations of phosphamidon were + 0.539 which might indicate probable absence of feeding deterrence ability in phosphamidon. The growth rate decreased, but the mean larval weight increased with increase of concentration and the correlation coefficient between larval weight and concentrations of phosphamidon was +0.930. Although variations were observed in effect of different concentration during larval period, in later stage of development, similarly with the effects of malathion, the percentage of pupation and adult emergence decreased significantly (Figure 1). The mechanism in which the two insecticides interfered with the insect's body physiology was probably different. In order to study the effect of LD40 dosages of malathion and phosphamidon on tissue weight and different bioconstituents, the early instar larvae were allowed to feed on treated leaves and grow till adult. Analysis in late fifth instar larvae before silk spinning revealed that the tissue weight decreased along with total lipid, protein, Glycogen (female) and cholesterol (male) (Figure 2).

INSECTICIDE MECHANISM OF ACTION

The major classes of synthetic pesticides are organochlorines, organophosphates, carbamates and pyrethroides. Preliminary survey revealed that organophosphates and pyrethroides are two of the most common pesticide classes used by common farmers against pests of paddy and other crops and vegetables in Assam. They also belong to the most commonly used pesticide groups in tea gardens. Organophosphates like malathion, phosphamidon and dimethoate even at sublethal dose have been reported to be highly toxic against the larvae of *A. assama* [6].Organophosphates (OPs) are known to cause inhibition of esterases in silk worms [18]. Carboxylesterases constitute a class of the metabolic enzymes involved in insecticide resistance to OPs, carbamates, and pyrethroids through gene amplification, upregulation and coding sequence mutations [19]. The major function of acetylcholine esterase (AChE) is hydrolysis of the neurotransmitter acetylcholine bounded at cholinergic synapses in the central nervous system of insects [20] and the latter confers target site for susceptibility to orgnophosphorous insecticides which in *Bombyx mori* is reported to act through inhibition of*Bm*AChE1 responsible for expression of acetylcholincholine esterase [19,21,22]. Like DDT, Pyrethroides are axonic poisons. But in contrast to the residual persistence and biomagnification effect of DDT for which its use in agricultural crop has been banned over the globe, pyrethroides are the fastest developing group of modern insecticides primarily because of their effectiveness and safety application [23]. The synthetic pyrethroid deltamethrin (Decis) although is effective against a notorious parasitoid of silk worm,*Exorista sorbillans* with LC50 value at 0.106%, the insecticide has been found to be more highly toxic (LC50= 9 x10^{-5})to the larvae of muga silk worm, *Antheraea assama* [24]. Aerial application of organophosphates to agricultural field and their drifting to nearby mulberry plantation was reported to influence food consumption and utilization of silk worms [25]. Lipophilic insecticides are mainly carried in blood [26] in protein and protein associated forms [27]. From blood, insecticides are redistributed to gastrointestinal system, adipose tissue and brain [28-30]. Toxic effect of insecticides and oil pollutants are reported to cause changes in levels of different bioconstituents and metabolic processes in insects[30-34] finally leading to growth inhibition (Figure.1& 2) [6,15,35].

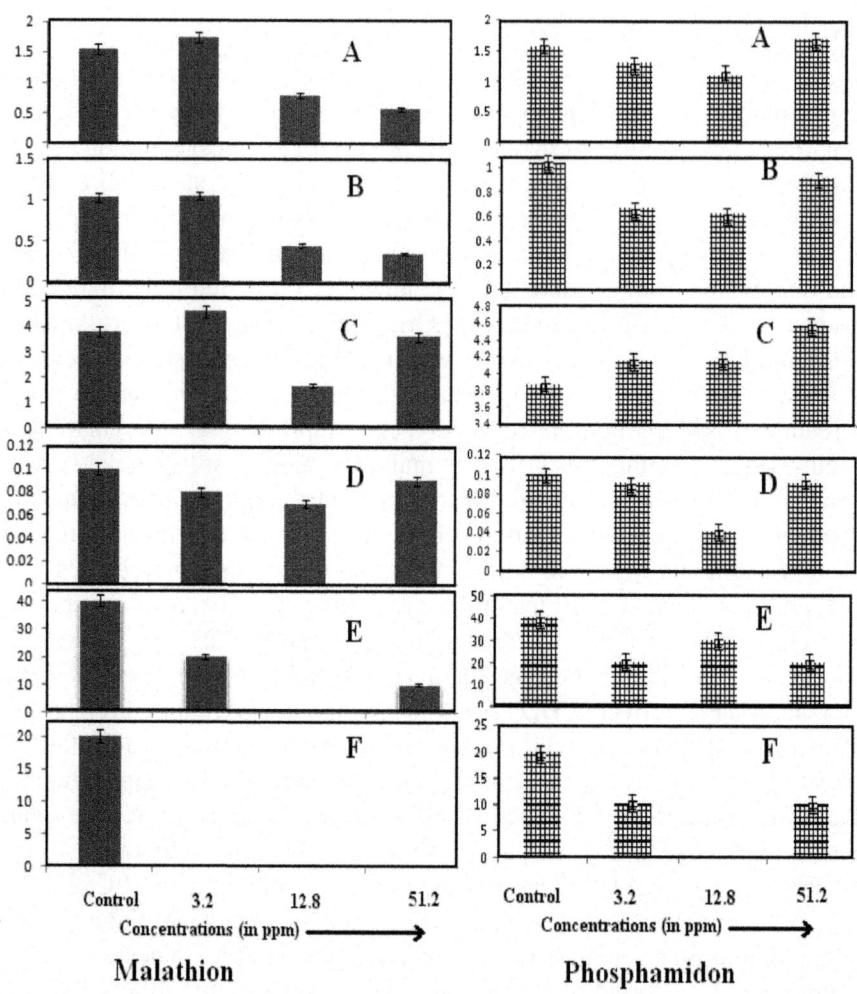

Figure 1. Effect of malathion and phosphamidon on developmental parameters of *A. assama*. A. Daily consumption, B. Daily utilization, C. Mean larval weight, D. Growth rate, E. Percentage of pupation, F. Percentage of emergence.

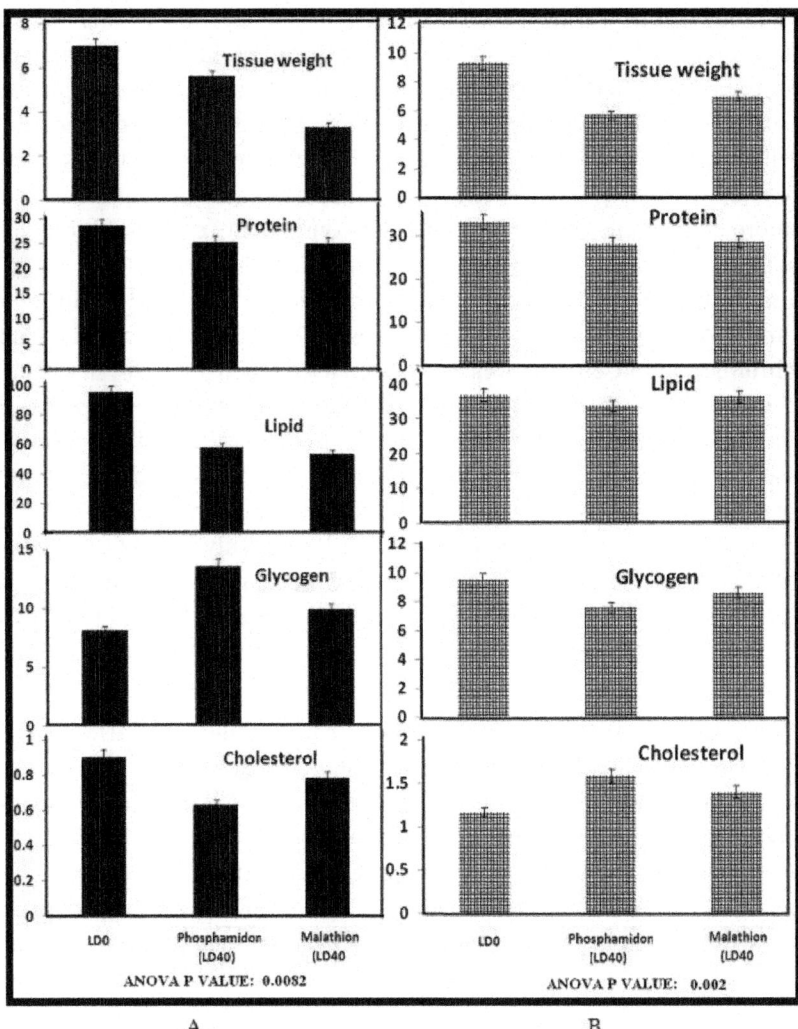

Figure 2. Effect of phosphamidon and malathion on tissue weight and level of bio-chemical constituents, Tissue weight (mg), Total Protein (mg/weight),Total lipid (mg/ml), Glycogen (mg/ml)and cholesterol (mg/ml). A. Male, B. Female.

NATURAL PRODUCTS

Natural products are excellent alternative to synthetic chemical pesticides. Plants exhibit enormous versatility in synthesizing complex materials which have no immediate obvious growth or metabolic functions. These complex materials referred to as secondary metabolites are produced as constitutive and induced defense as a result of co evolution arising out of millions of years of plant-herbivore interaction [36-41]. They may be exploited for the management of insect pests owing to their ability to act as toxicant, repellent, antifeedant and insect growth regulators. They are non phytotoxic, biodegradable and have little or no mammalian toxicity [42,43]. Plant extracts and essential oils come under the category of " Green pesticides" as they are safe, eco friendly and more compatible with environmental components compared to synthetic pesticides. 20th century witnessed an increasing trend in use of botanicals with more than 2000 bioactive plant species identified for their insecticidal and anti - pathogenic properties [39,40,44,45]. Natural insecticides such as pyrethrum, rotenone and nicotine have been used extensively for insect control [39]. Limnoids such as azadirachtin and gedunin, present in species from Meliaceae and Rutaceae are recognized for their toxic effects on insects and are used in several insecticide formulations in many parts of the world [46,47]. Neem formulations have been found effective against the mulberry hairy caterpillar, *Eupterotemollifera* Walker [48]. Pink mealy bug,*Maconellicoccus hirsutus* (Green) is one of the major pest which infests the mulberry plants and cause Tukra diseases leading to qualitative loss of leaves. Leaf extracts of *Andrographis* (99.0%), *Leucas* leaf extract,NSKE (99.0%), *Vitex* leaf and *Ocimum* (90.1%) have been reported to act as repellent against the bug[49].

Grasserie (viral), Flacherie (bacterial), Muscardine (fungal) and Pebrine (protozoan) are four common diseases of silkworm and they have been causing heavy loss to silkworm crops in silk producing countries like India and China. Herbal extracts have been tried for control of these diseases. Isaiarasu *et al.* [50] reported efficacy of aqueous and alcoholic crude extracts of *Acalypta indica, Ocimum sanctum*and *Tridax rocumbens* against flacherie and muscardine diseases in silkworm. The alcoholic extract of the plant *Tridax procumbens* were reported most effective followed by alcoholic extract of *Ocimum sanctum* and *Acalypta indica.*

The reported zone of inhibition (area mm2) of *Acalypta indica, Ocimum sanctum* and *Tridax procumbens* at a dose of 50 µl against flacherie causing bacteria were 29.9± 9.7, 42.4±6.8, 74.6±6.6 respectively and against muscardine causing fungus were 13.1±6.3, 23.6±3.4, 23.1±9.1 respectively. Aqueous extracts of thirteen plants were tested against cytoplasmic polyhedrosis virus (AmCPV) in tasar silkworm, *Antheraea mylitta* and out of them 2% concentrations of *Aloe barbedensis* (AKP 3), *Psoralea corylifoilia* (AKP13) and *Bougainvillea spectabilis* (AKP 9) were reported effective in suppressing virosis. They reduced mortality of larvae due to virus infection of 66.17%, 64.47% and 57.19% respectively. Total haemocyte count and haemolymph protein were also reported to increase in treated larvae which is considered as expression of immune response against the attack of pathogen [51]. Extracts of *Terminalia chebua* has been recommended against a potent bacterial strain, *Pseudomonas aerugenasa* (strain AC3) causing flacherie disease in *Antheraea assama*[52].

Plants of Ethnic Importance

Indigenous knowledge (IK) is unique to a particular culture and society. IK is embedded in community practices, institutions, relationships and rituals [53]. It forms the basis for local decision-making in agriculture, health, natural resource management and other activities and constitutes an important component in the global knowledge system. In most cases, IK is an underutilized resource in the development processes [53]. Learning from indigenous knowledge of specific communities used for generations after generations can improve the understanding of their local conditions and saves time, effort and money besides constituting the foundation for activities designed to address regional and global problems. Thus, the natural products based on the indigenous use of botanicals could be one way of mitigating the problems associated with inappropriate use of synthetic chemicals [54].

Botanicals in Sericulture Field

In this chapter we restrict our discussions to candidate plants for being used against a parasitoids of silk worm. In sericulture field, farmers of Assam traditionally sprinkle extracts of tulsi (*Ocimum sanctum*) leaf over egg bunches (Kharika) of *A. assama*. No other plants were recorded to be used against pests and pathogens of this silkworm in a survey carried out in Upper Brahmaputra valley of Assam during 2007-2011. Uzi fly, *Exorista sorbillans* (Diptera: Tachinidae) is a parasitoid of silkworm and a serious threat to sericulture industry.

The mated adult female fly lays eggs on the integument of third to fifth instar larva of *Antheraea assama*. The maggots after hatching pierce through the integument and grow inside the body of the silk worm by feeding on the fat body. The matured maggot pierce through the shell of the silk cocoon and crawl away from the site of the cocoonage in search of suitable place for pupation (Plate 2).

The piercing of the cocoon shell renders the silk cocoon unreelable. In this way, the fly is reported to cause 20-80% loss of seed crop of *A. assama*. The fly infestation is reported in all the commercial silkworm varieties and from almost all silk producing countries of the world [55,56]. A total of ten plants possessing insecticidal and medicinal values including *O. sanctum* were bioassayed against *E. sorbillans* (Table 1). Leaves of the plants were shade dried, ground to fine powder and extracts were prepared by using ethanol, water and hydroalcohol (50:50). Ethanolic extract of three plants namely *Catheranthus roseus, Ocimum sanctum and Ageratum conyzoides* proved effective by causing 53.33 %, 22.17%, 57.41% mortality after 24h at 10% concentration. The order of toxicity was*Ageratum conyzoides > Catheranthus roseus> Ocimum sanctum> Melia azedarach> Paederia foetida> Eupatorium odoratum> Polygonum hydropiper> Vitex negundo=Leucas aspera.*

\Further fractionation of ethanolic extract of the effective plants using an eleutropic series of solvents viz. petroleum ether, chloroform, butanol and water followed by subsequent bioassay showed that the petroleum ether extract of *Catheranthus roseus, Ocimum sanctum* and *Ageratum conyzoides* caused 46 %, 6.67%, 86.21% respectively after 24h and 100%, 33.79%, 100% mortality respectively after 48h at 5% concentration. Results of other solvent fractions were found negligible. The LC50 value of the most effective petroleum ether extract of *Ageratum conyzoides* was recorded as 0.74 percent [57]. Qualitative phytochemical studies of petroleum ether extract of the plants showed positive results for the presence of flavonoids, alkaloids, phenols and terpenoids. *Ocimum sanctum* has been reported to be effective against other Dipterans and many other pests and microorganisms[58-61].In addition to *Ocimum* the other two plants found effective were *Ageratum conyzoides* (Asteraceae) and *Catharanthus roseus*(Apocynaceae). Extracts of different species belonging to Asteraceae were earlier found effective against mosquitoes. *Ageratum conyzoides* with a long history in medicinal use is a common weed species and naturally grows in abundance in Assam. It is widely distributed in tropics and subtropics and has been used in various parts of Africa, Asia and in South Americafor curing various diseases like purulent ophthalmia, ulcers, wound caused by burns, asthma, dyspnea, pneumonia and also as purgative, febrifuge, antiinflammaory, analgesic, anti-diarrheic etc. Both essential oil and the major component precocene have antijuvenile hormone activity on a variety of insects [62]. In addition to their medicinal importance *Catharanthusroseus* has been reported effective against many insect pests including Dipterans. Volatile components obtained by hydrodistillation was found to contain 76 compounds [63].

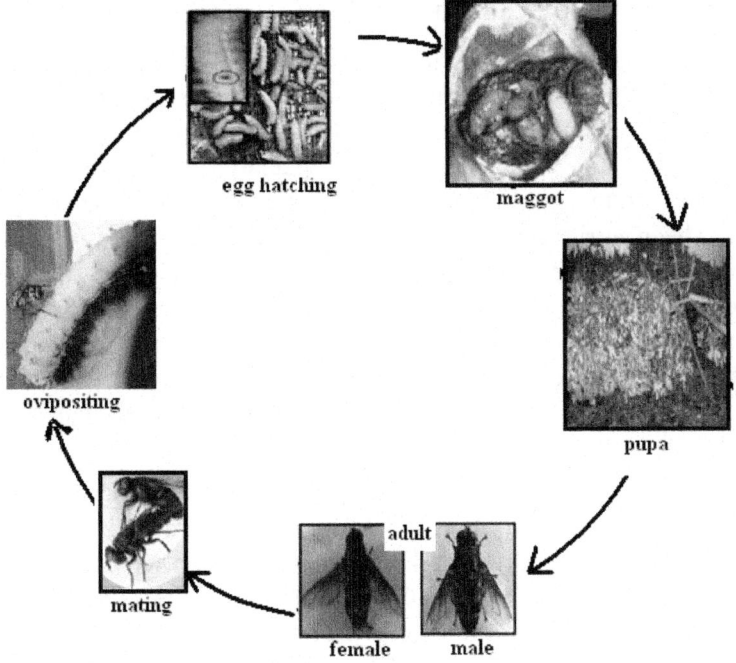

egg hatching

maggot

ovipositing

pupa

mating

adult

female   male

Plate 2. Life Cycle of *Exorista sorbillans* (Uzi fly).

Table 1. Comparative toxicity of crude extracts of plants (10%) to *E.sorbillans* (contact residual film technique)

Sl. No.	Plant	Cold water (Mean±SE)	Hot water (Mean±SE)	Hydroalcohol (Mean±SE)	Ethanol (Mean±SE)
1.	*O.sanctum*	0±0	3.33±3.33	3.33±3.33	22.17±4.54
2.	*O.gratissimum*	0±0	13.33±3.33	6.67±3.33	15±5.01
3.	*A.conyzoides*	0±0	10±5.78	13.70±3.17	57.41±4.31
4.	*C.roseus*	0±0	3.33±3.34	3.33±3.34	53.33±21.88
5.	*E.odoratum*	0±0	3.33±3.34	0±0	3.33±3.34
6.	*L. aspera*	0±0	0±0	0±0	0±0
7.	*M.azedarach*	0±0	0±0	0±0	6.67±3.3
8.	*P.foetida*	0±0	10±0	0±0	3.33±3.34
9.	*P.hydropiper*	0±0	3.33±3.34	0±0	0±0
10.	*V.negundo*	0±0	0±0	0±0	0±0

Botanicals in Paddy Cultivation

Farmers in some locations of Brahmaputra valley of Assam use plant materials in rice fields for prevention of pest infestation. For instance, robub tenga(*Citrus maxima*), chuka tenga (*Citrus medica*), jora tenga (*Citrus spp.*), leaf of baghdhaka (*Chromolaena odorata*), patharua bihlongoni (*Polygonum hydropiper*), chirata tita (*Andrographis paniculata*) etc. But, no systematic study has been carried out so far to ascertain if farmers are deriving any benefit or there is any scientific basis behind the beliefs of the farmers. A survey done by interactions with 200 farmers in 44 villages revealed that 23% farmers used synthetic chemical pesticides, 37% used traditionally used plant parts, 27% used both synthetic chemicals and traditionally used plant parts,while 13% adopted neither synthetic chemicals nor traditionally used plant materials (Table 2) [64].

Similarly with the discussions in context with pests and their control in sericulture field, here we limit our discussions regarding traditional use of plants and their scientific validation in context with a representative pest of paddy. *Nymphula depunctalis* (Guenée) [=*Parapoynx stagnalis* (Zeller)] (Lepidoptera; Pyralidae) is a serious pest of paddy that attacks young rice plants in waterlogged paddy fields of the Oriental region [65-70].It also thrives on various other grasses [71]. A number of related species occur in Asia but *N. depunctalis* is most widely distributed occurring in South and South East Asia, China, Japan, Australia, South America and Africa [72-74]. In many parts of Asia, *N. depunctalis*has been reported as a major pest [75-79]. In India, it is one of the serious pests in the states of Andhra Pradesh, Assam, Bihar, Karnataka, Orissa, Tamil Nadu, Manipur and Kerala. Damage to the plant occurs mainly due to defoliation by scraping of the green tissues by the larvae leaving only the white papery epidermis behind. Upon hatching, the first instar larvae climb onto a leaf and begin feeding on the green tissues by scraping the leaf surface. They then move to the leaf tip and cut a slit on the margin at a location 2–3 cm below the leaf tip. Then they make another cut about 1 cm below the first cut. Due to a lack of turgor pressure, the cut leaf segment rolls around a feeding larva to form a tubular case that is secured by silk spun by the larva. The inner surface of the leaf case is lined with silk to hold a thin film of water that is essential for respiration and to prevent desiccation of the larva[80]. Pupal period is spent within a closed case. (Plate 3).

Twenty two plants were found to be used traditionally in paddy fields against *N. depunctalis*and out of them extracts of 13 plants were selected for bioassay against *N. depunctalis*. The methods used for bioassay were residual film technique [81] and case dip technique [82]. The selected plant parts were shade dried, ground to powder and used to prepare extracts using water, hydro-alcohol (50:50) and ethanol. While hot water extract was the least toxic, both hydro-alcohol and ethanolic extracts were highly toxic. Out of thirteen plants tested, ethanol extract of seven plants were found highly effective (90 -100% mortality), five plant extracts were found moderately effective (50 – 89% mortality) and one was found the least effective (mortality less than 50%). The highlyeffective ones were leaves of*Calotropis procera*, root-bark of *Zanthoxylum nitidum*, leaves of *Zanthoxylum rhesta*, stem bark of*Crataeva nurvala* and leaves of *Croton tiglium*, *Vitex negundo* and *Chromolaena odorata*. The moderately effective ones were leaves of *Melia azedarach*, *Dryopteris filix-mass*, *Polygonum hydropiper* and *Tephrosia candida*, roots of *P. hydropiper*. Leaf of *Premna latifolia* was found the least effective. The degree of toxicity of the effective plants were *Calotropis procera*>*Zanthoxylum nitidum*>*Zanthoxylum rhesta* using residual film technique.

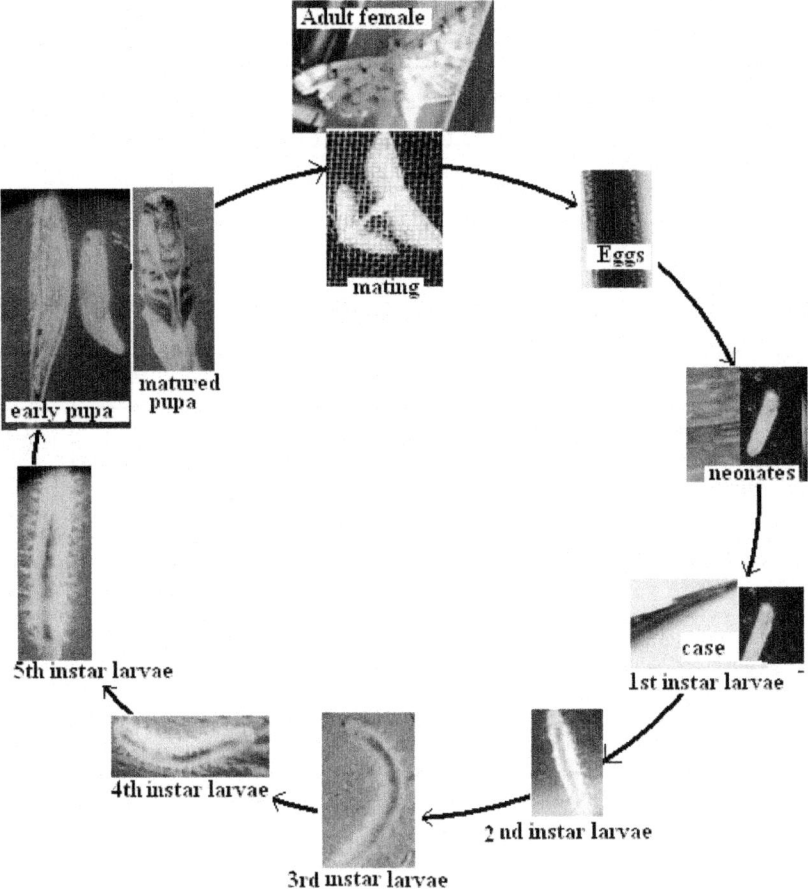

Plate 3. Life cycle of *Nymphula depunctalis*.

The degree of toxicity in bioassay using case dip technique was *Calotropis procera>Zanthoxylum nitidum >Zanthoxylum rhesta>Crataeva nurvala> Crotontiglium > Vitex negundo>Chromolaena odorata*. Phytochemical analysis for alkaloid, flavonoid, tannin, saponin, sterol and glycosides using crude ethanol extract showed the presence of flavonoid, tannin, saponin and glycoside in leaf extract of *Calotropis procera*; flavonoid, saponin, sterol and glycoside in the leaf extract of *Chromolaena odorata*; alkaloid, flavonoid, tannin and sterol in the stem-bark extract of *Crataeva nurvala*; flavonoid, tannin and sterol in the leaf extract of *Croton tiglium*; tannin and sterol in the leaf extract of *Vitex negundo*; alkaloid and sterol in the root-bark extract of *Zanthoxylum nitidum* and flavonoid, tannin, saponin and glycoside in *Zanthoxylum rhesta*. (Figure 1and 2)[59].

Table 2. List of plants used by the farmers against insect pests of paddy in North East India

Scientific Name	Common Name	Family	Part Used	Farmers in %
Ageratum conyzoides	Gundhua ban	Asteraceae	Leaf (LF)*	01.0
Andrographis paniculata	Chirata	Acanthaceae	Leaf	03.0
Azadirachta indica	Mahaneem	Meliaceae	Leaf and Seed	05.5
Calotropis procera	Akan	Asclepiadaceae	Leaf and stem	04.0
Chromolaena odorata	Germany ban	Asteraceae	Leaf	42.0
Citrus grandis, C. medica	Rebab and Bor Tenga	Rutaceae	Fruits and Fruit peel	11.5
Colocasia spp.	Kachu	Araceae	Stem	01.5
Crataeva nurvala	Varun	Capparidaceae	Stem-Bark (SB)*	03.5
Croton tiglium	Koni bih	Euphorbiaceae	Leaf and Seed	03.0
Cympogon nardus	Citronella	Gramineae	Leaf	01.5
Dryopteris filix-mas	Dhekia bi-lohngoni	Aspidiaceae	Leaf	03.0
Euphorbia neriifolia	Siju	Cactaceae	Stem	00.5

Jatropha gossypifolia	Bhoot ara	Euphorbiaceae	Leaf and Stem-Bark	03.0
Melia azedarach	Ghora neem	Meliaceae	Leaf	05.0
Murraya koeningii	Narasinga	Rutaceae	Leaf	00.5
Polygonum hydropiper	Patharuabihlongoni	Polygonaceae	Leaf, Stem, Root (RT)*	07.5
Premna latifolia	Pitha	Lamiaceae	Leaf	21.00
Tephrosia candida	Kuku mah	Fabaceae	Stem-Bark and Leaf	06.0
Thevetia nerrifolia	Rakta karabi	Apocynaceae	Stem-Bark, Seed	02.00
Vitex negundo	Pasatia	Verbenaceae	Leaf	13.00
Zanthoxylum nitidum	Ricom, Tezmui	Rutaceae	Root-Bark (RB)*	03.00
Zanthoxylum rhesta	Ongare	Rutaceae	Leaf and Stem Bark	02.00

Botanicals in Tea Gardens

Tea, *Camellia sinensis* (L.) O. Kuntze is grown in a monoculture system and about 1031 arthropod species are reported to be associated with the crop [83]. Cultivation and production of Assam tea was started by Assam company under the British regime around 1840 and the company monopolized the industy till 1860. With the advent of synthetic pesticides and their tremendous potentiality for controlling insect pests, during 50's tea industries started applying insecticides against tea pests which boosted production [83].With the growing concern over pesticidal residue in tea leaves and environmental impact of pesticides, various international regulatory bodies have fixed maximum permissible limit of residues in tea leaves. Synthetic pesticides are still being used in tea plantations although in limited quantity. These pesticides however do not remain confined to the tea plantations only and contaminate the host plants of silkworms grown in sericulture fields lying adjacent to the tea gardens[6]. Literature reveals that several plants have been found to be effective against pests of tea [83-86].

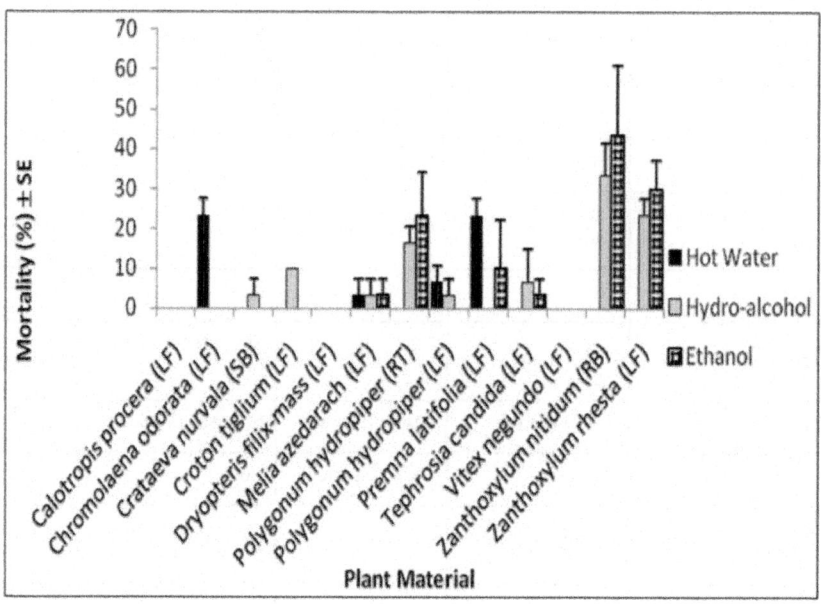

Figure 3. Effects of plant extracts on percent mortality of larvae of *N. depunctalis* in residual film technique.

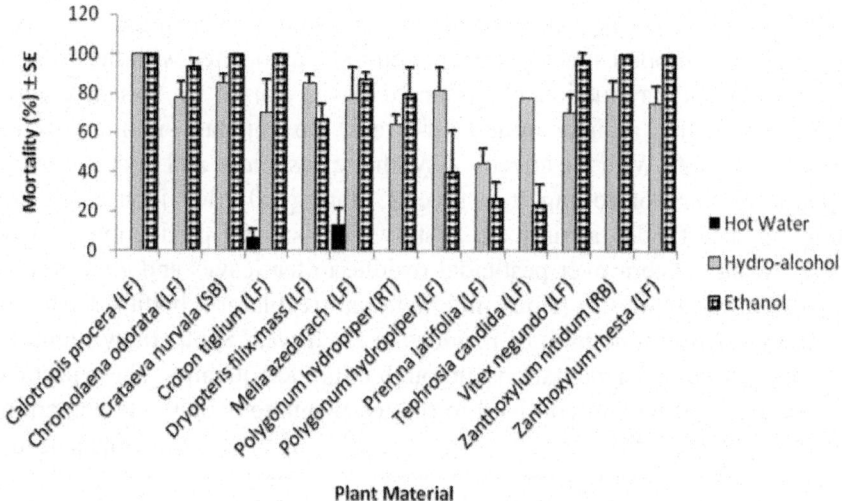

Figure 4. Effects of plant extracts on percent mortality of larvae of *N. depunctalis* in case dip technique.

SELECTIVE TOXICITY OF PLANT PRODUCTS

Integrated pest management emphasizes on use of pesticides having selective toxicity as against the use of broad spectrum pesticides. This reminds us of classical examples of using different insecticides for controlling aphids and thrips in potted chrysenthemum in glass houses of Southern Britain in 70's at the advent of development of the concept of integrated pest management [87]. Apart from the aspects why plant products are considered better options for pest control as discussed above, if they can be categorized for having differential efficacy against different insects, the latter can give an added value to plant products for being used in pest control. However the efficacy also depends not only on the compound(s) present in the plant preparation but also on the ability of the insect to defend against the compounds they are subjected to.Less tissue susceptibility, presence of detoxifying enzyme, high immune response, development of alternative physiological pathways, reduced penetration of pesticides through the cuticle and intestine, lower transport of pesticides to the target sites, storage of pesticides in fat body or other inert organs, genetically determined modified behavior in response to pesticide etc. may be factors responsible for conferring resistance to certain insects against high susceptibility of other insect species [23, 88, 89].

Silk worms have a bitrophic relationship with the Dipteran, *Exorista sorbillans* (Widemann) (Uzi fly) as the later is an endoparasitoid of the silk worm. As silkworms are beneficial insects, before recommending use of any botanical it is desirable to investigate whether the recommended plant or its product found effective against pest and pathogens in seri-ecosystem possess selective toxicity or not. An investigation carried out to evaluate selective efficacy of petroleum ether extract of certain plants against the component insects of the bitrophic system, *Exorista sorbillans and larvae of A. assama*revealed no mortality of late instar larvae of *A. assama* till 48h at 10% concentration while they were highly toxic to the parasitoid *Exorista sorbillans.*

Essential oils are volatile mixtures of hydrocarbons with diverse functional groups. Essential oils are defined as any volatile oil(s) that have strong aromatic components and that give distinctive odour, flavor or scent to a plant. These are the byproducts of plant metabolism and are commonly referred to as volatile plant secondary metabolites. Essential oils are found in glandular hairs or secretory cavities of plant cell wall and are present as droplets of fluid in the leaves, stems, bark, flowers, roots or fruits in different plants. The aromatic characteristic of essential oils attract or repel insect, protect plant from heat or cold and the chemical constituents of the oil act as defense material. Their probable diverse mode of action and use in food and pharmaceutical industry justify their status as possible environmentally benign candidate for pest management. *O.sanctum* and *A.conyzoides* are plants with rich source of essential oils. The toxicity of these oils were evaluated by both contact residual and topical application method against the parasitoid of silkworm, *E.sorbillans*. These oils were found more effective than petroleum ether extract of the respective plants and the calculated LC50 value of essential oil of *O. sanctum* and *A. conyzoides* in contact residual film method were recorded as 0.15% and 0.05% respectively. But, topical application of the oil showed a different order of toxicity and reflected the fact that degree of toxicity depended on method of application. The Lethal time for *O. sanctum* oil was 4±0.58 minutes and that of *A. conyzoides* oil was 118±12.45 minutes after topical application of 1μl of oil on thorax of the fly and hence degree of toxicity of oil on topical application was *O. sanctum* >*A. conyzoides*.When toxicity of these oils were compared against the larvae of *A. assama*, it was found that late instar larvae could survive till 48h after application of 0.5% concentrations of oil of the plants. The same concentration of oil of *A. conyzoides* while were appliedagainst early instar larvae 100% mortality was caused with LC50 value at 0.19% concentration(Table 3)[81]. But toxicity of oil of *O. sanctum*against *A.assama* larvae was lower and at 0.5 percent concentration, it could only cause a maximum of 5.78% mortality of early instar larvae at 48h of treatment and could cause no mortality in late instar larvae till 48h of observation period (Figure 5). This may suggest less tissue sensitivity of *Antheraea assama* against the action of extract and oil of *O. sanctum* applied [90].The plant *Ocimum spp.*is known to comprise more than thirty species, but only a few have been subjected tophytochemical studies (Grayer et al. 1996).The whole plant of *Ocimum* is rich in essential oil and based on composition of volatile principles of essential oil,intraspecific chemotypes of several species of *Ocimum* have been described[91-93]. The plant is a part of Indian tradition as a holy substance and its essential oil is larvicidal against both *Aedes* and *Culex* mosquitoes [94].

Table 3. Efficacy of extract & oil of *A. conyzoides* and *O. sanctum* as compared with Deltamethrin [(_) in cells indicate zero mortality].

Plant material	*Exorista sorbillans*			*Antheraea assama*		
	Regression equation	LC50 (%)	95% Fiducial limit	Regression equation	LC50 (%)	95% Fiducial Limit
*A. conyzoides*oil	Y=8.72+2.87X	0.05	2.603-3.075	Y=7.10663+2.92014X	0.19	2.394-2.837
*O.sanctum*oil	Y=9.15452+4.97895X	0.146	8.929-11.172	–	–	–

A .conyzoides-petroleum ether	Y=5.17654+1.27665X	0.72	1.220-1.352	Y=3.62422+0.625415X	154.88	0.311-0.937
Deltamethrin	Y=5.90148+904456x	0.103	1.688-1.969	Y=7.46583+.612381x	.00009	0.572-0.656

Identification of Haemocytes of *A. Assama* by Light Microscopic and Transmission Electron Microscopic Studies (TEM) Studies

In order to understand the possible effects of the extracts of *O. sanctum* and *A. conyzoides* at cellular level electron microscopic studies of haemocytes were carried out after topical application of the extracts on thoracic surface of fifth instar larvae of *A. assama.* Five types of haemocytes of *A.assama* were identified under light microscopic and transmission electron microscopic studies. The identified cells were-1. Plasmatocyte 2.Granulocyte 3. Spherulocyte 4. Prohaemocyte 5. Oenocyte (Plate-4-9)

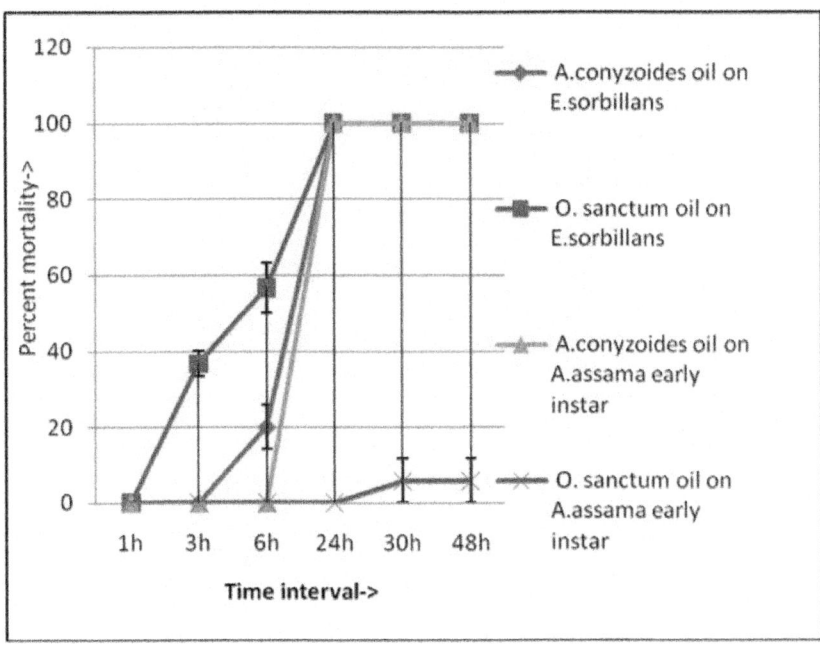

Figure 5. Comparative percent mortality of *A. assama* and *E. sorbillans* at 0.5% concentration of oil of *O. sanctum* and *A. conyzoides.*

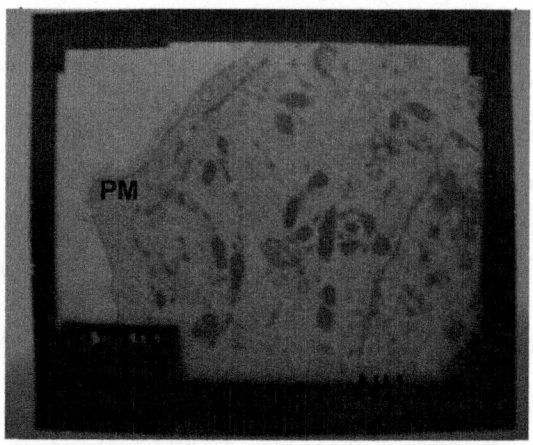

Plate 4. Plasmatocyte in *A. assama.*

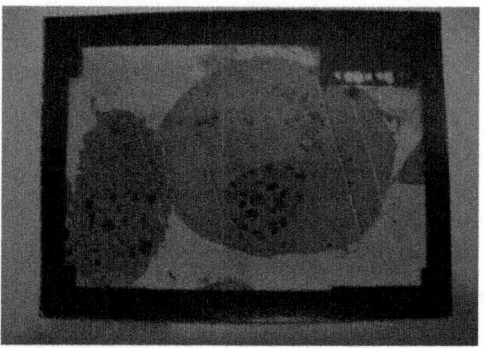

Plate 5. Plasmatocyte and Oeonocyte.

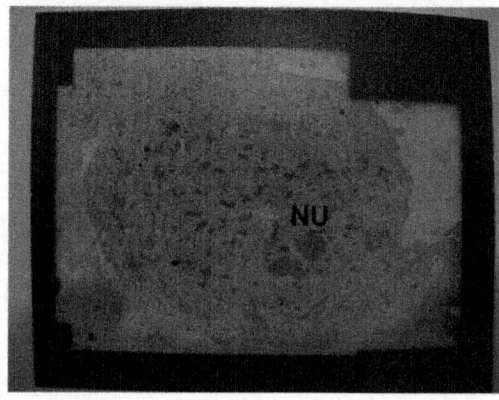

Plate 6. Prohaemocyte of *A. assama.*

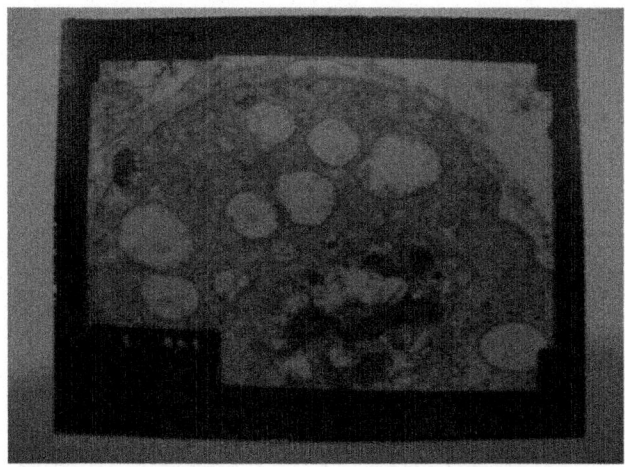

Plate 7. Spherulocyte of *A.assama.*

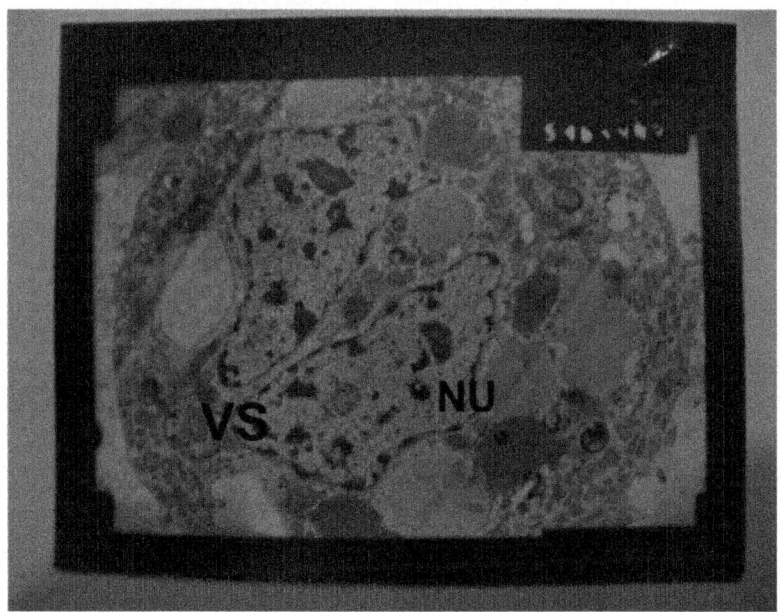

Plate 8. Granulocyte of *A. assama* under TEM.

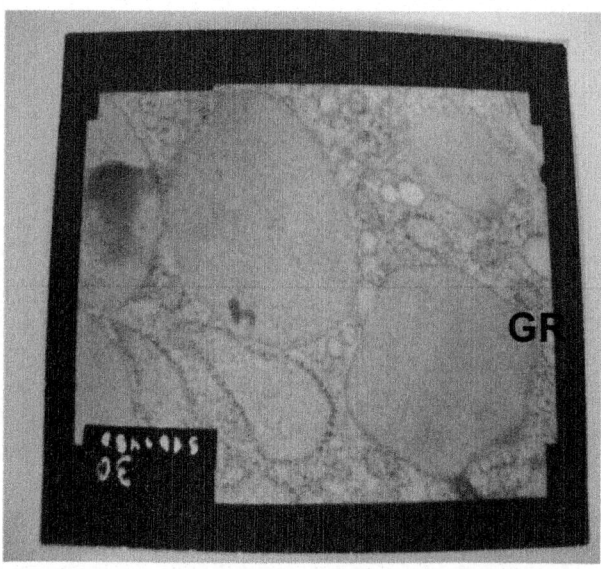

Plate 9. Structured granule.

Plasmatocyte

Plasmatoytes are polymorphic cells with round, ovoid,elongate and spindle shaped structures with filopodia. The cells are small to moderate size. The long axis is 25 μm and the short axis is 14.74 μm.Surface of PLs is not smooth and have ridges much flatter than those of granulocytes. Nucleus is dense, round, ovoid or elongated with distinct membrane. The cytoplasm is generally abundant, granular and agranular. Cytoplasm contains a good number of mitochondria and extensive rough endoplasmic reticulum. Golgi body recorded was in less number. Vacuole is rare in normal cell. The nucleus is mostly centrally placed, prominent, usually round but changes shape with the change in shape of the cell.

Granulocyte

They are rounded or oval in shape with long and short filopodia. The long axis is 8.382 μm. Two types of membrane bound granules were observed, electron dense and structured. Cytoplasm contains large number of mitochondria, endoplasmic reticulum, free ribosomes, lysosomes and small number of vacuoles. Binucleate granulocyte was also recorded. Nucleus generally small takes various shapes, round, ovoid or irregular. The cell surface contains many elevations and depressions giving a conspicuous pattern.

Spherulocyte

These are spherical or ovoid or elongated cells with long axis 7.21 μm. The cell surface contains pits.Wavy edges show presence of spherules. Nucleus is mostly centrally placed, sometimes obscured by spherules. Mitochondria, rough endoplasmic reticulum and golgi complex were observed in cytoplasm when spherule number was less.

Prohaemocyte

They are round, oval in shape and long axis is 7.94 μm. The nucleus occupies major area of the cell. Plasma membrane is almost smooth.Cytoplasm is thin, almost homogeneous, contains few number of mitochondria, Golgi bodies and endoplasmc reticulum. SEM shows a spherical surface.

Oenocytoid

These are large rounded or ovoid cells with eccentric nucleus. Nucleus is small, nucleoli were prominent. The plasma membrane is smooth and regular. Cytoplasm uniformly distributed and contains extensive rough endoplasmic reticulum. Filopods are generally absent. Other cellular organelles like Golgi bodies and mitochondria are less in number.

Tem Studies on Effect of Petroleum Ether Extract on Haemocytes of Late Instar Larvae

Transmission electron microscopic studies of haemocytes of fifth instar silk worm larvae was done after application of petroleum ether extracts of *O. sanctum* and *A. conyzoides* at different hours of treatment (Plates 10-21).

Application of petroleum ether extract of *O.sanctum* at 0.25 h caused less damage of PL and GR. Mitochondria and nucleus was almost intact in both types of cells. Nuclear membrane and cell membrane damage was negligible. Cell attachment was less in number. Release of material from GR was observed in some points. Golgi body was in formative phase in GR. PL cells in dividing stage and dividing nucleus in GR was recorded.

Cell membrane breakdown of PL was observed at 1h of treatment while mitochondria was intact. Filopods are less. But cell attachment was marked. GR cell was less affected and filopod like extension was many toward cell attachment site. Granules were released from the area of damage of the cell membrane. Dissolution of nuclear membrane was observed at certain points. OE, SP, PRO was less affected. At 6h of treatment PLwas less affected. Golgi body was observed at formative phase. Filopods are less in number. Export of material from the cell was observed. GR cell was also less affected. Cell membrane, mitochondria were well preserved. Nucleus in dividing stage was recorded. In some GR, granules release from the cell. Some GR attach to one another. Other cells are less affected. But at 12 h of treatment period, cell membrane breakage, mitochondrial damage vacuolization of PL cell was observed while GR cell was less affected. But again at 48h no effect on cell membrane, nuclear membrane, mitochondria was observed. Golgi was found to be in formative stage. Release of material was not recorded. GR cell was looking intact where cell membrane, mitochondria and nuclear membrane were not affected. Aggregation of cell was less. OE cell membrane was ruptured and released some cytoplasmic materials but mitochondria, nuclear membrane were well preserved. Attachment of OE with other cells was observed. SP and PRO were less affected.

At 0.25hof *A.conyzoides* petroleum ether extract treatment, heavy loss of the cell membrane of PL cell was noticed but mitochondrial structure was found almost in good condition. ER long while many Golgi bodies were found in formative stage casting off numerous vesicles. Cell attachment was very pronounced with release of flocculent material. Many vesicles were observed in cytoplasm. Extensive damage of the cell membrane of GR cell was observed. Some of the mitochondria was found in a state of completely dissolved mitochondrial membrane. But nuclear membrane was in good condition. Cell attachment was observed while ER and GB was looking normal. OE was less affected.

At 1h of treatment of *A.conyzoides* petroleum ether extract, damage of cell membrane of PL in some parts were noticed but mitochondria was unaffected. Nuclear membrane was damaged and Golgi body was in formative stage. Filopods present and large vesicles in the cytoplasm was observed. GR cell membrane was found totally damaged state and some of the mitochondria was also affected. But nuclear membrane was less affected. ER short and GB was not observed. Attachment of the cells were observed and numerous vesicles were seen. Other cells were less affected. At 12h of treatment also, damaged PL cell with ruptured cell membrane, affected mitochondria, large vesicles and filopod were recorded. GR cell was also affected with mitochondrial damage, cell membrane rupture, heavy vacuolization. Release of cytoplasmic material to outside of the cell and import of material was noticed.

At 1h of acetone treatment as control almost intact cell membrane, mitochondria, ER, nuclear membrane was observed. Cell attachment was less with few filopods in PL cell. Granules and vesicles number were less and release of cytoplasmic material were not observed in PL cell. But GR cell was affected and plasma membrane rupture and damage in mitochondria was observed. Plenty number of ER but few vesicles were observed. Both types of granules structured and electron dense were observed. Cell aggregation with release of flocculent material was noticed.OE was less affected than PL and GR. But at 6h of acetone treatment PL and GR were found intact with intact plasma membrane, nuclear membrane, normal mitochondria, few vesicles and very less cell attachment. A medium number of filopods were observed but release of cytoplasmic material was not observed. OE cell was also found intact with unaffected mitochondria. After 48h treatment of acetone also, the nuclear membrane and mitochondria of both PL and GR were found intact with less cell attachment. Cell membrane damage of PLwas less. Golgi body was found in formative phase and few number of filopods and vesicles were observed but release of flocculent material was not recorded. Nuclear division of PL cell was observed. Cell membrane rupture of some of the GR cell was recorded with discharge of granule while other cells were intact with almost absence of filopods. Plenty number of ER and large sized vesicles were seen. Both types of granules structured and electron dense were noticed.

A comparative assessment of the effects made showed that the rupture of plasma membrane (PM) in plasmatocyte (PLs) was maximum in treatment with *A. conyzoides*while it was comparatively less in treatment with *Ocimumsanctum*. Symptoms of immune response i.e. cell attachment between PLs and GRs and among the plasmatocytes along with release of cytoplasmic and granular material from granulocytes were observed in case of treatment with all the plant extracts. Toxic symptoms like highly affected mitochondria with break down of nuclear membrane, dissolution of cristae became evident in case of treatment with *A. conyzoides*. In case of treatment with *Ocimum sanctum* the affects were comparatively less and at 48h the PLs were exactly similar with those of control insects. Similar effects on ultrastructure PLs and GRs were reported to be caused in lepidopteran insect by neem gold [95,96]. Plant based pesticides were earlier reported to decrease prohaemocyte, plasmatocyte and spherulocyte and increase granunlocyte and oenocyte population [95,97]. Gupta(1998) reported on involvement of plasmatocytes and granulocyte in cell mediated immune response. In our investigation also cell lysis, release of granular material, vacuolization and clumping of PLs and GRs were observed after treatment with the plant extracts.

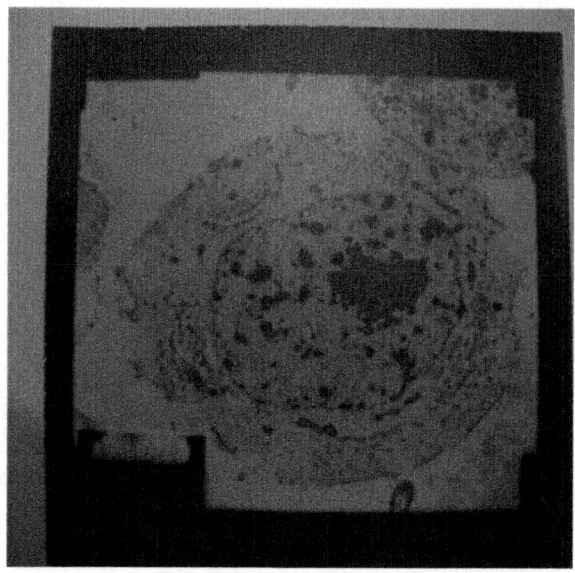

Plate 10. Plasmamembrane rupture in *A. conyzoides* PL under TEM.

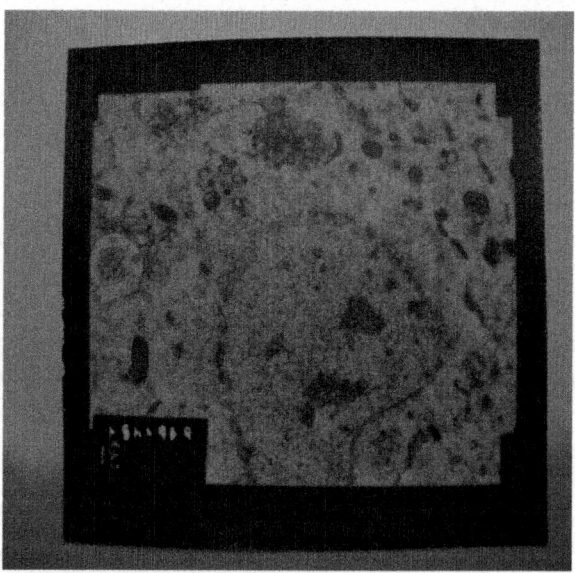

Plate 11. Nuclear membrane rupture A. *conyzoides* treated PL under TEM.

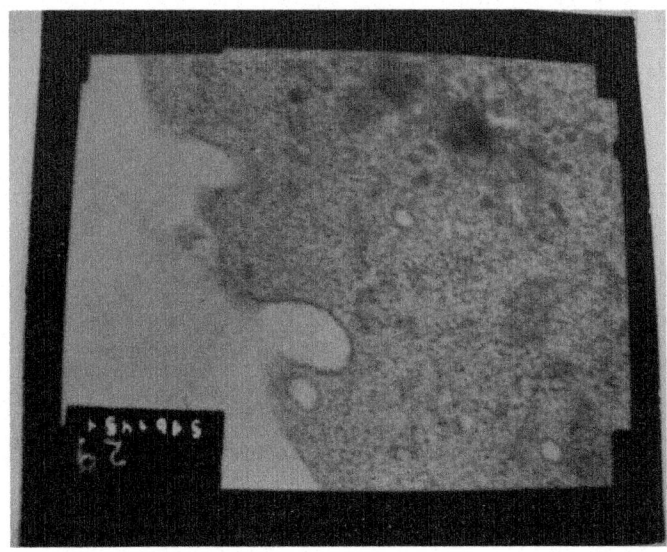

Plate 12. Pinocytic vesicle formation of PL at 12 h of treatment with *A. conyzoides*

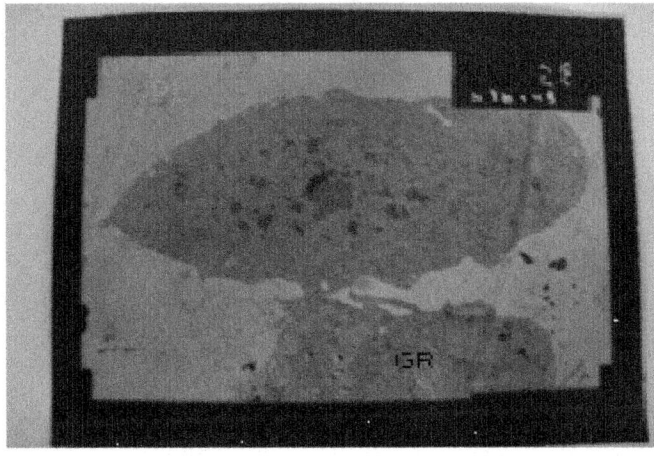

Plate 13. Cell clumping between PLs and GRs at 12 h of treatment with *A. conyzoides*

Plate 14. PM rupture at 12 hr after treatment with *O. sanctum.*

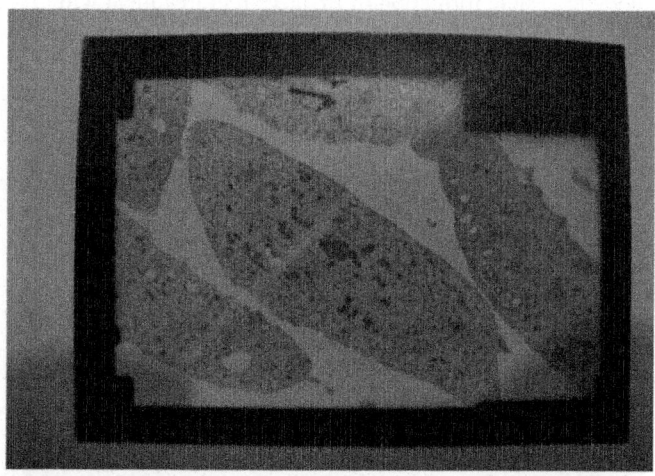

Plate 15. Normal PLs at 48hr of treatment with *O. sanctum* under TEM.

Plate 16. GR at 1hr of *O.sanctum* treatment under TEM.

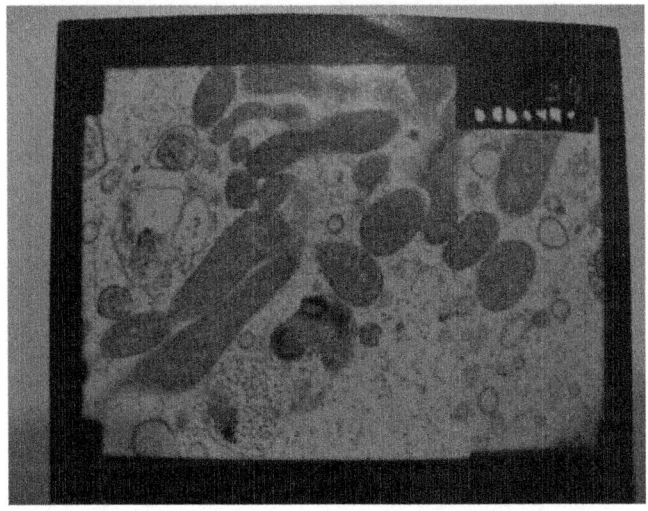

Plate 17. Mitochondria at 15min of *O. sanctum* treatment.

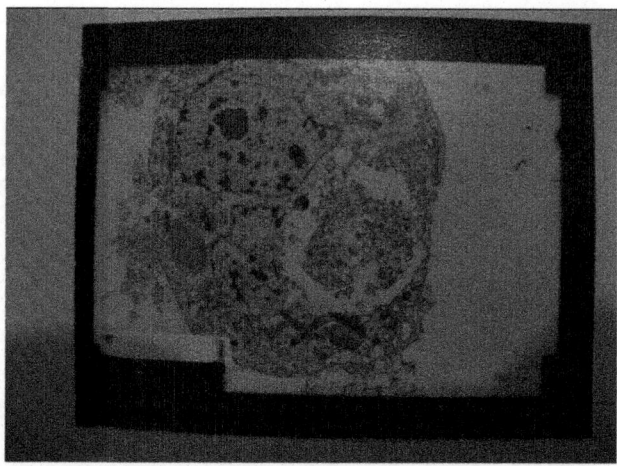

Plate 18. Granulocyte damage at 15min of *O. sanctum.*

Plate 19. Granular material release at 12hr of treatment in *O. sanctum.*

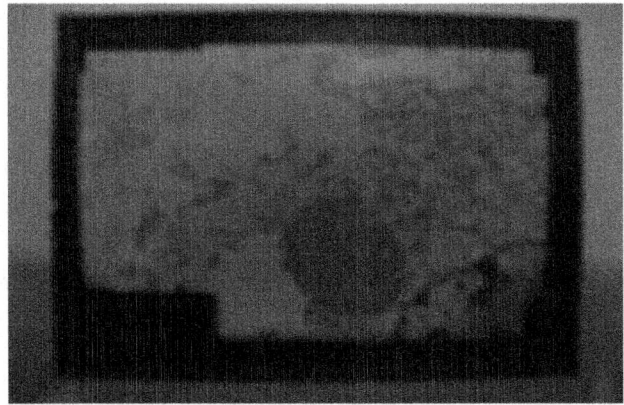

Plate 20. GR breakdown at 1hrof treatment in *A. conyzoides A. conyzoides* under TEM.

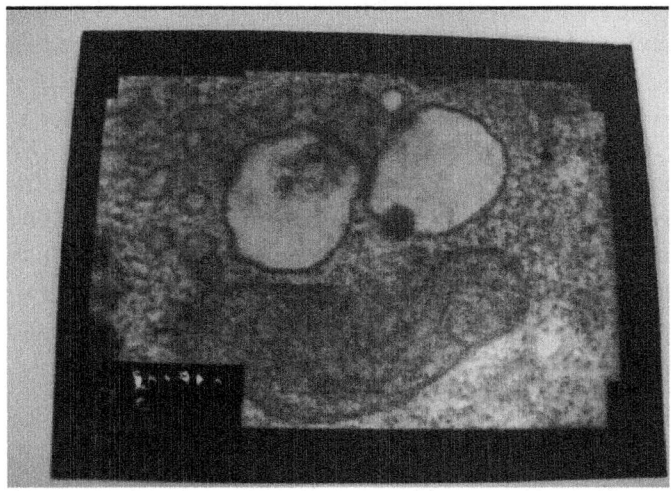

Plate 21. GR mitochondria damage at 12hr of treatment in *A. conyzoides.*

Effect of Essential Oil on Economic Characters of Silk Wrom

Study on impact of essential oils of *O. sanctum* and *A. conyzoides* after topical application at the dosages of 0.2, 0.1, 0.02 and 0.01 μlg^{-1} on 2nd day old *A. assama* fifth instar larvae showed that there was no significant difference on developmental period of fifth instar larvae from that of control in case of treatment of essential oil of *O. sanctum*. A significant decrease in developmental period was caused by essential oil of *A.conyzoides* at the highest dose. Early spinning of larvae after application of essential oil of *A. conyzoides* might be due to its effect on endocrine system. Precocene found in essential oil of*Ageratum* species is known to cause precocious metamorphosis through its action on corpora allata [23,99].

Economic characters of silkworm like cocoon weight and pupal weightof treated larvae did not vary significantly from that of control larvae. But significant decrease of shell weightwas recorded after application of essential oil of *A. conyzoides* at all dosages [24]. Reduction in shell weight might be due to reduction of spinning period causing precocious metamorphosis which might becaused by precocene II found in *A. conyzoides* [99]. Thus in comparison to *A. conyzoides*, *O. sanctum* is a preferable candidates for being used against *E. sorbillan*. However, nutritional indices like efficiency of conversion of ingested food into body matter (ECI), efficiency of conversion of digested food into body matter (ECD), approximate digestibility (AD), consumption index (CI), relative growth rate (RGR) were not found to differ significantly from that of control at lower dosages (0.02, 0.01 µl/g)after treatment of both essential oils of *O.sanctum* and *A.conyzoides* [24]. There are also reports stating beneficial effects of plant products on silk worms which is supportive of use of plant products in sericulture field. Botanicals like *Curcuma longa*, *Phyllanthus ambilica*, *Asparagus racemosus*, *Aegle marmelos*, *Boerhavia diffusa*, *Allium sativum* and *Ocimum basilicum*applied against flacherie infested*Bombyx mori* larvae increasedlarval weight, cocoon weight, shell weight, silk ratio [100]. Similar studies on beneficialeffects of plant products on growth, development and economic characters of silk worms have been carried out by several workers [101-103]. Synergistic actions of plant chemicals might be responsible for growth enhancement of certain insects [104].

Table 4. Variation of cocoon weight of *A. assama* late instar larvae after treatment with essential oil of selected plants (P<0.05)

Table: Variation of cocoon weight of *A. assama* late instar larvae after treatment with essential oil of plants (P<0.05).				
Dose→	0.2µl/g	0.1µl/g	0.02µl/g	0.01µl/g
Plant(oil)↓	Cocoon weight (Mean±SE)	Cocoon weight (Mean±SE)	Cocoon weight (Mean±SE)	Cocoon weight (Mean±SE)
O.sanctum	5.15±.45	4.85±.44	5.13±.44	4.85±.44
A.conyzoides	4.81±.31	5.30±.29	4.81±.31	5.3±.29
Control	5.15±.66	5.15±.66	5.15±.66	5.15±.66
Table: Variation of pupa weight of *A. assama* late instar larvae after treatment with essential oil of plants (P<0.05).				
Dose→	0.2µl/g	0.1µl/g	0.02µl/g	0.01µl/g
Plant(oil)↓	Pupal weight (Mean±SE)	Pupal weight (Mean±SE	Pupal weight (Mean±SE	Pupal weight (Mean±SE
O.sanctum	4.69±0.42	4.47±0.46	4.73±0.41	4.47±0.46
A.conyzoides	4.50±0.31	4.96±0.26	4.50±0.31	4.96±0.26
Control	4.75±0.64	4.75±0.64	4.75±0.64	4.75±0.64
Table: Variation of shell weight of *A. assama* after treatment with essential oil of plants (P<0.05).				
Dose→	0.2µl/g	0.1µl/g	0.02µl/g	0.01µl/g
Plant(oil)↓	Shell weight (Mean±SE)	Shell weight (Mean±SE)	Shell weight (Mean±SE)	Shell weight (Mean±SE)
O.sanctum	0.34±0.01	0.34±0.04	0.33±0.01	0.34±0.04
A.conyzoides	0.26±0.04	0.29±0.02	0.28±0.01	0.29±0.02
Control	0.32±0.02	0.32±0.02	0.32±0.02	0.32±0.02
Table: Variation of developmental period after treatment with essential oil of plants (P<0.05).				
Dose→	0.2µl/g	0.1µl/g	0.02µl/g	0.01µl/g
Plant(oil)↓	Dev period (Mean±SE)	Dev period (Mean±SE)	Dev period (Mean±SE)	Dev period (Mean±SE)
O.sanctum	11.67±.33	11.67±.33	11.33±.33	11.67±.33
A.conyzoides	10±0*	10.67±.33*	11.00±.58*	11.67±.33
Control	11.67±.33	11.67±.33	11.67±.33	11.67±.33

SE→ Standard error; *→ significant at 95% Confidence Interval

MECHANISM OF ACTION OF BOTANICALS

The efficacy of plant products depend on presence of specific organic compound (s) which may interfere with the body physiology of the target organism. The compounds may belong to different chemical groups of secondary metabolites of plants viz. Alkaloids, phenolics, flavonoids and terpenoids. The ethanolic extract of *O. sanctum* is reported to have greater and broader spectrum of activity against tested organisms [106]. Literature reveals that in comparison to essential oil less works have been carried out to isolate and characterize bioactive compound of *Ocimum sp.* having insecticidal potency through solvent extraction and hence knowledge regarding the use of solvent extracts of*Ocimum* is comparatively less. Petroleum ether and acetone extract of *A. conyzoides* earlier was reported to have juvenile hormone activity against *Culex quinqufasciatus, Aedes aegpti* and *Aedes stephensi.* Hexane extract was found effective against *Musca domestica* and methanolic extract caused abnormal development and suppression of population of *Anopheles stephensi* [99,106,107]. Thiophene derivative, a class of compound found in many Asteraceae species has been attributed for the toxic effect of the plant. The hexane fraction of acetone extract of *C. roseus* containing alpha-amyrin is postulated to be active IGR against many pests [63]. Moreira et al. [107] showed the hexane extract of*Ageratum conyzoides* to have insecticidal activity and purified the compounds using IR, 1H NMR, 13 c NMR, HMBC. The compounds were5,6,7,8,3/,4/,5/ heptamethoxy flavone, 5,6,7,8,3/ pentamethoxy-4/,5/-methylene dioxyflavone and coumarin out of which only coumarin showed insecticidal activity against dictyopteran, lepidopteran and dipteran.

The mechanism of toxic effect of essential oil and oil compounds on insect at present is not well known. Insects vary enormously in their response to different essential oils and oil compounds. Essential oils are liquid in room temperature and get easily transformed from a liquid to a gaseous state at room or slightly higher temperature without undergoing decomposition. The quantity of essential oil found in most plants is 1 to 2% but can contain amounts ranging from 0.01 to 10%. Most essential oils comprise of monoterpenes with 10 carbon atoms, sesuiterpenes with 15 carbon atoms and rarely diterpenes or higher terpenes. The most predominant groups are cyclic compounds with saturated and unsaturated hexacyclic or an aromatic system. Bicyclic and acyclic components are also present[2]. Monoterpenes are common essential oil constituents and several hundred naturally occurring monoerpenes are reported. They are biosynthesized from geranyl pyrophosphate of the isoprenoid pathway. These can be classified into two major groups- monoterpene hydrocarbons and oxygenated monoterpenes. Monoterpene hydrocarbons include acyclic aliphatic, monocyclic aliphatic and dicyclic aliphatic while oxygenated monoterpenes include acyclic monoterpenoid, monocycic monoterpenoid and dicyclic monoterpenoids [41].Some of the major oil constituents of *O. sanctum* are methyl eugenol, caryophyllene, germacrene D, β-elemene, eugenol, caryophyllene epoxide and α-cadinol [24,108,109]. Oil of *A. conyzoides* plant collected from South-East Asian countries mostly contains sesquiterpene Beta-caryophyllene, demethoxyageratochromene and many monoterpenes [24,110]. The monoterpenes in Indian oil of *A. conyzoides* have been reported to be mainly Ocimene (5.3%), alpha-pinene (6.6%), eugenol (4.4%), methyl eugenol (1.8%) and the sesquiterpenes are Beta- caryophyllene(1.9%),Delta- cadinene(4.3%), sesquiphellandrene (1.2%)

and caryophyllene epoxide (0.5%) [62]. Eugenol was earlier reported to be effective against mosquitoes in Michigan [60]. Compounds like coumarin and mainly furanocoumarins can alter the detoxication capability of an organism by reversibly or irreversibly inhibiting cytochrome P450 detoxication enzymes [111-112]. They are also reported to have neurotoxic mode of action probably through binding with different types of octapamine receptors and interference with octapamine activity [113-115] or interference with GABA-gated chloride channels [2,115] or inhibition of achetylcholine esterase [116-118]. Such activity of the oil may be attributed to action of a single major compound or synergistic actions of group of compounds [119].

The root-bark extract of *Zanthoxylum nitidum* having efficacy against rice pest contains several alkaloids [120,121] out of which benzo[c]phenanthridine alkaloids (sanguinarine and chelerythrine) are reported to have the molluscicidal activity [122]. Alkaloid is a large group of plant secondary metabolites and its occurrence has been reported in approximately 20% of all plant species. Most of the alkaloids exert their effect on different region of the nervous system. Different neuroreceptors like alpha, serotonin, muscarinic, nicotinic acetylcholine receptors, adrenergic receptor and enzymes involved in synaptic transmission are some of the target sites of alkaloids. Basic molecular targets are DNA intercalation, protein biosynthesis and membrane stability [123].

The structures of phenolic allelochemicals and their mode of action are diverse [124]. In lepidopteran larvae, phenolic toxicity might occur in the form of oxidative stress [125]. However, *Heliothis virescens*larvae after feeding with high phenolic foliage exhibited improved total Trolox equivalent antioxidant capacity (TEAC) in haemolymph [126]. Therefore it has been proposed that the elevated foliar phenolics in some plants might have beneficial antioxidant properties for herbivorous insects. Flavonoids are known to inhibit cholinesterases [127] and might be responsible for insecticidal action [128]. Various forms of saponins and sterols derived from different plants have been reported to act as insect growth regulator [129].

THE FUTURE OF PROSPECTS OF BOTANICALS IN SERI-ECOSYSTEM

An important essence of integrated pest management is to consider the whole ecosystem as the management unit. A seri-ecosystem should necessarily include not only the sericulture field or the silkworm culture units, but also the agricultural field in the neighbouring areas. A consensus approach is to keep the natural ecosystem largely intact. While attempts are made to control pests and pathogens in the neighbouring agricultural fields, issues regarding their impact on sericulture activities must be taken into consideration or vice-versa. Emphasis is to be given on studies associated with efficacy of plant and plant products against pests and pathogens in the whole management unit, their mode and mechanism of action, identification of target sites and use of resistant strains. Life scientist and chemist need to act coherently to make effective use of botanicals. With the knowledge in hand we hypothesize that holistic approach involving in depth research for using botanical in both sericulture field and the other crop systems in the vicinity might be able to provide a green environment to the silk worms.

ACKNOWLEDGEMENT

The authors are grateful to UGC, India for funding part of the work. Thanks to Mr. B.Deka for helping in statistical analysis.

REFERENCES

1.   R. . Carson, 1962The Silent spring. Houghton Mifflin. US.

2.   O. Koul, S. Walia, G. S. Dhaliwal, 2008Essential oils as green pesticides: potential and constraints. Biopesticides Int. 416384

3.   I. Afshan, Murthy, 2005Price transmission in silk industry of Karnataka. In: Conference Papers-The 20th Congress of the International Sericultural Commission.36257261

4.   Teotia RS, Sen SK1994Mulberry disease in India and their control. Sericologia. 34118

5.   S. C. Datta, R. Datta, 2007Increased silk production by effective treatment of naturally infected root-knot and black leaf spot diseases of mulberry with acaciasides. Journal of Environmental & Sociobiology. 4209214

6.   Bora DS1998Effect of environmental stress with special reference to photoperiod and insecticide on muga worms. Antheraea assama Westwood. Ph.D. Thesis, Dibrugarh University, Dibrugarh, Assam, India.

7.   B. Li, Y. Wang, H. Liu, Y. Xu, Z. Wei, Y. Chen, W. Shen, 2010Resistance comparison of domesticated silkworm (Bombyx mori L.) and wild silkworm (Bombyx mandariana M.) to phoxim insecticide. African Journal of Biotechnology. 91217711775

8.   Wiayanthi N,Subramanyam MV2002Effect of fenvalerate-20EC on sericigenous insects. I. Food utilization in the late-age larva of the silkworm, Bombyx mori L. Ecotoxicol Environ Saf. 532206211

9.   A. Bhagyalakshmi, R. S. Venkata, R. Ramamuthi, P. S. Reddy, 1995Studies on the effect of hexachlorocyclohexane on the growth and silk qualities of silkworm, Bombyx mori L. Chemistry and Ecology. 11297104

10. Naik PK, Delvi MR1984Effects of insecticide, permethrin, on dietary water utilization in eri-silkworm Philosamia ricini. Proc. Indian Acad. Sci. (Animal Sci.). 935497504

11. Wilkinson CF1976Insecticide Biochemistry and Physiology. Plenum Press. New York.

12. Zabik MJ1985Photochemistry of pesticides. In : Comprehensive Insect Physiology, Biochemistry and Pharmacology. Gilbert LI and Kerkut GA editors. Pergamon Press, Oxford. 12776801

13. A. Chukudebe, R. B. Othman, M. Fukuto, T. R. , 1989Formation of trialkyl phosphorothioate esters from organophosphorus insecticides after exposure to either ultraviolet light or sunlight. J. Agric. Food. Chem. 372539545

14. Haynes KF1988Sublethal effects of neurotoxic insecticides on insect behaviour. Ann. Rev. of Entomology. 33149168

15. Kariappa BK, Narasimhanna MN1978Effect of insecticides in controlling mulberry thrips and their effect on rearing silk worm Bombyx mori L. Indian J. Seric. 17714

16. Troitskaya EN, Chichigina IP1980The effect of combined insecticidal preparations on silkworm larvae. Uzbekshii Biologicheskii Zhurnal. (3): 50-53.

17. Waldbauer GP1968The consumption and utilization of food by insects. Adv. Insect Physiol. 5233282

18. Y. Tazima, 1978The Silkworm: an important laboratory tool. Kodansha Ltd.

19. Q. U. Yu, C. Lu, W. L. Li, Z. H. Xiang, Z. . Zhang, 2009Annotation and expression of carboxylesterases in the silkworm, Bombyx mori. BMC Genomics.10:553doi:10.1186/1471-2164-10-553.

20. Toutant JP (1989Insect acetylcholinesterase: catalytic properties, tissue distribution and molecular forms. Prog Neurobiol. 32423446

21. J. G. Oakeshott, C. Claudianos, P. M. Campbell, R. D. Newcomb, R. J. . Russell, 200, 2005Biochemical genetics and genomics of insect esterases. In :Comprehensive molecular insect science. Gilbert LI, Iatrou K, Gill SS editors. London: Elsevier. 5309361

22. J. Y. Shang, Y. M. Shao, G. J. Lan, G. Yuan, Z. H. Tang, C. X. . Zhang, 2007Expression of two types of acetylcholinesterase gene from the silkworm, Bombyx mori, in insect cells. Insect Sci. 14443449

23. Pedigo LP2002Entomology and Pest Management. Pearson Education Inc.

24. B. Khanikor, 2011Evaluation of extracts and essential oils of Ocimum and Ageratum against uzi fly, Exorista sorbillans (Wiedemann), a parasitoid of Antheraea assama Westwood. Ph.D. Thesis. Dibrugarh University. Dibrugarh. India.

25. Suresh. A. Nath, Kumar. R. P. S. Verma, 1997Changes in protein metabolism in haemolymph and fat body of the silk worm, Bombyx mori (Lepidoptera : Bombycidae) in response to OP insecticidal toxicity. Ecotoxicology and Environmental safety.362169173

26. Burt PE, Lord KA1977In: Biochemical Insect Control. Quarshi S. Wiley Interscience Publication, New York.183.

27. Morgan DP, Roan, CC, Paschal EH1976In : Toxicology of insecticides. Matsumura F. editor. Plenum Press, New York.

28. Burt PE, Lord KA, Forrest JM, Goodchild RE1971The spread of topically applied pyrethrin-I from the cuticle to the central nervous system of cockroach, Periplaneta Americana. Entomol. Exp. Appl. 14255269

29. F. Morairity, M. C. French, 1971The uptake of dieldrin from the cuticular surface of Periplaneta amaricana L. Pesticide Biochemistry Physiology. 19286292

30. Saxena PN1990Organophosphate intoxication in insect body: penetration, kinetics and target site interaction In : Environmental impact on Biosystems. Dalela RC editor. Academy Environmental Biology, India. 4767

31. D. Bora, R. Handique, 1996Insecticide induced biochemical changes in Antheraea assama Westwood. Proc. Acad. Environ. Biol. 52221226

32. Tiwari SK, Bhatt RS1996Methoxychlor and dimethoate induced changes in biochemical components in the haemolymph and fatbody of the larva of rice moth, Corcyra cephalonica Staint. (Lepidoptera:Pyralidae). Uttar Pradesh J. Zool. 16118

33. Patel IS, Yadav DN, Shukla YM1996Biocemical basis of mechanism in a monocrothphos resistant strain of the predator, Green lace wing, Chrysopa scelestes Banks. Annal. Plant Protec. Sci. India. 66 (B): 2.

34. R. Handique, D. Bora, P. Chetia, D. Kotoky, 2002Effect of diesel exhaust on cholesterol uptake by muga silkworms. Indian J. Sericulture. 4112933

35. D. Bora, R. Handique, 1992Effect of several synthetic insecticides on growth, development, food consumption and food utilization of Antheraea assama Westwood larvae. Proc. Acad. Environ. Biol. 12195203

36. T. Swain, 1977Secondary compounds as protective agents. Annu. Rev. Plant. Physiol.28479501

37. G. A. Rosenthal, D. . Janzen, 1979Herbivores: Their interaction withsecondary plant metabolites. NY: Academic.718

38. Dethier VG (1982Mechanism of host plant recognition. Ent exp appl Ned Entoml Ver Amsterdam. 314956

39. Balandrin MF (1985Natural Plant Chemicals: Sources of Industrial and Medicinal Materials. Science. 22811541160

40. K. Sukumar, Boobar. L. R. Perich, 1991Botanical derivatives in mosquito control: A review. J. Amer. Mosq. Control Ass. 7210237

41. HarborneJB(1998Phytochemical Methods. A guide to modern techniques of plant analysis, Third Edition. Chapman and Hall, London. 302.

42. Mishra AK, Dubey NK1994Evaluation of some essential oils for their toxicity against fungi causing deterioration of stored food commodities. Applied and Environmental Microbiology.6011011105

43. B. Subramanyam, R. Roesli, 2000Inert dusts. In: Alternatives to Pesticides in Stored-Product IPM, Subramanyam BH and Hagstrum DW editors. Kluwer Academic Publishers, Dordrecht. 321380

44. Rawls RL1986Experts Probe Issuses, Chemistry of light activated pesticides. Chem. Eng. News. Sept. 22: 2124.

45. D. Yadav, 2011Bio-pesticides and bio-fertilizer for sustainable sericulture. The Silkworm. 28th August.

46. Dura VK, Nagpal BN, Sharma VP1995Repellent action of Neem cream against mosquitoes. Indian J. Malariol. 324753

47. B. N. Nagpal, A. Srivastava, V. P. Sharma, 1996Control of mosquito breeding using scrapings treated with Neem Oil. Indian J. Malariol. 326469

48. Kumar GS2012Experimental study to find the effect of different neem (Azadirachta indica) based products against moringa hairy caterpillar (Eupterotemollifera Walker). International journal of Biology, Pharmacy and Allied sciences. 112228

49. V. Sathyaseelan, V. Bhaskaran, 2010Efficacy of some native botanical extracts on the repellency property against the pink mealy bug, Maconellicoccus hirsutus (green) in mulberry crop. Recent Research in Science and Technology. 2103538

50. L. Isaiarasu, N. Sakthivel, J. Ravikumar, P. Samuthiravelu, 1969Effect of herbal extracts on the microbial pathogens causing flacherie and muscardine diseases in the mulberry silkworm, Bombyx mori L. Journal of Biopesticides. 42150155

51. Kumar KPK, Singh GP, Sinha AK, Madhusudhan KN, Prasad

BC2012Antibacterial action of certain medicinal plants against AmCPV and their effect on cellular and biochemical changes in tasar silkworm, Antheraea mylitta D. Research Journal of Medicinal Plants. 619299

52. B. Unni, P. Dowarah, S. Wann, A. Gangadharrao, 2011Muga heal-Terminalia chebula based bioformulation as an antiflacherie agent and a silk fiber enhancer. Science and Culture. 456460

53. R. Woytek, 1998Indigenous knowledge for development: a framework for development.Knowledge and Learning Centre, Africa Region, World Bank.

54. Amaugo GO, Emosairue SO2003The efficacy of some indigenous medicinal plant extracts for the control of upland rice stem borers in Nigeria. Tropical and subtropical agroecosystems. 2121127

55. J. O'Hara, 1992Exorista sorbillans-a serious tachinid pest of silkworms. The Tachinid Times: 5117

56. M. Sahu, A. K. Sahu, B. B. Bindroo, 2008Biological control of uzi fly Infestation in Muga. Indian silk.46101819

57. Khanikor. B. Bora, M. Konwar, 2010Plant extracts for management of UZI fly Exorista sorbillans Wiedemann (Diptera: Tachinidae). In: Bioresources For Rural Livelihood. Kulkarni GK, Pandey BN and Joshi BD, editors. Narendra Publishing House. New Delhi, India. 217224

58. D. Obeng-Ofori, C. Reichmuth, 1997Bioactivity of eugenol, a major component of essential oil of Ocimum suave (Wild) against four species of store product Coleoptera. Intl. J. of Pest Manag.4318994

59. Kelm MA, Nair MG1998Mosquitocidal compounds and a triglyceride, 1, 3-dilinoleneoyl-2-palmitin from Ocimum sanctum. J. Ag. Food Chem. 4630923094

60. Kelm MA1999Bioactive compounds from Ocimum sanctum Linn. Lamiaceae. Ph.D.Thesis, Michigan State University. 95.

61. C. Kamaraj, Bagavan. A. Rahuman, Elango. G. Zahir, P. Kandan, G. Rajakumar, S. Marimuthu, T. Santhoshkumar, 2010Larvicidal efficacy of medicinal plant extracts against Anopheles stephensi and Culex quinquefasciatus (Diptera: Culicidae). Tropical Biomedicine. 272211219

62. Okunade AL2002Ageratum conyzoidesL. (Asteraceae). Fitotherapia 73116

63. G. Brun, J. M. Bessiere, F. M. G. Dijoux, B. David, A. M. Mariotte, 2001Volatile components of Catheranthus roseus(L.) G.Don (Apocynaceae). Flavour and Fragrance Journal. 162116119

64. H. Gogoi, Bora, 2012Bio-efficacy of extracts of some plants

of ethnic importance against Nymphula depunctalis Guenee (Lepidoptera:Pyralidae).North Bengal University Journal of Animal Sciences. (Press).

65. Shroff KD1920Notes on miscellaneous pests of Burma. In Proceedings of the Third Entomological Meeting, Dept. Agric. Calcutta.Pusa, India. 351353

66. P. Sison, 1938Some observations in the life history, habits and control of the rice caseworm Nymphula depunctalis Guenee. Philipp. J. Agric. 9273301

67. Alum AZ1967Insect pest of rice in East Pakistan. John Hopkins press. 633655

68. Grist DH, Lever RJAW1969Pest of rice. Longmans, Greens & Co. Ltd. London

69. Chi TTN, Tam BTT, Dau HX, Khoa, NT, Lan NTP, Paris TR1995Current status of rice pest management by farmers in direct-seeded rice and transplanted rice area. Omonrice. 44250

70. N. Vromant, A. J. Rothuis, N. T. T. Cuc, F. Ollevier, 1998The effect of fish on the abundance of rice caseworm Nymphula depunctalis (Guenee) (Lepidoptera: Pyralidae) in direct-seeded, concurrent rice-fish fields. Biocontrol Science and Technology. 8539546

71. P. Patgiri, 1997Bioecological studies of Nymphula depunctalis (Guenee) (Pyralidae: Lepidoptera) and its management. Dissertation, Ph.D. Assam Agricultural University, Jorhat, India.

72. Pathak, MD1975Insect pests of rice. Manila (Philippines): International Rice Research Institute.

73. W. H. Reissig, E. A. Heinrichs, J. A. Litsinger, K. Moody, L. Fiedler, T. W. Mew, A. T. Barrion, 1985Illustrated guide to integrated pest management in rice in tropical Asia. Los Banos, the Philippines: IRRI.411.

74. Hill DS1987Agriculturalinsect pests of the tropics and their control (2nd ed.). Cambridge University Press, Cambridge.

75. Lefroy HM1908Indian insect life. Jhacker and Co. Creed Lane, London. 516.

76. R. L. M. Ghosh, M. B. Ghatge, V. Subramanayan, 1956Rice in India. ICAR publication. New Delhi.

77. Joseph KV1969Incidence of rice caseworm Nymphula depunctalis (Guenee) as a major pest in Kerela. J. Bombay Nat. Hist. Soc. 66395396

78. P. S. Prakashrao, G. Padhi, 1984Varietal susceptibility to rice caseworm Nymphula depunctalis Guenee and its behaviour. Oryza 21157162

79. M. Haq, N. M. M. Haque, A. N. M. R. Karim, 2006Incidence pattern of rice caseworm (Nymphula sp). J. Agric. Rural Dev. 4 (1 & 2): 7-81.

80. E. A. Heinrichs, V. D. Viajante, 1987Yield loss in rice caused by the rice caseworm Nymphula depunctalis (Guenee) (Lepidoptera: Pyralidae). Current Science. 47928929

81. Busvine JR1971A critical review of the techniques for testing insecticides. Commonwealth Agricultural Buereux: London.345

82. Morse JG, Bellows TS, IwataY(1986Technique for evaluating residual toxicity of pesticides to motile insects. J. Econ. Entomol. 791281283

83. L. K. Hazarika, M. Bhuyan, B. N. Hazarika, 2009Insect pests of tea and their management. Annual Review of Entomology. 54267284

84. M. K. Deka, K. Singh, R. Handique, 1999Antifeedant and oviposition deterrent effect of Melia azadirach L. and Adhatoda vasica L. against tea mosquito bug. Annal. Of Plant Protec. 712629

85. Patil GS, Patil MG, Mendki PS, Maheswari VL, Kothari RM2000Study of antimicrobial and pesticidal activity of Nerium indicum. Pestology. XXIV:5

86. I. Rahman, 2010Insecticidal potential from some locally available plants against bunch caterpillar, Andraca bipunctata Waker. (Lepidoptera: Bombycidae). Bulletin of Life Sciences. 161117

87. Emden HFV1989Pest contol. Cambridge University Press. London. 82105

88. Hsin CY, Coats JR1986Metabolism of isofenphos in southern corn rootworm. Pesticide Biochemistry and Physiology. 253336345

89. J. Stenersen, 2004Chemical Pesticides- Mode of Action and Toxicology. CRC Press, USA.

90. Khanikor. B. Bora, 2011Selective toxicity of Ageratum conyzoides and Ocimum sanctum against Exorista sorbillans(Diptera:Tachinidae) and Antheraea assama (Lepidoptera: Saturniidae). Natl. Acad. Sci. Lett. 34914

91. Lawrence BM1992Labiatae oils: mother nature's chemical factor.. Eds. Lawrence BM, editors. Essential oils 19881991Allured Publ. Stream, IL. 188-206.

92. Grayer RJ, HarborneJB(1994A survey of antifungal compounds from higher plants 1982-1993. Phytochemistry. 371942

93. Vieira RF, SimonJE(2006Chemical characterization of basil (Ocimum spp.) based on volatile oils. Flavour and Fragrance Journal.21214221

94. S. Ananth, P. Thangamahi, S. Pazhanisami, S. Meena, 2009Larvicidal

activity of three medicinal plants against filarial mosquito Culexquinqufasciatus(Say). Journal of Basic and Applied Biolog. 3(1&2):53-58.

95. Sharma PR, Sharma OP, Saxena BP2003Effect of neem gold on haemocytes of the tobacco armyworm, Spodoptera litura (Fabricius) (Lepidoptera: Noctuidae). Curr.Sci.845690695

96. Sharma PR, Sharma OP, Saxena BP2008Effect of sweet flag rhizome oil (Acorus calamus) on hemogram and ultrastructure of hemocytes of tobacco armyworm, Spodoptera litura (Lepidoptera: Noctuidae). Micron. 39544551

97. J. A. Rajkumar, B. Subramaniyam, Devakumar, 2000Growth regulatory activity of silver fern extract on the cocoon bollworm Helicoverpa armigera (Hubner). Insect Science Appl.20295302

98. Gupta AP1991Insect immunocytes and other hemocytes: roles in cellular and humoral immunity. In: Immunology of Insects and Other Arthropods. Gupta AP, editor. CRC Press. Florida. USA. 119p.

99. J. Calle, A. Rivera, J. G. Luis, Z. Aguiar, H. M. Nimeyer, P. J. Nathan, 1990Insecticidal activity of petroleum ether extract of Ageratum conyzoides L. Rev Colomb QUIM.199196

100. P. Pachiappan, M. C. Aruchamy, S. K. Ramanna, 2009Evaluation of antibacterial efficacy of certain botanicals against bacterial pathogen Bacillus sp. of silkworm, Bombyx mori L. Int. J. Indust Entomol. 1814952

101. R. Rajasekaragouda, M. Gopalan, N. Jeyaraj, Natarajan, 1997Field performance of plant extracts on mulberry silkworm, Bombyx mori L. Entomon. 22(3 & 4) : 235- 238.

102. K. Murugan, D. Jeyabalan, N. Senthikumar, S. Senthilnathan, N. Sivaprakasam, 1998GrowthPromoting effect of plant products on silkworm. A Biotechnological Approach. Journal of Scientific andIndustrial Research57740745

103. M. Manimuthu, L. Isaiarasu, 2010Herbal tonic on silkworm Bombyx mori. J. of Biopesticides. 33567572

104. J. Lawless, J. Allen, 2000Aloe vera- Natural Wonder Cure. Harper Collins Publishers, London.

105. K. Mahmood, U. Yaqoob, R. Bajwa, 2008Antibacterial activity of essential oil of Ocimum sanctum L. Mycopath. 66365

106. R. C. Saxena, O. P. Dixit, P. Sukumaran, 1992Laboratory assessment of indigenous plant extracts for antijuvenile hormone activity in Culex quinquefasciatum. Indian J Med Res 95204206

107. Moreira MD, Picanco MC, Barbosa LC, Guedes RNC, Borros EC, Campos MR2007Compounds from Ageratum conzyoides: isolation, structural elucidation and insecticidal activity. Pest Management Science. 636615621

108. P. Prakash, N. Gupta, 2005Therapeutic uses of Ocimum sanctum Linn (Tulsi) with a note on eugenol and its pharmacological actions: a short review. Ind. J. Physiol. Pharmacol. 492125131

109. A. Kicel, A. Kurowska, D. Kalemba, 2005Composition of essential oil of Ocimum sanctum L. grown in Poland during vegetation. J. of Essential Oil Res. 172217219

110. Rana VS, Blazquez MA2003Chemical composition of the volatile oil of Ageratum conyzoides aerial parts. Intl. J.of Aromatherapy. 134203206

111. J. J. Neal, D. Wu, 1994Inhibition of insect cytochromes 450by furanocoumarins. Pesticide Biochemistry and Physiology. 50: 43-50.

112. E. Enan, 2001Insecticidal activity of essential oils : octopaminergic sites of action. Comparative Biochemistry and Physiology. 130325337

113. M. Kostyukovsky, A. Rafaeli, C. Gileadi, N. Demchenko, E. Shaaya, 2002Activation of octopaminergic receptors by essential oil constituents isolated from aromatic plants: possible mode of action against insect pests. Pest Management Science. 581111011106

114. Price DN, Berry MS2006Comparison of effects of octopamine and insecticidal essential oils on activity in the nerve cord, foregut, and dorsal unpaired median neurons of cockroaches. J. Insect Physiol. 523309319

115. Priestley CM, EM Williamson, KA Wafford, DB Sattelle2003Thymol, a constituent of thyme essential oil, is a positive allosteric modulator of human GABA receptors and a homo-oligomeric GABA receptors from Drosophila melanogaster. Br. J. Pharmacol., 14013631372

116. M. F. Ryan, O. Byrne, 1988Plant insect coevolution and inhibition of acetylcholinesterase. J. of Chem. Ecol.1419651975

117. Coats JR, Karr LL, Drewes CD1991Toxic and neurotoxic effects of monoterpenoids in insects and earthworms. In: Natural Occuring Pest Bioregulators. Hedin P editor. Am.Chem. Soc. Symposium series. 449305316

118. Abdelgaleil SAM, Mohamed MIE, Badawy MEI, El-arami SAA2009Fumigant and contact toxicities of monoterpenes to Sitophilus oryzae (L.) and Tribbolium castaneum (Herbst) and their inhibitory effects on acetylcholinesterase activity. J Chem Ecol.35518525

119. L. A. Gallindo, A. M. Pultrini, M. Costa, 2010Biological effects of

Ocimum gratissimum L. are due to synergic action among multiple compounds present in essential oil. J Nat Med. 644436441

120. D. Geng, D. X. Li, Y. Shi, J. Y. Liang, Z. D. Min, 2009A new Benzophenanthridine alkaloid from Zanthoxylum nitidum. Chinese J. Nat. Med. 74274277

121. Chen JJ, Lin YH, Day SH, Hwang TL, Chen IS2011New benzenoids and anti-inflammatory constituents from Zanthoxylum nitidum.Food Chemistry. 1252282287

122. Z. Ming, Jian. Gui-Yin-Guo, Z. Li, Z. Ke-Long, H. Jin-Ming, S. Xiao, L. , W. Wang-Yuan, 2011Evaluation of molluscicidal activities of benzo[c] phenanthridine alkaloids from Macleaya cordata (Willd) R. Br. on snail hosts of Schistosoma japonicum. Journal of Medicinal Plants Research. 54521526

123. M. Wink, T. Schmeller, B. Latz-Bruning, 1998Modes of action of allelochemical alkaloids: Interaction with neuroreceptor, DNA and other molecular targets. Journal of Chemical Ecology. 24(11). 18811937

124. Z. H. Li, Q. Wang, X. Ruan, D. Pan, D. A. Jiang, 2010Phenolics and Plant Allelopathy. Molecules. 1589338952

125. Summers CB and Felton GW1994Prooxidant effects of phenolic acids on the generalist herbivore Helicoverpa zea (Lepidoptera: Noctuidae): Potential mode of action for phenolic compounds in plant antiherbivore chemistry. Insect Biochemistry and Molecular Biology. 249943953

126. Johnson KS and Felton GW2001Plant phenolics as dietary antioxidant for herbivore insects: a test with genetically modified tobacco. Journal of Chemical Ecology. 271225792597

127. M. Jung, M. Park, 2007Acetylcholinesterase inhibition by flavonoids from Agrimonia pilosa. Molecules. 1221302139

128. M. Kuroyanagi, T. Arakawa, Y. Hirayama, T. Hayashi, 1999Antibacterial and anti androgen flavonoids from Sophera flavescens. J Nat Prod. 6215951599

129. Moore SJ and Lenglet AD2004An overview of plants used as insect repellents. In: Traditional medicinal plants and malaria, eds. Wilcox M., Rasoanaivo P., Bodeker G., CRC press, Boca Ratan, FL, USA. 343363

Chapter 7

PLANTS AS POTENTIAL SOURCES OF PESTICIDAL AGENTS: A REVIEW

Simon Koma Okwute

Department of Chemistry, University of Abuja, Gwagwalada, Federal Capital Territory, Abuja,, Nigeria

INTRODUCTION

For global food security, the agricultural sector of the world economy must achieve a production level that ensures adequate food supply to feed the increasing population as well as provides raw materials for the industries. This is particularly so as the energy sector is vigorously pursuing research into the use of grains and root crops as sources of starch for conversion into bio-fuels. Coincidentally, these crops (maize, rice, millet, *sorgum*, soybeans, cowpeas, sugarcane, groundnuts, e.t.c.) are the stable foods in most parts of the developing countries of the world such as Africa, South America and Asia. In addition to the above new development in the industrial utilization of these crops, they are frequently and vigorously attacked by *herbivorous insects* and other pests such as *phytopathogens* and mollusks. In fact the loss due to pests and diseases is about 35% on the field and 14% in storage, giving a total loss of about 50% of agricultural crops annually. Thus the world food production is adversely affected by insects and pests during crop growth, harvest and storage [1]. Apart from the farm environment insects and pests constitute serious menace in the home, gardens and bodies of water, and transmit a number of diseases by acting as hosts to some disease-causing parasites. Thus elimination of these insects and pests or mitigation of their activities will go a long way in reducing world food crisis as well as improve human and animal health.

Insects and other pests have been in existence since the creation of the universe, and of cause man. The threat of insects and other pests such as mosquitoes, cockroaches, rodents, parasitic worms, pathogens and snails, has been well known and challenged by man. The ancient man had deployed different methods of control, including prayers, magic spells, cultivation systems, mechanical practices as well as application of organic and inorganic substances to protect his crops from the attack of weeds, diseases and insect pests [1].

Between 500 BC and the 19th century a number of substances classified as pesticides and defined as" any substances or mixture of substances intended for preventing, destroying, repelling or mitigating any pest" were used to control pests. They included sulphur, arsenic, lead and mercury [2]. In 1874 DDT (dichlorodiphenyltrichloroethane) was synthesized and during the second half of World War II its insecticidal activity was discovered and was effectively used to control malaria and typhus diseases among the troops. It became the first synthetic organic pesticide and was used after the war for agricultural purposes [3].There is no doubt that the use of insecticides has contributed immensely to the increase in agricultural productivity and to the improvement in human health, particularly the eradication of malaria in the developed countries of the world in the 20th century and beyond [4]. However, it has been established that use of synthetic organic pesticides, particularly the chlorinated hydrocarbons such as DDT and derivatives has led to serious environmental pollution (water, air and soil), affecting human health and causing death of non-target organisms (animals, plants, and fish). This situation led to the Stockholm Convention in 2001 and the eventual ban of DDT in 2004[5]. Before the ban efforts were already made by researchers for alternative sources of pesticides due to other reasons including (a) non-selectivity/specificity, (b) ineffectiveness, (c) not many of the synthetic compounds have been successfully marketed due to lack of interest by potential users, (d) high cost of synthetic chemicals and (e) development of resistance[6-7]. Natural products from plants have attracted researchers in recent years as potential sources of new pesticides. The folkloric use of higher terrestrial plants by the natives of various parts of the world as *pesticidal* and antimicrobial materials has been well known [8-9]. Perhaps, one of the early plants so recorded as *pesticidal* material was tobacco (*Nicotiana tabacum*). The use of tobacco leaf infusion to kill aphids led to the isolation of the alkaloid, *nicotine*, while the chemical investigation of the Japanese plant, *Roh-ten (Rhododendron hortense)* in 1902 showed *rotenone*, as the active constituent [10]. In this class of age-old *pesticidal* plants are species belonging to the genus *Chrysanthemum* found in Kenya and other highlands in Africa, which are the sources of the all purpose and very successful insecticidal extract, *pyrethrum,* and the active constituents, the *pyrethrins* [11].

DDT

Nicotine

Rotenone

Pyrethrins

Tens of thousands of natural products have been identified from plants and hundreds of thousands are yet to be isolated and screened for their bioactivities. This large reservoir of organic chemicals is largely untapped or under-tapped for use as pesticides. In this chapter the traditional applications of native plants as pesticidal agents and the results of biological and chemical studies on these plants in the past few decades are examined with a view to assessing their potential use in agriculture and related fields. The factors influencing efficacy, the advantages of and problems associated with the use of plant-based*pesticidal* products are also discussed.

The pesticidal agents that will be dealt with will include insecticides (insect killers including adults, ova, and larvae) insect *repellents, antifeedants, molluscicides, fungicides* and *phytotoxins* (herbicides). It must however be stated at this stage that although much work has been done in the past decades to show that indeed plants have the potentials to provide alternative and safe pesticides to replace the synthetic ones not enough work has been done in the area of identifying the active components. Whether or not it is very necessary to utilize pure constituents will be discussed later from the point of view of safety, cost and effectiveness (synergism). It is equally important to note that this review will be restricted to those plant-based pesticides that have the potential to be used as extracts (solutions), smoke or dust that have the potential of killing pests or their hosts or mitigating their effects. Consequently, although plant materials that act against worms that destroy crops of economic importance may be discussed, *anthelmintics* for intestinal worms in humans and other animals will not be included in the discussion.

PESTICIDAL PLANTS

There is no doubt that a number of plants possess pesticidal activity and investigations by various research groups in different parts of the world have confirmed this. One of the most recent studies was the survey by Mwine *et al.* which established that thirty-four species belonging to eighteen families are used in traditional agricultural practices in Southern Uganda [12]. Also, Rajapake and Ratnasekera studied the toxicity of the ethanol extracts of the leaves of twenty plant species from different families to *Callosobruchus maculatus* and *Callosobruchus chinensis*. It was observed that mortality reached a maximum level in 72 hours of exposure to the leaves oils which indicated a high level of lethality [13]. Similarly, Lajide *et al.* and Fatope *et al.* have investigated the protectant effectiveness of some plants native to Nigeria against the maize weevil, *Sitophilus zeamais* Motsch, and the cowpea weevil, *Callosobruchus maculatus* F, respectively[14-15]. On the basis of the results of pesticidal screenings it has been established that a number of plants have broad pesticidal activity and those commonly used in traditional agricultural applications in many parts of the developing countries, particularly in the tropical areas, are shown in Table 1 which are only representative but not exhaustive of the thousands of plants so far screened [13-16]. From various investigations it has been established that activity is usually distributed in most cases among the various parts of the same plant though the lethality and quantities of the active components may vary [13].

Having provided a background to the potential use of plant materials as pesticides we shall now look at efforts made in the last few decades by researchers to give us hope that if we return to the ways of our ancestors in combating pests by applying science and technology the terrestrial environment which is our home will be protected against the harmful effects of synthetic pesticides.

Insecticidal Plants

In the past decades, apart from the *pyrethrum* which has attained international and commercial acclaim due to its high effectiveness and broad spectrum insecticidal activity (repels and kills insects depending on concentration) very few natural insecticides have been developed. Of particular economic significance among the plants in common use today is the tropical plant *Azadirachta indica*, popularly known as the neem tree. In India as well as in Nigeria the plant is effectively used to control over 25 different species of insect pests. The activity has been associated with the presence of *azadirachtin*, which is said to be highest in the kernel than in the leaves and other tissues of the plant [1,13].The effectiveness of nine insecticidal species of Chineese origin has been compared with synthetic insecticides against 40 species of insects. Three of the plants *Milletia pachycarpa* Benth, *Trpterygium Forrestii Loes* and *Rhododendron molle* G. Don were studied in detail. The finely ground powder when applied as spray in suspension or as dust were highly active against *aphids, pentatomids* and leaf-beetles as well as against *caterpillars*, body lice and plant lice. Among the plants *R. molle* displayed specific toxicity against certain species of *lepidopterous* larvae, *pentatomids* and leaf-beetles. The three plants were shown to contain *rotenone*, [17].

Investigation of the Sri Lankan plants showed that extracts of three plants, *Plearostylia opposita* (Wall) Alston (Celastraceae), *Aegle marmelos Correa* (Rutaceae) and *Excoecaria agallocha* (Euphorbiaceae) were insecticidal. For the first time three compounds possessing the *daphnane orthoester* skeleton, which are constituents of the ethyl acetate extract of *E. agallocha*, were found to be insecticidal[18-19].

Table 1. Species, families, parts used and evaluated

Species	Families	Parts
Abrus precatorius L	Fabaceae	L, S
Allium sativum L	Alliaceae	L
*Anacardium occidentale*L	Anarcadiaceae	L
*Annona senegalensis*Pers.	Asteraceae	S, B
Artemisia annua L	Asteraceae	L, B
Azadirachta indica A. Juss	Meliaceae	L,B,R, F
*Balanites aegyptiaca*Linn Bel.	Zypophyllaceae	R
Bidens pilosa L	Asteraceae	L
Cannabis sativa L	Cannabaceae	L, S, F
Capsicum frutescens L	Solanceae	F
Carica papaya L	Caricaceae	R, B
Chrysanthemum coccineum Wild	Asteraceae	L, F
Clausena anisata	Rutaceae	L, R

Dalbergia saxatilis	Fabaceae	L, B
Dannettia tripetala	Annonaceae	L
Eucalyptus globulesLabill	Myrtaceae	L, B
Gmelina arborea Juss.	Verbenaceae	L
Hyptis sauvcolens Poit.	Labiate	shoot
Jatropha curcas L	Euphorbiaceae	sap, F, S, B
Khaya senegalensis A. Juss	Meliaceae	S, B
Lannea acida	Anacardiaceae	B
Lawsonia inermis	Lythraceae	L
Melia azadarach L	Meliaceae	L, R, B
Mitracarpus scaberZucc	Rubiaceae	shoot
Nicotiana tabacum L	Solanaceae	L
Ocimum gratissimum L	Liminaceae	L
Parkia clappentonianaKeay.	Mimosaceae	S, B
Phytolacca dodecandraL'Herit	Phytolaceae	L, F
Piper guineense Schum &Thonn	Piperaceae	F
Piliostigma thonningii	Caesalpiniaceae	R,B
Prosopis africana Linn.	Mimosaceae	S, B
Spenoclea zeylanicaGearth	Sphenocleceae	shoot
Tagetes minuta L	Asteraceae	L
Tephrosa vogelii Hook	Fabaceae	L
Vernovia amygdalina L	Asteraceae	L

Azadirachtin

Daphnane orthoester

The pawpaw tree, *Asimina tribola (Annonaceae)*, a plant found in various traditional communities, particularly in Africa and the Americas, has been investigated and found to possess *antitumor,*pesticidal, and *anti-feedant* properties. The pesticidal activity is known to reside in the seeds and bark and the focus has been on *asimicin*, which is the major bioactive component. It is active against blowfly larvae, *Calliphora vicina* Meig, the spotted spider mite, the *melon aphid*, the mosquito larvae *(A. aegypti)*, the Mexican *bean beetle, nematodes*, and many pests of agricultural concerns [20].

Asimicin

Some of the investigations have revealed the mode of action of some of the plant products. N. M. Ba and co-workers have studied *Cassia nigracans* V., *Cymbopogon schoenanthus* S. and *Cleome viscose* L from Burkina Faso for their insecticidal potentials and established that they were reasonably active and that they were most effective by inhalation. Consequently, such plants are not suitable for field applications. The plants however showed potent stomach and contact toxicity on 1st instars larvae irrespective of the crude extract and therefore good for cowpea protection in storage [21]. Similarly, Okwute *et al.* have demonstrated the *protectant* property of the powdered dry leaves of *Dalbergia saxatilis* against the cowpea *bruchid, Callosobruchus maculatus* and established that *oviposition* and damage to seeds was less and mortality higher with *D. saxatilis* as a contact poison than as a respiratory poison (Table2)[22]. It was also shown that the treated seeds were quite viable after the treatment with over 70% germination rate after 5 days exposure to planting (moist) conditions [22]. For *Dalbergia* spp. the insecticidal activity against adult mosquitoes and houseflies has been demonstrated (Figure 1) and the activity has been attributed to the presence of *cinnamylphenols*, [23].

Cinnamylphenol

The genus *Piper* (family Piperaceae) is probably one of the most studied. With over 1000 species, about 112 genera have been screened for *pesticidal activity* and over 611 active compounds have been isolated and identified from various parts of the species [24]. Perhaps, of great significance are extractives from *Piper guineense, Piper longum,* and *Piper retrofractum* which are known to be active against *Callosobruchus maculatus,* the garden insect, *Zonocerus variegatus* L, and the mosquito larvae causing 96-100% mortality rate in 48 hours mostly as solution sprays [25-26]. From the chloroform and petroleum extracts of *P. guineense* fruits were isolated two *Piper amides, guineensine,* and *piperine,* having terminal isobutyl and piperidyl basic moieties, respectively. In these experiments *piperine* was shown to be a synergist rather than an insecticide in the crude extracts. The significance of this co-occurrence in the efficacy and efficiency of crude drugs and bio-pesticides will be discussed later. In an effort to enhance the insecticidal activity of the *piperine amides* some workers have embarked on structure- activity relationships (SAR) studies and have come to the conclusion that the pipenonyl group does not influence activity and that the isobutyl group does not confer any special advantage as previously reported [27]. However, using *piperine* (95% mortality) as a template pesticide and replacing the piperidyl group gave a higher insecticidal activity (97.5% mortality) with N-diethyl moity than the isopropyl analogue (95% mortality) against *Aedes monuste erseis* [28].

Table 2. Evaluation of *protectant* potentials of *Dalbergia saxatilis*

Treatment (gm)	No. of eggs laid on seeds	No. of damaged seeds	Insect mortality	% Germination
2.00	3.2 ±0.84	0±0.00	10.0±0.00	83.3
1s.75	3.4±0.55	0±0.00	10.0±0.00	91.7
1.50	3.6±0.95	0±0.00	10.0±0.00	83.3
1.25	3.8±1.00	0±0.00	10.0±0.00	75.0
1.00	4.8±1.64	0±0.00	10.6±0.89	100
0.75	5.8±1.48	0±0.00	12.0±0.71	100
0.50	10.6±2.70	9.8±1.30	14.0±1.92	100
0.25	16.4±3.11	10.8±1.95	0.0±0.00	91.7
0.0g(Control)	17.0±3.16	32.0±3.49	0.0±0.00	91.7

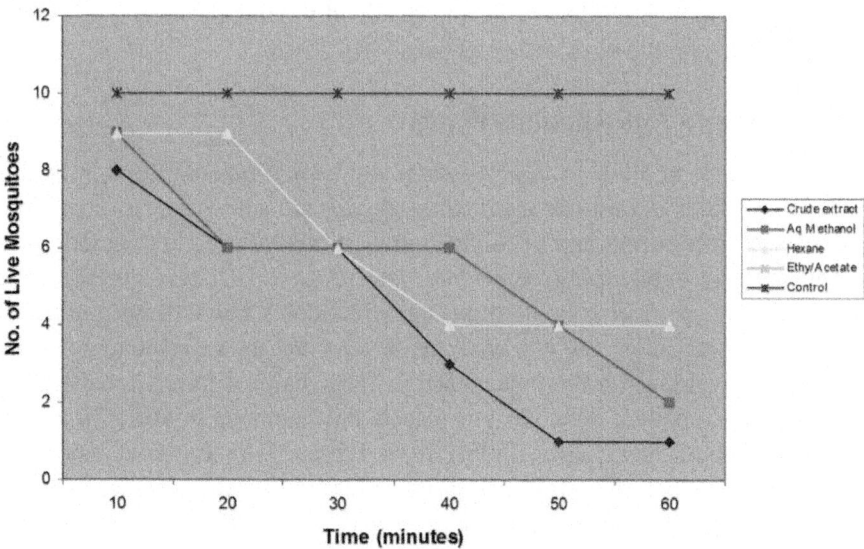

Figure 1. Mortality rate of mosquitoes exposed to 0.2% solutions of the crude extract and fractions of *Dalbergia saxatilis*.

Guineensine

Piperine

A new class of insecticides was recently discovered by Beltsville researchers led by Puterka in the U.S.A. that offers a safe and effective alternative to commercial insecticides. They are polyesters of sugars and include *sucrose* and *sorbitol octanoates*. They were isolated from the poisonous hairs on the tobacco leaves which hitherto were assumed to contain nicotine, a popular insecticide. When insects were contaminated by rubbing they caused death of the insects by a dehydration process, and rapidly degraded to harmless sugars and fatty acids. These polyesters are known to be effective against a variety of farm and domestic insect pests and the deadly parasitic *Varroa* mite which usually settles on the back of honey bees [29].

Repellent and Anti-Feedant Plants

Closely related to the insecticidal agents and sometimes used in combination with insecticides in pest management strategies are some classes of pesticidal agents with interesting and peculiar biological activities. They include insect *repellents, anti-feedants* or *deterrents*, and *attractants*.These classes are far less common in plant sources than the insecticides but will be given some attention. Sometimes, a given insecticide may act as an insecticide or as a *repellent* depending on the concentration. The major difference between the two is that a repellent does not kill insects but keeps them away by exuding pungent vapours or exhibits slightly toxic effects [13]. By these activities a *repellent* prevents insects from perching or landing on the surfaces of targets. Thus *repellents* can be used to prevent and control the outbreak of insect borne diseases such as malaria. The insects of interests in this regard include mosquito, flea, fly, and the *arachnid tick*.[30]. The use of plant materials as *insect repellents* is increasingly receiving attention, particularly in the developing countries. For example Seyoun *et al.*reported that in Western Kenya the natives employ direct burning of the species *Ocimum americana* L,*Lantana camara* L, *Tagetes minuta*, and *Azadirachta indica* A. Juss against the malaria vector,*Anopheles gambiae* S.S.Giles[31]. Some recent studies on *repellent* plants have led to the isolation and characterization of some active components. Prominent among these compounds are *callicarpenal*, and*intermedeol,* from the species *Cymbopogon nardus* which showed promising alternative in the control of infestations by *Amblyomma cajennense*[32]; *nepetalactone*, a *catnip* compound for the control of the Asian adult male and female Lady beetle as well as cockroaches, flies, termites and mosquitoes[33-34]; and *geraniol,* and *p-menthane-3,8-diol*(PMD), monoterpenoid alcohols from the *citronella* and *lemon oils*, respectively[35]. Some researchers have found that products containing 40% *lemon eucalyptus oil*are as effective as products containing high concentrations of *DEET* and that *neem oil* can give up to 12 hours protection against mosquitoes in cage experiments[36-37].

Literature on the direct production of chemicals with specific activity to act as insect *anti-feedants* is very scanty probably as *anti-feedancy* and *repellency* are closely related bioactivities. However, a number of plants produce *polyphenols* called tannins which confer astringency or bitter taste on such plants and consequently herbivores stay away from eating such plants [38]. Among the few plants studied for feeding *deterrency or anti-feedancy* the species *Xylopia aethiopica* is very significant. The hexane and methanol extracts of the fruits and seeds have been shown to possess strong termite *anti-feedant* activity and *ent-kauranes* and some *phenolic amides* have been implicated. Among the *ent-kauranes* the activity was significantly dependent on the structures and that *(-)-kau-16-en-19-oic acid,*had the strongest anti-feedant activity [39]. Another species with promise is *Jatropha podagrica*cultivated in West Africa. The organic extracts showed reasonable *anti-feedant* activity against *Chilo partellus*, the maize stem borer, at concentrations of 100 %/leaf disc, the chloroform extract being the strongest. The most active compound isolated was *15-epi-4E-jatrogrossidentadione,*[40].

Attractants are *semio-chemicals* produced usually by some insects with effect on other insects as a communication tool and can be used to determine or control insect populations, particularly by disrupting their mating patterns. Rarely do plants produce chemicals that attract insects that are natural enemies of other insects that feed on the plants except the tea tree [41]. Thus field application of this phenomenon is not common and therefore will not be discussed further.

Callicarpenal

Intermedeol

Geraniol

Nepetalactone

p-Menthane-3,8-diol(PMD)

(-)Kau-16-en-19-oic acid

15-epi-4E-jatrogrossiden
-tadione

Fungitoxic Plants

Plant diseases, particularly fungal infections, contribute significantly to agricultural crop losses globally. Research has been on to utilize botanicals in plant disease control worldwide and extracts from many plant species have been found to be active against many *phyto-pathogenic fungi* without imposing ill side effects[42]. In some cases the active components have been identified and tested directly. The results so far are quite encouraging and some are discussed in this chapter.

Many plants produce *essential oils* as secondary metabolites but their exact role in the life processes of the plants has been unknown. There is however no doubt as revealed in this survey that the results of various investigations have overwhelmingly implicated essential oils of many species as possessing *fungitoxic activity*. They are therefore agents of protection in plants against diseases. Consequently, since the leaves, resins, and latices of plants contain essential oils more commonly than other parts of plants they have been more commonly investigated for *fungitoxicity* [43-44].

Typical studies included the investigation of the *essential oil* of the leaves of *Phenopodium ambrosioides* which has been shown to exhibit strong *fungitoxic activity* against mycelia growth of *Phizoctonia solani*, the causative organism of *damping off disease* of seedlings, at 1000 ppm without any *phytotoxicity* on the germination and seedling growth of *Phaseolus aureus*[45]; the activity of the steam-distillate and hot-water extracts of fresh leaves of *Cymbopogon, Ocimum gratissimum, Chromoleana odorata* and fruits of *Xylopia aethiopica* against *Utilago maydis, Ustilaginoidea virens, Curvularia lunata,* and *Phizopus spp*, reducing growth by 10-60%[46]; the screening of the leaves of 30 angiospermic taxa against *Pythium aphanideratum,P. debaryanum* with *Hyptis suaveolens*(Labiatae),*Murraya knoenigii*(Rutaceae), and *Ocimum canum*(Labiatae) which displayed strong toxicity at 43-86% inhibition in soils infected with *P. debaryanum* [47]; the use of Ocimum gratissimum and *Eucalyptus globules* water extracts to control cowpea seedling *wilting i*nduced by *Sclerotium rolfsil* from 39.6% for untreated to 4-12% for treated[48]; and the tomato fruit rot, which is commonly observed in local markets in many parts of Africa, can be significantly reduced with the extracts of a number of local plants such as *Cassia alata, Alchornea cordifolia and Moringa oleifera* as post-harvest agents[49]. Of particular interest and importance is the availability of some species such as the popular neem tree (*Azadirachta indica*) and the pawpaw leaves extracts which are known to act against the yam rot. Yam is an important tuberous food crop of the tropical South America and Africa where both plants are found commonly around villages and within family compounds.

The pawpaw leaves extract at the various concentrations of 20, 40, 60, and 80% were found to be more active than the neem against*Alternaria solani* [50].

Efforts have been made by some researchers to investigate the active constituents of some of the fungitoxic plants. The constituents of the essential oils of 9 Turkish species including *Thymbra spicata*were investigated using GC-FID technique. At least 20 components were identified and the activity was attributable to the presence of *phenolic agents* such as *thymol,* and *carvacol,* [51].

Thymol   **Carvacol**

In other studies the *fungitoxic* chloroform extract of the underground parts (bulbs) of *Eleutherine bulbosa*(Miller) Urban*(*Iridaceae) gave 4 compouds of which three *naphthaquinones,* incuding *eleutherinone,* were active at 100μg/ spot*(bioautography).* The fourth compound, *eleutherol,* which lacked the *quinone* moity was not active, showing the strategic role of this group in the bioactivity of the series[52]. The relationship between *fungitoxicity* and the *quinone*skeleton is also exhibited in the broad spectrum *fungitoxicity* of *lawsone,* isolated from the leaves of *Lawsonia inermis,* against eight different *phytopathogenic fungi*(Table 3) [53]. Working on the rhizomes of *Zingiber cassumunar* N. Kishore and R.S. Dwivedi isolated the *fungitoxic* and *non-phytotoxic monocyclic sesquiterpene, zerumbone,* which was active at 1000 ppm against *Rhizoctonia solani,a damping-off pathogen*[54][55].

Eleutherinone

Lawsone

Zerumbone

Eleutherol

Table 3. *Fungitoxicity* measured as % inhibition of *lawsone* against eight different fungi (% inhibition)

Fungi	PPM of lawsone		
	1000	2000	4000
	Inhibition %		
Alternasia solani	60	100	100
Alternasia tenuis	100	100	100
Aspergilus niger	65	100	100
Aspergilus wanti	100	100	100
Absidia ramosa	100	100	100
Absidia orymbifera	100	100	100
Absidia crophlalophora fusispora	100	100	100
Circinella umbellate	84	100	100

Molluscicidal Plants

Biharzia affects millions of people, particularly children who play or swim in infected freshwaters in the developing countries of Africa, Asia and Latin America. The disease was discovered in 1851 by Theodor Bilharz as the cause of urinary *schistosomiasis*.It is associated with certain species of aquatic snails of the genera *Biomphalaria, Bulinus and Oncomelania*.Therefore, one way of attacking the disease is to eliminate the host snails[56-57]. Chemicals that kill snails are called *molluscicidal agents.*Most of the *molluscicidal agents* in use today are synthetic and like most synthetic pesticides are harmful to man and the environment. *Molluscicidal agents* of natural origin are important in the widespread control of *Schistosomiasis*. Mirazid, an Egyptian drug from *myrrh* was being developed as an oral drug until 2005 when it was found to be only 8 times as effective as *praziguantel,* a synthetic chemical, and has therefore not been recommended by WHO. However, other plants have been studied and some have demonstrated potential activity which may provide leads for future drugs but more importantly are the searches for molluscicidal agents from plants to eliminate the host snails. This is the focus of the presentation in this chapter.

Adesina and Adewunmi, and Kloos and McCullough, have separately investigated the species *Clausena anisata* and found it to possess *molluscicidal activity* which is distributed among the root, leaves, bark and stem in a decreasing order of potency [58-59]. Adedotun and Alexander evaluated the *molluscicidal activity* of the aqueous and ethanolic extracts of fruits and roots of *Dalbergia sissoo* against the egg mass and adults of *Biomphalaria pfeifferi* and found that only the ethanolic extracts showed significant activities. Thus ethanol extracts the active constituents more than water[60]. Similar observations have been recorded for *Clausena anisata* parts and *Tetrapleura tetraptera* fruits, particularly when the active components are *glycosides*(Table 4)[58,61].

Table 4. Molluscicidal activity of Clausena anisata and Tetrapleura tetraptera

extracts			
Plant parts	**Concentration**	**% Mortality**	**Solvent**
Clausena anisata			
Root	6-10 ppm	100	Methanol
Leaves	1000 ppm	53.3	Water
Stem	1000 ppm	7	Water
Bark	1000 ppm	40	Water
Tetrapleura tetrptera			
Fruit	100%	100	Water
Fruit	10%	100	Methanol

Investigation of the extracts of the Argentine collection of the fern *Elaphoglossum piloselloides* led to the isolation of two new *bicyclic phloroglucinols* which showed acute *molluscicidal activity* against the*Schistosomiasis* vector, *Biomphalaria peregrina*[62]. Other phytochemical and biological investigations have implicated *jatrophone,* as one of the *molluscicidal agents* in the active crude ethanol extract of *Jatropha elliptica,* while *a monodesmosidic saponin,* and *thujone,* have also been identified as the active constituents of the bark powder of *Saraca asoca* and the leaf powder of *Thuja orientalis,* respectively, against the freshwater snail *Lymnaea acuminata*[63-64]. Finally, the *molluscicidal properties* of the leaves of *Alternanthera sessellis,* a plant found in West Africa, have been investigated and confirmed. The effect of heat on the stability of the product has been determined by comparing the activities of the unevaporated and evaporated aqueous extracts which showed that the unevaporated has higher activity than the evaporated and the fresh leaves higher than the dry leaves(Table 5)[65].

Table 5. Expected effective lethal concentrations of *A. sesselis* extracts of dry and fresh leaves on adult *B. globosus.*

*Mo*lluscicide	*LC50 and limits(mg/ml)*
Crude unevaporated fresh leaves extract	32.57(25.15-39.08)
Crude unevaporated dry leaves extract	40.42(35.15-46-47)
Crude evaporated fresh leaves extract	43. 57(38.38-49.46)
Crude evaporated dry leaves extract	48.07(42.81-54.28)

Phloroglucinol I, R=H

Phloroglucinol II, R=Et

Jatrophone

Monodesmosidic saponin

Thujone

Herbicidal Plants

Apart from insects and diseases disturbing crop plants and animals on the farm and in the environment weeds need also to be controlled because they retard plant growth and therefore reduce crop yields. *Herbicides,* also commonly known as *weed-killers,* are pesticides used to kill unwanted plants. *Selective herbicides* kill specific targets while leaving the desired crop relatively unharmed. Some of these act by interfering with the growth of the weed. Some plants produce *natural herbicides,* and such action of *natural herbicides* (interfering) is called *allelopathy.* Herbicides are widely used in agriculture and in landscape turf management.

The plant *Centaurea maculisa* provides a good example of *allelopathy.* The root secretes (+) and (-)*catechins,* but it is *(-) catechin* which is phytotoxic and accounts for the invasive behavior in the rhizosphere [66-67]. The phenolic root exudate of Buckwheat(*Fagopyrum esculentum*) has been studied using HPLC and GC-MS and *palmitic acid methyl ester* and a *gallic acid derivative* have been implicated as the active constituents[68]. *Allelopathic properties* have also been found among some*terpenoids.* Investigation of the aerial part of *Eupatorium adenophorum* led to the isolation of eleven components of which *5,6-dihydroxycadinan-3-ene-2,7-dione,* was the only active herbicidal compound[69]. It has been observed that certain varieties of common fescue lawn grass come equipped with their *natural broad-spectrum herbicide* that inhibits the growth of weeds and other plants around them. A group of Cornell researchers led by Frank Schroeder has identified the *natural herbicide* to be the *amino acid, m-tyrosine,* and that the grass exudes the compound from the roots. The compound is toxic to plants but not to fungi, mammals or bacteria. The major drawback is its high solubility in water, making it ineffective if applied directly as a herbicide in the field [70]. On the other hand the *spotted knapweed plant* spreads over large areas because it releases *catechin* through its roots into the soil that kills the surrounding plants. Unlike *m-tyrosine,* *catechin* is quite stable and does not kill certain species of grass and grass-like plants like wheat. Therefore, it can be sprayed or added to soil to maintain lawns and wheat fields and is environmentally friendly [71]. Thus it has great potentials as experiments have shown that it is as effective as 2,4-D against weeds, kills weeds within one week of application and ordinary tapping of the leaves activates the plants' chemical response[71].

(-) Catechin

5,6-dihydroxycadinan-
3-ene-2,7-dione

m-Tyrosine

COMMERCIAL BOTANIC PESTICIDES

Plant-based pesticides (botanic pesticides or botanicals) have been in use as pesticides for over 150 years. It was only very recently that the synthetic insecticides effectively became the prominent agrochemicals for controlling all forms of agricultural pests and have assumed a very important position in the marketplace. However, in the past three decades so much has been reported in literature in respect of natural products that were identified with potent pesticidal activity such as feeding deterrency and toxicity to insects in laboratory assays. In spite of the above success not many of the isolated compounds or the crude material (extract or dust) have really found the market due to regulatory procedures associated with product development, particularly in the United States of America.For example, in the past twenty years probably only very few new sources of botanicals have been developed to the commercial status. Thus the four major commercial products today include the*pyrethrum/pyrethrins, rotenone, neem* and the *essential oils.* Three others in limited use or importance include *ryania, nicotine* and *sabadilla,* while *garlic oil* and *Capsicum* oleoresin are relatively new extracts [72]. *Botanical pesticides* are processed in various ways, principally (a) as crude plant material in the form of powder or dust; (b) as extracts from plant resins, formulated into liquid concentrates; and (c) as pure isolated constituents by extraction/chromatographic techniques or hydro-distillation of the plant tissue, particularly the leaves. For a pesticide to be considered safe for use and be registered as a commercial product, the $LD_{50,}$ the term used to describe the lethal dose required to kill 50% of the test animals expressed in milligrams(mg) per kilogram (kg) of body weight, must be determined. Technically, the lower the value the more toxic the sample is to mammals. Although botanical pesticides are generally considered safer than their synthetic counterparts, some have much lower LD_{50}than standard synthetic insecticides like *carbaryl* and *malathion.* The *pesticidal* characteristics of some of the current *commercial botanicals* are outlined below and summarized in Table 5. [73-74].

The *pyrethrins,* account for about 80% of global use of botanicals. Kenya is the major supplier, followed by Tanzania and the Botanical Resources Australia. The material is highly degradable under sunlight, oxygen and moisture. It therefore requires frequent applications. Its activity is usually enhanced by incorporating *piperonyl butoxide*(PBO) as a synergist. It acts against a wide range of pest.*Rotenone,* on the other hand, is available mainly from *Lonchocarpus* spp and the *Derris* spp. found in East Indies, Malaya and South America (Venezuela and Peru). It is obtained by solvent extraction to yield resins containing about 45% *rotenoids* of which the major component, *rotenone,* is 44%. Its activity and persistence are comparable to DDT and it is used to protect lettuce and tomato crops. It is slower acting than any of the *botanicals* currently in use and yet readily degradable, taking several days to kill insects. The *neem* product has become popular commercially in recent time because of its broad spectrum activity and low mammalian toxicity. There are two neem-based products, the first being the neem oil from the cold pressing of seeds for the management of *phytopathogens* while the other product is medium- polarity extract containing the potent compound azadirachtin,5 (0.2-0.6% of seed/weight). The actual commercial product is a 10-50% concentration using solvents. Although it has a half-life of about 20 hours its systemic action on foliage ensures reasonable persistence in field applications.

The *essential oils* are products of steam distillation of aromatic plants, mostly of the family*Lamiliaceae*, giving *monoterpenoid phenols* and *sesquiterpenes*. Examples are *thymol* and *carvacol*. They possess high volatility and therefore are not suitable for field applications but appropriate for stored grains. The essential oils are components of many commercial foods and beverages and are therefore more readily approved for use without going through the rigorous regulatory procedures even in the USA.

The other botanicals, though not very important commercially have some advantages in applications [74]. They include the following:

- *Ryania* has a low mammalian toxicity but has the longest residual activity, providing up to two weeks of control after an initial application. It works best on caterpillars and worms but also kills a number of other pests with the exception of spider mite.

- *Nicotine*, a constituent of *Nicotiana tabaccum,* is the most toxic of all botanicals and extremely harmful to humans. It is a very fast-acting nerve toxin and is most effective on soft-bodied insects and mites.

- *Sabadilla* is available from the seeds of the plant *Schoenocaulon officinale* which is cultivated in Venezuela. It is one of the least toxic of the botanicals. It is toxic to honeybees, caterpillars and leafhoppers as well as beetles.

From the above it is clear that a serious drawback to the commercialization of botanicals is the high cost of processing plant materials to meet World Health Organisation(WHO) and Food and Agriculture Organisation(FAO) safety standards.

Table 6. List of some commercial botanical pesticides

Plant Name	Product/Trade Name	Group/Mode of Action	Targets
1. Lonchocarpus spp Derris eliptica	Rotenone	Insecticidal	Aphids, bean leaf beetle, cucumber beetles, leafhopper, red spider mite

2. *Chrysanthemum ciner-ariaefolium*	Pyrethrum/Pyrethrins	Insecticidal	Crawling and flyingin-sects such as cockroach-es, ants, mosquitoes, termites
3. *Nicotiana tabaccum*	Nicotine	Insecticidal antifungal	Aphids, thrips, mites, bugs, fungus gnat, leafhoppers
4. *Azadirachta indica[Dogonya*ro (Ni-geria)]	Azadirachtin/Neem oil Neem cake Neem powder Bionimbecidine (GreenGold)	Repellent Antifeedant Nematocide sterilant Anti-fungal	Dandruffs(shampoos) eczema, nematodes, sucking and chewing insects(caterpillars, aphids, thrips, maize weevils)
5. Citrus trees	d-Limonene, Linalool	Contact poison	Fleas, aphids, mites, paper wasp, house cricket, dips for pets
6.*Shoenocaulon officinale*	Sabadilla dust	Insecticidal	Bugs, blister beetles flies, caterpillars, potato leafhopper
7. *Ryania speciosa*	Ryania	Insecticidal	Caterpillars, thrips, beetles, bugs, aphids
8. *Adenium obesum(Heliotis sp)*	Chacals Baobab(Senegal)	Insecticidal	Cotton pests, par-ticularly the larvae of ballworm

Given the large number of plants traditionally used as *pesticidal agents* by various local communities globally, particularly in the developing countries, the number of plants so far investigated and the products developed from them, the impact on agricultural production from this source is very insignificant. Therefore, there is need for more plants to be harnessed for use in agriculture and related fields. However, there is need to examine the modalities for their utilization, particularly with respect to consistency of constituents as well as efficacy and quality of the products, vis-à-vis the production of bioactive plant-based products using western models or utilize the plants according to traditional procedures that eliminate purification. For example, *the anti-sickle cell anaemia* drug, *NICOSAN* has been found to be less potent and more toxic on separation into individual components [75].Thus, there are some advantages in the traditional procedures of preparing herbal products in a manner that preserves the constituents of the plants and hence enhances synergism and potency. However, while appreciating the low cost of production of *botanic products* by eliminating sophisticated purification and formulation procedures, a middle of the road approach that ensures consistency of active constituents and enhances efficacy and safe delivery is necessary. This may be achieved by using bioassay-guided fractionation which has been shown by some workers to ensure that bioactive compounds of the same chemical class in a crude plant extract are consistently pooled together. The procedure has been shown to improve activity dramatically and has been used to obtain active compounds from plants that were previously considered to be inactive [76].

Thus, cheap plant-based bioactive products may be prepared with improved efficacy if processed using bioassay-guided fractionation of the crude extract and classified as *orphan pesticide* as is sometimes done in drug development. The content of the identified components can be used to standardize the crude pesticide as *gedunin* has been proposed for crude *neem-based antimalarial drugs* [77]. It is only in this way will the abundant plant-based natural resources of the developing countries be readily and cheaply made available for agricultural production without polluting the environment.

CONCLUSION

The results of *pesticidal* and *phytochemical screenings* of a number of higher plants based on traditional knowledge strongly indicate that plants are endowed with *pesticidal* properties that can be harnessed cheaply for use in agriculture and related fields. The need to use plant-based products arises from the fact that the synthetic pesticides are harmful to humans, and the entire ecosystem due to high toxicity and persistence. Also, they are too expensive for the poor farmers in the developing countries of the world. On the other hand, plant- based products are cheap and bio-degradable and are therefore environmentally friendly. However, an agricultural programme that depends essentially on plant-based materials must be backed-up by a vigorous research programme into new plant sources. As revealed in this review traditional knowledge has so far guided studies on possible active plants and the results have overwhelmingly confirmed the activity of a reasonable percentage of the plants. The results have equally established that plants belonging to certain families of plants are more likely to possess*pesticidal* activity. Thus, these results will serve as useful guides in the collection of plants for laboratory and field research studies.

One area of difficulty in laboratory research studies is the *bioassay* of plant extracts. It has been established that certain crude extracts contain active components but may appear inactive in primary screens due to *antagonistic actions* of the constituents. Such problems may be overcome, particularly in screening against plant pathogens, by the application of *bio-autographic techniques.* Associated with the detection and the determination of level of activity in crude extracts is the appropriateness of the solvent for the extraction of plant materials. Use of a less desirable solvent can lead to low extract activity due to low concentration of the active principle. For example, aqueous alcoholic extracts have been found to be more active than aqueous extracts as most of the active compounds are *lipophilic* in character and are therefore more readily extracted into an organic medium.

For the poor countries it would be more expensive to use the plant extracts or the pure constituents than the plant powder or dust in large-scale field applications. Crude extracts can however be cheaply used if a readily available solvent such as water is the solvent of choice. Use of extracts also allows for easy dosage calculation and spraying applications which need to be done repeatedly due to high volatility of plant-based pesticidal products. The efficacy of such products can be enhanced by *bioassay-guided fractionation* which is known to concentrate activity and promote synergism between structurally related constituents.

Obviously, in large-scale field utilization of botanic agricultural chemicals there must be adequate and constant supply of candidate plants to the areas in need. This means that since plants grow well usually in areas of natural habitat effort should be made to invest in large-scale cultivation of such plants in their various localities as is the practice in China, Japan and Kenya. This will be of great economic advantage in the developing countries as such programmes can lead to economic empowerment of the poor farmers and ultimately improve the economies of these countries.

ACKNOWLEDGEMENT

The author would like to thank Dr. Egharevba, Henry Omoregie of the National Institute for Pharmaceutical Research and Development, Idu, Abuja, Nigeria, my former Ph.D student, for the preparation of the structures for this chapter.

REFERENCES

1.   Plant-based pesticides for control of Helicoverpa amigera on Cucumis; by Jitendra Kulkarni, Nitin Kapse and D.K Kulkarni; Asian Agric. History,13 (4),2009

2.   US Environmental (July 24,2007What is a pesticide? Epa.gov. Retrieved on September 15, 2007.

3.   WHO1979Environmental Health Criteria 9: DDT and its derivatives.

4.   van Emden HF, PealallDB(1996Beyond Silent Spring. Chapman and Hall, London, 322 pages.

5.   UNEP2005Ridding theWorld of POPs. In: A guide to the Stockholm Convention on Persistent Organic Pollutants.

6.   O. Stephen, (1990. Duke, 1990Natural pesticides from Plants. In: Advances in new crops. Timber Press, Portland, OR. J. Janick and J. E.

Simon (eds), 511517

7. E. F. Abraham, E. Chain, 1940An enzyme from bacteria able to destroy penicillin. Nature 146: 837.

8. J. M. Dalziel, (1937, 1937The useful plants of West Tropical Africa. Crown Agents for Overseas Governments, London.

9. S. Ayensu, 1978Medicinal Plants of West Africa. Reference Publications Inc., Algonac Michigan, U.S.A.

10. P. Tooley, (1971, 1971Crop Protection. In: Food and Drugs.Chemistry in Industry series.Chapter 3. John Murray Albermark Street, London.

11. S. David, Siegler, 2005Plant-derived insecticides. In: Plants and their uses. Department of Biology, University of Illinois,Urbana, Illinois.

12. J. Mwine, et al.2011Ethnobotanical survey of pesticidal plants used in South Uganda: Case study of Masaka district; Journal of Medicinal Plants Research 5711551163

13. R. H. S. Rajapake, D. Ratnaseka, 2008Pesticidal potential of some selected tropical plant extracts against Callosobruchus maculates F. and Callosobruchus chinensis L., Tropical Agricultural Research and Extension, 116971

14. Lajide,L(1988Insecticidal activity of powders of some Nigerian plants against the maize weevil (Sitophilus zeamais Motsch). In: Entomology in the Nigerian Economy. Research focus in the 21st Century, 227235

15. M. O. Fatope, et al. (1995, 1995Cowpea weevil bioassay: a simple prescreen for plants with grain protectant effects; International Journal of Pest Management, 41208486

16. S. K. Okwute, 1992Plant-derived Pesticidal and Antimicrobial Agents for use in Agriculture: A Review of Phytochemical and Biological Studies on some Nigerian Plants. Journal of Agric. Sci. and Technol. 216270

17. Shin-Foon Chiu et al.1950Effectiveness of Chineese insecticidal plants with reference to the comparative toxicity of botanical and synthetic insecticides. Journal of the Science of Food and Agriculture; 19276286

18. Samarasekera Javaneththi1997Insecticidal Natural Products from Sri Lanka. Ph.D Thesis, Open University.

19. A. Paul, Wender, 2011Gateway synthesis of daphnane congeners and their C affinities and cell-growth activities; Nature Chemistry,3615617

20. S. Ratnayake, et al.1993Evaluation of the Pawpaw tree, Asimina triloba(Annonaceae) as a commercial source of the pesticidal annonaceous acetogenins. In: New crops. J. Janick and J. E. Simon (eds). Publishers: Wiley, New York. 644648

21. N. M. et, al, 2009Insecticidal activity of three plant extracts on the cowpea pod sucking bug, Clavigralla tomentosicollis, STAL. Pakistan J. of Biological Sciences; 1213201324

22. S. K. Okwute, et al.2009Protectant, insecticidal and antimicrobial potentials of Dalbergia saxatilis Hook f (Fabaceae). African Journal of Biotechnology; 82365566560

23. M. Gregson, et al.1978Violastyrene and isoviolastyrene cinnamylphenols from Dalbergia miscolobium. Phytochemistry; 1713751377

24. C. O. Ogobegwu, 1973Studies of the insecticidal activities of some Nigerian plant extracts. Dissertation for the award of B. Sc. Agric.Biol., University of Ibadan, Nigeria.

25. M. F. Ivbijaro, M. Agbaje, (1986, 1986Insecticidal activities of Piper guineense Schum and Thonn and Capsicum species on the cowpea bruchid Callosobruchus maculates F. Insect Science and its Application; 74521524

26. L. A. Dyer, J. Richard, C. D. Dodson, (2004, 2004Isolation, synthesis and evolutionary ecology of Piper species. In: A model genus for studies of phytochemistry, ecology, and evolution. L. A. Dyer and A.O.N. Palmer (eds).Publishers: Kluwer /Plenum New York 117139

27. I. K. Park, et al.2002Larvicidal activity of isobutylamides identified in Piper nigrum fruits against three mosquito species. J. Agric. Food Chem; 507186670

28. M. Miyakado, et al. (1985, 1985The Piperaceae amides-6. Chemistry and insecticidal activities of Piperaceae amides and their synthetic analogues. J. Pesticide Science; 101117

29. Rosaline Marion Bliss2005Death by desiccation: Sugar esters dry out insect pests of flowers and ornaments. US Agricultural Research Services (ARS)Report.

30. M. F. Maia, S. J. Moore, 2011Plant-based insect repellents: A review of the efficacy, development and testing. Malaria Journal 10 (1).

31. et. Seyoun, al, 2002Traditional use of mosquito-repellent plants in western Kenya and their evaluation in semi-field experimental huts against Anopheles gambiae: Ethnobotanical studies and application by thermal expulsion and direct burning. Transactions of the Royal Society of Tropical Medicine and Hygiene; 963225231

32. Sara Fernandes Soares et al.,2010Repellent activity of plant derived compounds against Amblyomma cajennense(Ascari: Ixodidae) nymphs. Vet. Parasitology; 16716773

33. JanSuszkiw(2009Catnip compounds curb Asian Lady Beetles.US Department of Agriculture, Agric. Research Services(ARS)Report.

34. M. Samuel, et. Mc Elvain, (1941. al, 1941The constituents of the volatile oil of catnip.1. Nepetalic acid, Nepetalactone, and related compounds. Journal of American Chemical Society;63615581563

35. D. R. Bernad, R. Xue, 2004Laboratory evaluation of mosquito repellents against Aedes albopictus, Culex nigripalpus and Ochlerototus triseriatus(Diptera Culicidae). J. Med. Entomology; 414726730

36. S. P. Carroll, J. Loye, 2006Journal of the American Mosquito Control Association; 223507514

37. A. K. Mishra, N. Singh, V. P. Sharma, 1995Use of neem oil as a mosquito repellent in tribal villages of mandla district, Madhya Pradesh. Indian J. Malario; 32399103

38. S. K. Okwute, A. A. Nduji, (1992, 1992Isolation of schimperiin: A new Gallotannin from the leaves of Anogeissus schimperii (Combretaceae). Proceedings of the Nigerian Academy of Science; 43641

39. Lajide. Labunmi, et al.1995Termite anti-feedant activity in Xylopia aethiopica(Annonaceae). Phytochemistry 40411051112

40. O. Olapeju, et. Aiyelaagbe, al, 2011Insect anti-feedant and growth regulatory activities of Jatropha podagrica Hook. Abstract 008 at 1st PACN/RSC Congress on Agricutural Productivity, Accra, Ghana,2123November, 2011.

41. R. Weinzierl, et al.2009Insect attractants and traps. In: Alternatives to Insect Management. Office of the Agricultural Entomology,University of Illinois at Urbana-Campaign.

42. G. D. Lyon, T. Beglinski, A. C. Newton, 1995Novel disease control compounds: the potential to "immunize" plants against infection. Plant Pathology 44407427

43. Rai et al,1999In vitro susceptibility of opportunistic Fusarium spp to essential oils. Mycoses 42(1,2): 97-101.

44. M. Rai, D. Acharya, 2000Search for fungitoxic potential in essential oils of Asteraceous plants. Compositae Letters 351823

45. N. K. Dubey, et al.1983Fungitoxicity of some higher plants against Phizoctonia solani. Plant and Soil 729194

46. R.T.Awuah(1989Fungitoxic effects of extracts from some West African plants. Ann. Appl. Biol. 115451453

47. V.N.Pandey and N.K. Dubey1994Antifungal potentials of leaves and essential oils from higher plants against soil phytopathogens. Soil

Biology 261014171421

48. D. A. Alabi, et al.2005Fungitoxic and Phytotoxic effects of Vernonia amygdalina(L), Baryophylum pinnantus Kurz, Ocimum gratissimum(Closium)L and Eucalyptus globules(Caliptos) Labill water extracts on cowpea and cowpea seedling pathogens in Ago-Iwoye, South-Western Nigeria. World Journal of Agricultural Sciences 11070

49. O. A. Enikuomehin, E. O. Oyedeji(2010, Oyedeji(2010Fungitoxic effects of some plant extracts against tomato fruit rot pathogens. Archives of Phytopathology and Plant Protection 433233240

50. M. N. Suleiman, 2010Fungitoxic activity of Neem and Pawpaw leaves extracts on Alternaria solani, casual organism of Yam Rots. Advances in Environmental Biology 42159161

51. Frank Mueller-Riebau, Bernard Berger, and OktayYegen(1995Chemical composition and fungitoxic properties to phytopathogenic fungi of essential oils of selected aromatic plants growing wild in Turkey. J. Agric. Food Chem. 43822622266

52. Tania Maria Almeida Alves, Helmut Kloos, Carlos Leomar Zani. Mem. Inst. OswaldoCruz 98(5).

53. R. D. Tripathi, H. S. Srivastava, S. N. Dixit, 1978A fungitoxic principle from the leaves of Lawsonia inermis Lam. Experientia 345152

54. N. Kishore, R. S. Dwivedi, a. Zerumbone, fungitoxic. potential, isolated. agent, Zingiber. from, Roxb. cassumunar, Mycopathologia 1203155159

55. [55] Sukh Dev1960Studies in Sesquiterpenes-XVI. Zerumbone, a monocyclic sesquiterpene ketone. Tetrahedron 8(3-4): 171-180.

56. B. [56], D. S. rown, 1980Freshwater snails of Africa and their medical importance.Publishers, London,Taylor and Francis. [57] Dalton, P. R. and Pole, D. (1978). Water-contact patterns in relation to Schistosoma haematobium infection. Bulletin of World Health Organisation 56417426

57. S. K. Adesina, C. O. Adewunmi, 1981The isolation of molluscicidal agents from the root of Clausena anisata (Wild) Oliv. In: Abstrats 4th International Symposium on Medicinal Plants, 4445Ile-Ife, Nigeria.

58. H. Kloos, F. S. Mc Cullough, 1987Plants with recognized molluscicidal activity.In: Plant Molluscicides. K. E. Mott (eds). Chapter 345108UNDP/ Wold Bank/WHO Special Programme for Research and Training in Tropical Diseases.

59. A. Adedotun, Ademsi, B. Alexander, Odaibo(2008Laboratory assessment of molluscicidal activity of crude aqueous and ethanolic extracts of Dalbergia sissoo plant parts against Biomphalaria pfeifferi. Travel Med.

Infect. Diseases 4219227

60. S. K. Adesina, C. O. Adewunmi, V. O. Marquis, 1980Phytochemical investigations of the molluscicidal properties of Tetrapleura tetraptera(Taub.). Journal of African Medicinal Plants 3715

61. Cecilia Socolsky et al.,2009Activity against the schistosomiasis vector snail Biomphalaria peregrina. J. Nat. Products 724787790

62. A. F. , dos Santos and A. E. Sant'Ana(1999Molluuscicidal activity of the diterpenoids jatrophone and jatropholones A and B from Jatropha elliptica(Pohl) Muel. Arg. Phytother Res. 138660664

63. Vet. Parasitology2009Publishers B. 164

64. B. A. Azare, S. K. Okwute, S. L. Kela, 2006Molluscicidal activity of crude leaf water extracts of Alternanthera sesselis on Bulinus(Phy) globosus. African Journal of Biotechnology 64441444

65. Harsh Pal Bais, et al.,2003Structure-dependent phytotoxicity of catechins and other flavonoids: Flavonoid conversions by cell-free protein extracts of Centaurea maculosa roots. J. Agric. Food Chem. 514897901

66. Jana Kalinova and Nadezda Vrchtova2009Levels of catechin, myricetin, quercetin and isoquercitrin in Buckwheat (Fagopyrum esculentum Moench), changes of their levels during vegetation and their effects on the Growth of selected weeds. J. Agric. Food Chem. 57727192725

67. Jana Kalinova, Nadezda Vrchotova and JanTriska(2007Exudation of allelopathic substances in Buckwheat (Fagopyrume sculentum Moench). J. Agric. Food Chem. 551664536459

68. Xu Zhao et. al.2009Terpenes from Eupatorium adenophorium and their allelopathic effects on Arabidopsis seeds germination. J. Agric. Food Chem. 572478482

69. Cornell University2007Fescue: A common lawn grass uses natural herbicide to control weeds. ScienceDaily, October 27, 2007.

70. Colorado State University2002Colorado University identifies natural plant-produced herbicide. scienceDaily,June 27, 2002.

71. Murray Isman,2006Botanical insecticides, deterrents, and repellents, in modern Agriculture and an increasingly regulated World. Annu. Rev. Entomol. 514566

72. of. University, I. F. A. S. Florida, Extension, E. N. Y-275 -http//edis.ifa. ufl.edu, I. N08, products. Natural, insect. for, Eileen. A. management, Buss and Sydney G.Park-Brown.

73. Illinois pesticide review,2004Raymond A. Clyod; Natural indeed: Are natural insecticides safer and better than conventional insecticides.

74. XeCHEM2006Personal communication with Dr.Pandey/MD of XeCHEM.

75. A. L. Mitscher, S. Drake, S. R. Gollapudi, S. K. Okwute, 1987A Modern Look at Folkloric Use of Anti-infective Agents; Journal of Natural Products,50610251040

76. Kinnon. S. Mac, et al. (1997, 1997Antimicrobial Activity of Tropical Meliaceae Extracts and Gedunin Derivatives. Journal of Natural Products, 60 (4), 336-341.

Chapter 8

EXPOSURE TO PESTICIDES IN TOMATO CROP FARMERS IN MERCED, COLOMBIA: EFFECTS ON HEALTH AND THE ENVIRONMENT

Marcela Varona Uribe[1], Sonia Mireya Díaz[1], Andrés Monroy[1], René A. Castro[2], Edwin Barbosa[2], and Martha Isabel Páez[3]

Marcela Varona Uribe[1], Sonia Mireya Díaz[1], Andrés Monroy[1], René A. Castro[2], Edwin Barbosa[2], and Martha Isabel Páez[3]

[1]Environmental and Occupational Health Group, Research Department, National Institute of Health, Bogotá,, Colombia

[2]National Agricultural Inputs Laboratory, Agricultural Protection Associate Management, Analysis and Diagnosis Associate Management, Colombian Agricultural Institute, Mosquera,, Colombia

[3]Environmental Research Group for Metals and Pesticides (GICAMP), Department of Chemistry, University of Valle,, Colombia

INTRODUCTION

The advance of chemicals in industry during the XX century gave rise to a number of highly aggressive compounds to human beings, and that altered the ecosystems balancing. Human population is inevitably exposed to environmental pollution through air-degraded products, water, the soil and food and their introduction into the food chain (Gomez et al, 2011).

The use of pesticides has been recognized and accepted as an essential ingredient in the modern agriculture for the control of pests, which damage crops and as a result, they produce a severe loss in food production. However, the extended use of pesticides, together with the inadequate behaviors of prevention and use of basic protection requisites will increase the probability of accidental intoxication in a notorious manner (Ntow et al, 2009), (Páez et al, 2011). The estimated worldwide pesticides application is about 4 million tons (Elersek and Filipic, 2011) and according to Instituto Colombiano Agropecuario (ICA) Colombia produce 16.999.216 litters of herbicides, 6.392.387 litters of insecticides, and 19.690.293 kilograms of fungicides (ICA, 2010) during 2010.

Approximately 1.8 billion people worldwide are engaged in agriculture and it has been estimated that up to 25 million agriculture laborers have suffered non-intentional intoxications every year (Alavanja, 2008). In developing countries, pesticides are the cause of up to one million cases of intoxication and up to 20.000 deaths a year (Duran-Nah and Colli-Quintal, 2000).

Among the different pesticides used, 85% are used for agriculture applications and the remaining 15% are used in homes, gardens, business applications, public health and veterinary (Idrovo, 2000) (CEPIS/PAHO, 2005).

Certain works like agriculture or pests killing represent the biggest risks of acute intoxication, while there is a latent danger for the population at large in their food chain (Ospina et al, 2009) (Thundiyil et al, 2008).

The agricultural development model in Colombia is mainly based upon the use of agro chemicals and according data reported by the Public Health Surveillance System of Colombia (SIVIGILA), there were in 2008 6.650 intoxication cases for the use of pesticides followed by 7.405 cases in 2009 and 8.016 cases in 2010, being the organophosporic and carbamate pesticides the principal reasons for intoxications (SIVIGILA, 2010). Such pesticides are widely used agricultural inputs, and they are esters of the phosphoric acid and the derivates there of, and they share in common as a pharmacological characteristic, the action of inhibiting enzymes having esteracic activities, and more specifically, the inhibition of the acetylcholinesterase. They are easily hydrolyze and they have a low capacity of remaining in the environment (Palacios and Moreno, 2004), (Chakraborty, 2009), (Ntow et al, 2009).

Other pesticides under study are the organochlorated, which are persistent, lypophilic and very steady. They can be accumulated in ecosystems, causing many toxic effects on reproduction, development and immunological functions of animals (Waliszewski et al, 2005). They have been universally reported in the adipose tissue and human serum (Rivas et al, 2007), (Côte et al, 2006).

This study determined the biomarkers the inner dosages, exposure and effect caused by the use of organophosphoric (OF), carbamates (C) and organochlorated (OC) pesticides. The levels for these pesticides were established in a sample of tomato and the good agriculture practices were implemented for the crops of tomato, which afforded to assure the crop sustainable management and the perception of hazards on the pesticides adequate usage and management.

MATERIALS AND METHODS

A descriptive cross section study was done including 132 laborers of the tomato crop in the location of la Merced – Caldas, during 2009 and 2010. This study considered three phases: the first was the diagnosis to determine the biomarkers for the chosen pesticides. An analysis was made on pesticides residues in tomatoes as well as the characterization of the present productive systems of the crop through a participating rural diagnosis. The second phase was intervention to guide the demonstration plots implementation wherein the good agricultural practices (GAPs) were shown which were compared to other plots managed under a traditional production system. The process of intervention was assessed during the last phase.

A questionnaire was applied including variables social, demographic, occupational, clinical, toxicological and tomatoes consumption habits. The pilot study was carried out on 10% of the total of the sample, although they did not make part of the research.

Following the criteria of inclusion, all laborers engaged in tomato planting entered into the study, provided that they were permanent residents in the community, who had used OF, C and OC pesticides at least during the six months previous to the study, and also who volunteered to participate in the study.

For the analysis of biological samples two blood samples were taken, one with 5 ml heparin for the determination of acetyl cholinesterase (AChE) and pseudo cholinesterase (PChE) by the technique of Michel and Aldrige (Vorhaus and Kark, 1953) and another of 5 ml without anticoagulant for the analysis of OC pesticides in serum. For this group 12 different pesticides (α-BHC, β-BHC, HCB, heptachlor, oxychlordane, α-chlordane, i-chlordane, α -endosulfan, β-endosulfan, 4,4-DDE, endosulfan and 2,4-DDT) were considered, which were determined by gas chromatography with electron micro capture (EPA, 1995) reporting the levels found.

With respect to the sampling of tomato it occurred at the beginning, on the highest peak and at the end of the production stage of tomato. For each sampling unit a zigzag path by the crop was followed, harvesting a tomato every three places along the zig-zag, and then based on a quartering system, obtaining a sample of 1 kg per plot. For determination of pesticide residues OF and OC the internal method for extraction AR-NE-03 was used, based on the multiresidue S-19 extraction method of the German Convention (DFG, 1987), followed by the gas chromatographic analysis with flame photometric detector FPD and ECD electron capture. Meanwhile, for the determination of residues of N-methyl carbamate an internal method based on W. Blass and C.

Philipowsky (Blass and Philipowsky, 1992) was used. The levels found for these pesticides were considered as contamination.

A descriptive analysis by frequency counting, central trends measures and dispersion was made for those continuous type variables, as some of the variables inherent to laborers, and environmental and biological measurements. Continuous variables were transformed to normalize them. We also explored possible relationships between some variables and they were crossed by constructing contingency tables. We used the Student›s t tests and chi square tests for the comparison of quantitative and categorical variables. Subsequently, we then performed a bivariate, stratified and logistic regression analysis. To compare results among laborers who worked in plots with GAP and traditional applications, the paired data test was used of Wilcoxon and Fisher for quantitative variables and the MacNemar test for qualitative variables. This study took into account Resolution 8430 of 1993 by the Ministry of Health, which classified this research as a minimal risk work. This study was approved by the Technical Committee of Research and by the Ethics Committee of the National Institute of Health.

RESULTS ANALYSIS OF THE TOTAL POPULATION INCLUDED IN THE STUDY

Social and Demographic Variables

A total of 132 agricultural laborers were registered for the study, which were occupationally exposed to pesticides in the location of La Merced, belonging to the urban area 12,1% (16) and to the rural area 87,9% (116). Some general characteristics of the population are shown in Table 1.

With respect to gender, 90,9% (120) were men and 9,1% (12) women. We found a statistically significant difference between ages by sex p <0,05. As for affiliation to the social security system, 99,2% (131) of individuals in the sample had some form of social health security.

Occupational History

At the time of the interview, 100,0% (132) of individuals reported to be engaged in agriculture, of them, five laborers were enrolled in a GAP program, which abolished the pesticide use as compared with other five who followed the traditional practices.

The time of exposure to pesticides ranged from three months to 35 years (Table 1), we found a statistically significant difference in the time of exposure to pesticides among men and women (p = 0,006).

Regarding the frequency of spraying 78,1% (104) of laborers reported applying at least once a week and 21,9% (28) used to apply pesticides every 15 days or more. Other variables related to the pesticide use are shown in Table 1.

Table 1. Population characteristics of location La Merced-Caldas, 2010.

Characteristics	Participants	Standard deviation	Range		Intervals of confidence
Age (years)					
Mean (SD) for both sexes	40,0	10,8	13	74	35,9-49,3
Men	39,7	10,8	18	69	37,7-41,6
Women	42,6	10,5	13	74	35,9-49,3
Time using pesticides (months)					
Men	109,2	88,1	3	420	94,8 - 127,2
Women	70,8	73,1	4	240	55,0 – 147,9
Hours a day of application					
Mean (SD)	5	2,5	------	------	--------
Smoking while using pesticides, N (%)					
Yes	10 (7,8%)	----	----	----	----
No	118 (92,2%)	----	----	----	----

In dealing about the storing of pesticides 95,5% (126) of laborers reported having an exclusive area and 19,5% (25) keep them indoors. On the use of personal protective equipment (PPE) a high percentage of laborers 96,2% (127) reported using some type of PPE, only 3,8% (5) did not use them, being the most frequent the use of the high heel boot (86,3%), and it is important to clarify that an employee may report using more than one element.

Laborers reported the greatest use of PPE related to protection of the body (120,5%), followed by protection of the lower limbs (93,9%), while the high boot the most common. 82,3% (105) of laborers said they were changing their work clothes at the end of the workday. The highest percentage of laborers 99,2% (131) washed their clothes at home and of these, 36 (27,3%) reported washing work clothes together with the rest of the family's clothes.

Table 2. Use and amount of pesticides studied and applied by agricultural laborers, in La Merced-Caldas, 2010.

Types and trade names of pesticides used,	Active ingredient	Control group	Toxicological category	Quantity of application per harvest	Application number of applications per harvest	Standard deviation	Median	Maximum	Minimo
Organophosphoric Compounds									
Lorsban	Chlorpyrifos	Insecticide	III	25.414 L	63	101,55	2	500.000	0,12
Tamaron	Methamidophos	Insecticide	I	2,81 L	36	4,91	2	30	0,5
Monitor	Methamidophos	Insecticide	I	8.334 L	30	45,64	1	250.000	0,12
Roxion	Dimethoate	Insecticide	II	1,54 L	29	1,02	1	4	0,12
Sistemin	Dimethoate	Insecticide	II	9.093 L	22	42,63	1.125	200.000	0,1
Carbamate compounds									
Furadan	Carbofuran	Insecticide	I	17.648 L	51	73,36	2	400.000	0,2
Roundup	Glyphosate	Herbicide	IV	3.850 L	13	13,86	3	50.000	1

An inquire was made on training on the safe handling of pesticides, being established that 74,2% (98) had never been trained, therefore, they had no knowledge about the use and handling of pesticides.

In relation to pesticide exposure, 74 (61,2%) reported having presented some symptoms at the time they were using them and of these 85,2% (63) did nothing with respect to this issue, or they self-medicated and only 14,8% (11) consulted a doctor. It was determined that Furadan was the pesticide causing most of intoxications to the population under study, being this toxicity category I (extremely toxic).

Use of Pesticides

Pesticides reported by individuals in the sample evidenced that the most commonly used were the insecticides in OF group, of these Lorsban (chlorpyrifos) was the most widely used for a crop of tomatoes. The most frequently used toxicological category was II (highly toxic), followed by I (extremely toxic) (Table 2).

Clinical Manifestations

The most frequently ailment reported was headache with 43,9%, followed by dizziness with 38,6%, weakness 36,4%, ocular burning 34,8% and redness of eyes with 31,8%. Grouped symptoms by systems most frequently found were in the central nervous system (95,5%) (Figure 1).

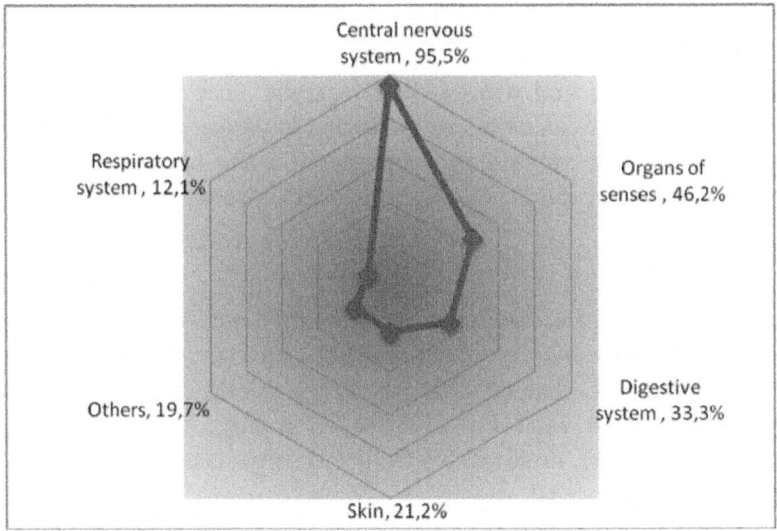

Figure 1. Distribution of systems, location of La Merced-Caldas 2010.

Biomarkers of Internal Dose, Exposure and Effect

Determinations were made for OC in 132 farm laborers, who showed an average level of 1,3 µg/L, with the highest values 4,4 DDE (mean = 3,4 µg/L, SD = 2,8, Minimum = 0,3, maximum = 13,5, IC = 2,9 to 4,0) and hepta chlorine (mean = 2,3 µg/L, SD = 1,0, = 6,1 Minimum, Maximum = 0,6, CI = 2,1 to 2,5). The average was obtained from the results of twelve OC pesticides, which perform the same mechanism of action.

Out of the total of workers, 45 (34,1%) showed inhibition of AChE enzyme in erythrocytes (mean = 0,84, SD = 0,020, IC = 0,80 to 0,88) and only one (0,8%) in plasma level (mean = 1,69, SD = 0,025, IC = 1,64 to 1,74). Of those with inhibition of the enzyme in erythrocytes, 14 (10,6%) individuals were below 25% with respect to the reference value.

ANALYSIS OF AGRICULTURAL LABORERS INCLUDED IN THE PROJECT DEMONSTRATION PLOTS

As specified in the methodology, 10 laborers were taken, 5 of which voluntarily agreed to participate in planting the tomato crop using the GAP, that is, pesticides were not used however a biological control was made. The remaining 5 laborers planted tomatoes as they were used to, using the pesticides frequently applied. The results shown below correspond to these 10 laborers who are involved in the project within the intervention phase.

It was seen that laborers of the parcels with GAP used more protection than those of conventional plots.

Plots were compared with GAP and traditional regarding the change of clothes at the completion of the workday and it was seen that laborers in both plots after receiving training in the proper use and handling of pesticides, successfully performed this activity, which became a protective factor.

Use of Pesticides

Laborers who worked in traditional plots reported the use of seven pesticides, all of them being insecticides, of which six belong to the chemical group of OF and one belongs to C. The most frequently used pesticides were Lorsban and Latigo with 28,6% (4) each. Two of the seven employees, (28,6%) belonged to the toxicity category I, three (42,8%) to the toxicity classes II and two (28,6%) to the toxicity category III.

Clinical Manifestations

With respect to manifestations grouped by systems, laborers who worked in the plots using GAP, showed more symptoms corresponding to organs of senses, while laborers in the traditional plot the referred symptomatology belonged to the central nervous system. In general, traditional plots laborers, had clinical manifestations more frequently in all systems, although there were no significant differences.

Biomarkers of Internal Dose, Exposure and Effect

With regards to biomarkers of exposure the presence of OC pesticide levels was found in serum, in laborers of both plots.

For OF and C the determination of the AChE enzyme showed some inhibition after the exposure in one of the five cases (20,0%) in conventional plots, while laborers belonging to GAP plots showed no inhibition. For plasma no inhibition of enzyme was seen in any worker.

Environmental Samples

With regards to the samples of tomato, in crops were traditional practices were implemented, pesticide residues were found belonging to the chemical groups of (chlorpyrifos and phenthoate) and n-methyl carbamates (carbofuran and 3-hydroxycarbofuran), presenting the highest concentration of residues in OF. For plots with GAP the presence of residues of same pesticides n-methyl carbamates and chlorpyrifos was seen, being the only difference the finding of dimethoate. These active molecules were not in the formulation of products recommended by the agronomist for the control of pests and diseases.

Tomato production, had a statistically significant difference between plots with GAP and the traditional plots (p = 0,020) (Figure 2).

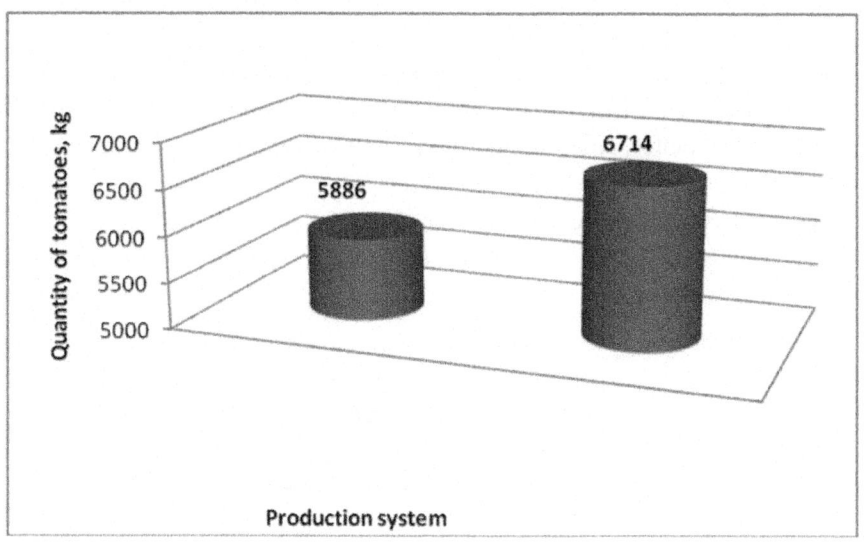

Figure 2. Comparison of production (kg of tomato) between production systems, in the location of La Merced-Caldas, 2010.

DISCUSSION

Pesticides have been of great help to developing countries in their efforts to eradicate insects, endemic diseases and to produce adequate food (Alavanja, 2009), (Ecobichon, 2001). There is a controversy about the world's dependence on these agents, due to their excessive use, volatility, long-range transport and eventual contamination of the environment (Ecobichon, 2001).

In Colombia, pesticide exposure has become a public health problem (Ministry of Social Protection, 2003) due to the higher demand in the use thereof and to the impact on the population health and the environment.

According to this study, all laborers who were hired, used to work in the agricultural sector and they were occupationally exposed to pesticides, most of which laborers came from the rural area, and 90,9% belonged were males, with a wide age range from 18 to 69, indicating that this is a working population and that young adults are the ones most commonly hired to carry out agricultural activities. Although laborers who qualified for the study, used to be informal workers, and many of them had no working contract, a high percentage (87,1%) of individuals in the sample belonged to the subsidized system of health, in other words, they had health coverage.

Concerning the time of exposure to pesticides, this exposure was considered as chronic, as workers had been exposed for an extended period of time, having an average of 9 years of exposure, which can result in long-term harmful effects, being men who have longer exposure times because they are mostly engaged to farming. This data is also supported by the spraying operations frequency, since about 70% of laborers applied the products at least once a week at an average of 5 hours a day.

In dealing with the storage of pesticides, a high percentage of laborers (95,0%) reported having an exclusive area, which reduces exposure to both the worker and his family.

For EPP, (personal protection elements) laborers used to perform agricultural activities wearing clothes for work exclusively, being the high-leg boot the most commonly used item, as well as the disposable mouth masks. It is Important to point out that EPPs used by laborers were not commensurate with the risk they were exposed to, as for example, wearing face masks is not a proper practice for their protection as they have to handle these harmful chemicals, as they allow the entry via inhalation of pesticides, and it was also found that only a minority of laborers use gloves, allowing the entry of these substances by the dermal route, especially of those products having the characteristic of being lipo soluble pesticides such as OF and C. The parts of the body that were mostly protected were the trunk and lower limbs, being the upper limbs the ocular and respiratory regions the least frequently protected.

Although laborers reported that they changed their work clothes at the end of the workday, they washed their clothes at home mixing it with the clothes of the rest of their family, exposing their family members to intoxication risk by such substances.

We inquired about the training courses that laborers have received at some point in their working lives on topics like the safe handling of pesticides and it was found that 74,2% had never been trained, so they did not have the necessary skills to handle such substances.

Only a small percentage reported to see a doctor when they showed some kind of symptoms when they used pesticides, while others took home remedies or they medicated themselves which results in an underreporting of cases of poisoning by such substances, as the intoxication cases are not reported to the system of public health surveillance in Colombia (SIVIGILA).

A high number of pesticides is used in Colombia primarily in categories I and II toxicological categories, and by chemical group the OF and C. This information is confirmed by other studies conducted in Colombia (Varona et al, 2007, 2009), (SIVIGILA, 2010), thus increasing the chances of triggering effects on health. Among the clinical manifestations reported by laborers, most of them are related to neurological and sensory organs disorders. Neurological disorders, may be related to pesticides OF and C, while manifestations of sense organs can be triggered by the use of multiple chemicals, including pesticides that are the subject of this study. The same occurs with the manifestations of the digestive system, which can onset by the ingestion of different chemicals, although they can also have a bacterial and viral origin, among other causes.

This study used biomarkers of internal dose, exposure and effect, which allowed setting the pesticides levels in 132 biological samples. OC pesticides which were most frequently found in biological samples were 4,4-DDE and endosulfan. It is important to point out that laborers did not report the use of these pesticides in the tomato crop, so the presence of these is explained by the environmental pollution and toxicokinetics inherent to this group of pesticides. Although OC were banned in the country since 1993 due to their high persistence, their ability of bio-magnification and their neurotoxic effects, they were still used for about 40 years.

The determination of enzyme AChE continues to be widely used to measure the exposure to OF and C, however, interpretations of results are highly variable, since there are genetic and physiological causes as well as associated pathologies, which can decrease the levels of this enzyme (Varona et al, 2007).

In addition, there is a significant variation within the same individual, therefore, the medical surveillance of laborers continuously exposed to these two groups of pesticides must also include not only the medical examination, but also the determination of enzyme AChE pre-exposure (baseline) and quarterly for the duration of the exposure (Varona et al, 2007).

This study reported that 34,1% of all laborers showed inhibition of this enzyme in erythrocytes, which confirms the fact that OF and C compounds are the most used by the laborers included in the study.

Within the research project and as stated earlier in the methodology, a second phase known as intervention was conducted. During this stage 10 individuals were included, who had some schooling, which allowed them to gain a better understanding of the concepts used in training on GAP and thus the implementation thereof was facilitated.

It was found that laborers used pesticides in plots with GAP, but his recommendation was not given by the agronomist, while in conventional plots they reported the use of pesticides inhibiting the AChE, which are classified as extremely and highly toxic causing a large exposure to this chemical group of pesticides, which in turn affect the central nervous system, a situation that is related to the clinical manifestations reported by laborers in the study of these plots (Idrovo, 2000), (Cassaret and Doull, 2005) (Goldfrank et al, 2006). Regarding the use of EPP, laborers of parcels with GAP used more protection, especially respiratory and eye protection than the conventional plots.

The analysis of biomarkers of exposure and effect for laborers who participated in the intervention phase showed the presence of OC pesticide levels for both traditional plots for GAP. None of the laborers reported the use of these pesticides in the tomato crop, as stated in the diagnostic phase, and their presence is due to their high persistence and their ability of bio magnification. Despite of the fact that their use is prohibited in Colombia, it is not uncommon to identify patients with acute and chronic effects resulting from the exposure to this type of pesticides (Varona et al, 2010), (Córdoba, 2006).

The determination of enzyme AChE showed the after exposure inhibition in a worker of conventional plots, however, this reduction did not require any treatment but a medical surveillance. It is noteworthy to say that there was not a greater number of laborers with inhibition of AChE enzyme considering that in conventional plots the use of OF and C pesticides was reported. This can be explained because this group of pesticides are easily hydrolyzed and excreted by renal way, and there is no bioaccumulation or bio magnification, but this is also due to the fact, that the specimen taken for the determination of the enzyme AChE, should take place within 24 hours after the exposure as a maximum. Since this enzyme subsides to the exposure, it starts to regenerate and therefore, it cannot show the true percentage of inhibition.

When comparing the production of tomato obtained by the two crop systems, it is seen that the range of average production of crops following the traditional practices were very similar to the crops following the GAP. However, when comparing the average tomato production, crops with GAP showed a statistically significant difference higher than the results attained by the traditional system. This behavior can be explained because the fertilization plan in plots using GAP satisfactorily met the needs of the soil.

Regarding the pesticides residues that followed traditional practices, we have found that OF showed the highest levels of residues in the initial stage of crops, where chlorpyrifos was the highly concentrated pesticide but it did not exceed the Codex maximum residue limits (MRLs). For crops established with GAP, we found that five crops presented residues of pesticides the active molecules of which were not recommended by the agronomist for the control of pests and diseases. It was also found that the highest concentration of residuals detected were due to the use of OF.

Although there was no evidence of commitment to health in terms of the effects assessed in this study, we detected significant correlation with respect to one of the traditional plots. It was reported for that plot that one of the individuals presented cholinesterase inhibition and coincidentally it is the plot where the residues in tomato reported the presence of chlorpyrifos. This is a situation of high concern, because this means that farmers in this area are significantly applying pesticides at this time.

The above implies that there is no risk perception by farmers, which makes it necessary to carry out educational campaigns to warn them about the need of at least meeting the rules and procedures laid down for each product. It is recommended to continue with the follow-up and support to this type of population through training and sensitization of laborers in an attempt to reduce the pesticide use and as a result to reduce the effects on their health from exposure to such substances.

It is necessary to strengthen farmers in the implementation of GAP and the advantages of this production system.

The use of chemical pesticides should be performed following the manufacturer's technical recommendations such as presence of pests, application dose, frequency of application and exhaust period in the context of an integrated pest management program which reduces the risk of finding concentrations exceeding the MRL which implies some weakness of competitiveness of crops.

FUNDING SOURCES

This research project was funded by Colciencias, the National Institute of Health, the Colombian Agricultural Institute, University of Valle and Caldas Territorial Health Department.

ACKNOWLEDGEMENT

We want to express our gratitude to the workers participating in this study. To Dr. Natalia Carvajal of University of Quindio for their contributions to this research. To Drs. Nelcy Rodriguez and Viviana Rodriguez, for their statistical analysis of information. To Hernán Correa and Jhon Jairo Gonzalez of Caldas Health Territorial Department, for their assistance to create the brochure as well as in field assistance and Jose Gabriel Muñoz, for his part in finding workers for the study.

REFERENCES

1. M. Alavanja, Pesticides use and exposure extensive worldwide. Rev Environ Health 2008 24 4 303309 .

2. W. Blass, C. Philipowsky, Determination of N-methyl carbamate residues using HPLC and on-line coupling of a post-column reactor in food of plant origin and soil. PflanzenschutzNachrichten Bayer 1992,45:277-318.

3. L. Cassaret, J. Doull, of. Fundamentals, First. Toxicology, Mexico. edition, Hill. Mc Graw, Interamericana, 2005

4. CEPIS / PAHO, Pan-American Center for Sanitary Engineering and Environmental Sciences. Self-taught course on diagnosis, treatment and prevention of acute pesticide poisonings, 2005

5. S. Chakraborty, S. Mukherjee, S. Roychoudhury, S. Siddique, T. Lahiri, M. Ray, Chronic exposures to cholinesterase-inhibiting pesticides adversely affect respiratory health of agricultural workers in India. Journal of Occupational Health 2009,51:488-97.

6. D. Córdoba, Ed. Toxicology, Manual. Modern, Bogotá, 2006

7. S. Côté, P. Ayotte, S. Dodin, C. Blanchet, G. Mulvad, H. Petersen, S. Gingras, et al. Plasma organochlorine concentrations and bone ultrasound measurements: a cross-sectional study in peri-and postmenopausal Inuit women from Greenland. Environmental Health: A Global Access Science Source, 2006

8. D. F. G. Deutsche, Forschungsgemeinschaft, Manual of Pesticide residue analysis.VCH Verlagsgesellscaft, Weinheim, Federal Republic of Germany (Method S-19).1987, 1: 383400

9. J. Duran-Nah, J. Colli-Quintal, pesticide. Acute, poisoning, Mexico Public Health 2000,42:53-55.

10. D. Ecobichon, Pesticide use in developing countries.Toxicology 2001 160:27-33.

11. T. Eleršek, M. Filipič, Pesticides. Book, impacts. The, pesticide. of, exposure, Chapter 12: Organophosphorus pesticides- Mechanism of their toxicity. Edited by INTECH, 2011

12. EPA, Environmental Protection Agency. EPA Method 5081 Determination of chlorinated pesticides, herbicides, and organohalides by liquid-solid extraction and electron capture gas chromatography. Accessed: March 9, 2010. Available in: http://www.caslab.com/EPA-Method-508_1/, 1995.

13. L. Goldfrank, N. Lewin, N. Flomenbaun, emergencies. U. S. A. Toxicological-New, The. York, Companies. Mc Graw-Hill, Inc., 2006

14. S. Goméz, C. Martínez, R. Villalobos, S. Waliszewski, Pesticides. Book, impacts. The, pesticide. of, exposure, Chapter 15: pesticides- genotoxic risk of occupational exposure. Edited by INTECH, 2011

15. ICA, the Colombian Agricultural Institute of Plant Production Deputy Manager, Technical Department of Food Safety and Agricultural Inputs. Technical Bulletin: Statistics marketing of chemical pesticides for agricultural use, 2010 Editorial produmedios.

16. A. Idrovo, Surveillance of pesticide poisonings in Colombia. Public health 2000,2(1):36-46.

17. Ministry of Social Welfare, National Institute of Health, Pan American Health Organization. Public Health Surveillance Protocol for acute and chronic poisoning by pesticides. Bogotá, 2003

18. W. Ntow, L. Tagoe, P. Drechsel, P. Kelderman, E. Nyarko, H. Gijzen, Occupational Exposure to Pesticides: Blood Cholinesterase Activity in a Farming Community in Ghana. Arch Environ Contam Toxicol 2009 56: 623-30.

19. J. Ospina, F. Manrique, N. Ariza, Educational intervention on knowledge and practices regarding work-related risks in potato farmers in Boyaca, Colombia. Public health 2009,11(2):182-190.

20. M. Páez, M. Varona, S. Díaz, R. Castro, E. Barbosa, N. y. Carvajal, A. Londoño, Human risk evaluation for pesticides in tomato cultivated with traditional systems and GAP (Good Agricultural Practice). Journal of Science, 15 2011 15366

21. PAHO, Pan American Health Organization. The health of the Americas. Scientific and Technical Publication 2007 (622).

22. M. Palacios, L. Moreno, Health differences in male and female migrant agricultural workers in Sinaloa, Mexico. Journal of Public Health of Mexico, 4 46 28693 2004

23. A. Rivas, I. Cerrillo, A. Granada, M. Mariscal, F. Olea, Pesticide exposure of two age groups of women and its relationship with their diet. Science of the Total Environment 2007,382:14-21.

24. SIVIGILA, National Institute of Health, Public Health Surveillance and Control Department, Environmental Risk Factors Group. Report of pesticide poisonings, 2010

25. J. Thundiyil, J. Stober, N. Besbelli, J. Pronczuk, Pesticide. Acute, a. poisoning, classification. proposed, tool, Bulletin of the World Health Organization 2008 86(3):205-9.

26. M. Varona, S. Diaz, Murcia. A. Lancheros, G. Henao, A. Idrovo,

Organochlorine pesticide exposure among agricultural workers from Colombian regions with illegal crops: an exploration to a hidden and dangerous world. International Journal of Environmental Health Research 2010,20(6):407-14.

27. M. Varona, G. Henao, S. Diaz, A. Lancheros, A. Murcia, N. Rodriguez, et al. Evaluation of the effects of glyphosate and other pesticides on human health in areas covered by the program to eradicate illicit crops. Biomedicine 2009 29 (3):456-75.

28. M. Varona, G. Henao, A. Lancheros, A. Murcia, S. Diaz, R. Morato, et al. Exposure factors organophosphorus pesticides and carbamates in the department of Putumayo. Biomedicine 2007 27 4009 .

29. L. Vorhaus, A. Kark, Serum Cholinesterase in health and disease.Am J 1953 1953; 14 70719 .

30. S. Waliszewski, M. Bermudez, R. Infanzon, C. Silva, O. Carvajal, P. Trujillo, S. Gomez, et al. Persistent organochlorine pesticide levels in breast adipose tissue in women with malignant and benign breast tumors. Bull. Environ. Contam. Toxicol, 75 7529 2005

Chapter 9

GREEN ASPECTS OF TECHNIQUES FOR THE DETERMINATION OF CURRENTLY USED PESTICIDES IN ENVIRONMENTAL SAMPLES

Jolanta Stocka, Maciej Tankiewicz, Marek Biziuk, and Jacek Namieśnik

Department of Analytical Chemistry, Chemical Faculty, Gdansk University of Technology, Narutowicza Street 11/12, Gdansk 80-233, Poland

ABSTRACT

Pesticides are among the most dangerous environmental pollutants because of their stability, mobility and long-term effects on living organisms. Their presence in the environment is a particular danger. It is therefore crucial to monitor pesticide residues using all available analytical methods. The analysis of environmental samples for the presence of pesticides is very difficult: the processes involved in sample preparation are labor-intensive and time-consuming. To date, it has been standard practice to use large quantities of organic solvents in the sample preparation process; but as these solvents are themselves hazardous, solvent-less and solvent-minimized techniques are becoming popular. The application of Green Chemistry principles to sample preparation is primarily leading to the miniaturization of procedures and the use of solvent-less techniques, and these are discussed in the paper.

INTRODUCTION

Pesticides are a numerous and diverse group of chemical compounds. They make it possible to control the quantities and quality of farm products and food, and they also help to limit diseases in humans transmitted by insects and rodents. They are very widely used not only in agriculture but also in public health, domestic and urban areas, for example as: insect repellents for personal use; rat and other rodent poisons; flea and tick sprays, powders, and pet collars; kitchen, laundry, and bath disinfectants and sanitizers; products that kill mold and mildew; some lawn and garden products, such as weed killers; some swimming pool chemicals [1].

Despite their many merits, pesticides are considered to be some of the most dangerous environmental contaminants because of their ability to accumulate, as well as their mobility and long-term effects on living organisms. The presence of pesticides in the environment is particularly hazardous and their fate and function are still largely unknown. They may cause humans and other living organisms to become more susceptible to diseases [2].

They can also participate in various physical, chemical and biological reactions, as a result of which even more toxic substances may be produced; by accumulating in living organisms, these can lead to irreversible, deleterious changes. The non-rational application of pesticides also adversely affects the environment and humans, increasing susceptibility to diseases and poisoning. Pesticides are a global risk because they move with the wind, rain and sea currents from other regions to places where they have never been used before.

CURRENTLY USED PESTICIDES

The range of applications of pesticides is continually expanding, hence their consumption is ever increasing and more of them are infiltrating into the environment. In 2009 sales of pesticides in Poland reached 49,760.8 tons, according to figures from the Ministry of Agriculture and Rural Development [3].

It is estimated that EU countries consume more than 300,000 tons of pesticides per annum on crop protection alone. The world market for pesticides is estimated at $33.59 billion, of which the Unites States represents the largest part, in terms of dollars (33%) and pounds of active ingredients (22%) [4]. Table 1 presents the World and U.S. amount of pesticides used in 2006 and 2007 [1].

Table 1. The World and U.S. amount of pesticides used in 2006 and 2007 (in millions of pounds).

Type of pesticide	World market	US market	US percentage of world market [%]
2006			
herbicides	2018	498	25
insecticides	955	99	10
fungicides	519	73	14
other	1705	457	27
total	5197	1127	22
2007			

herbicides	2096	531	25
insecticides	892	93	10
fungicides	518	70	14
other	1705	439	26
total	5211	1133	22

Currently, more than 800 pesticide active ingredients are present in a wide range of commercial products. These substances belong to more than 100 substance classes. Benzoylureas, carbamates, organophosphorous compounds, pyrethroids, sulfonylureas, or triazines are the most important groups. The chemical and physical properties of pesticides may differ considerably. There are several acidic pesticides; others are neutral or basic. Some compounds contain halogens, others phosphorous, sulfur, or nitrogen. These heteroatoms may have relevance for the detection of pesticides. A number of compounds are very volatile, but several do not evaporate at all. This diversity causes serious problems in the development of a "universal" residue analytical method, which should have the widest scope possible.

The choice of methodology for determining pesticides depends in large measure on the sample matrix and the structure and properties of the target analytes. In view of the numerous legal regulations laying down highest permissible levels of pesticides in various matrices, sensitive and selective analytical techniques are used, appropriate to the low concentrations at which the target analytes occur in them. In addition, each stage in the analytical procedure, as well as this process in its entirety should be validated [5,6].

Traditional methods for the determination of these pollutants are known and described by the EPA—United States Environmental Protection Agency [7] and standards, also Polish [8–12], but often they do not meet expectations, mainly due to the large time and effort required, the need of large amounts of organic solvents and, generally, hazardous and multi-step processes of isolation and enrichment of analytes which could be a source of further contamination and error. Moreover, there is a lack of research devoted to the issue of the new methodologies for the determination of currently used pesticides from different chemical groups. This is mainly due to the fact that these xenobiotics are present in environmental samples at very low concentration levels and the often complex matrix composition, which mandates the use of highly sensitive and selective instrumental techniques, preceded by the isolation and enrichment of analytes. Although reports appear on new analytical procedures for the determination of pesticides, they concern individual chemicals rather than classes of pesticides.

GREEN ANALYTICAL CHEMISTRY

Due to scientific and public concern about the environment pollution, environmentally-friendly practices have been introduced in different areas of society and research. Green Chemistry is the use of chemistry techniques and methodologies that reduce or eliminate the use or generation of feedstock's, products, by-products, solvents, reagents, *etc.*, that are hazardous to human health or the environment [13]. The adverse environmental impact of analytical methodologies has been reduced mainly in three different ways: reduction of the amount of solvents required in sample pre-treatment; reduction in the amount and the toxicity of solvents and reagents employed in the measurement step, especially by automatization and miniaturization; development of alternative direct analytical methodologies not requiring solvents or reagents [14].

The main different steps of the analytical process (sample collection, sample preparation, separation, detection, and data evaluation) make different contributions to environmental pollution and there are different potential ways to make them greener and closer to Green Chemistry principles. The trends in new sample-preparation methods that minimize the amount of reagents and organic solvents contribute to improving the environmentally-friendly features of those methodologies that cannot be applied directly to samples with no sample treatment. Nevertheless, when the use of reagents is unavoidable and their substitution is not feasible, the best alternative is minimization of their consumption. At this point, automation of analytical procedures by means of flow-injection (FI) methodologies plays an important role in the Green Chemistry context [15]. Miniaturization is one way to avoid side effects of analytical methods. In this respect, combination of modern analytical techniques with breakthroughs in microelectronics and miniaturization allows development of powerful analytical devices for effective control of processes and pollution. Combining miniaturization in analytical systems with advances in chemometrics is very important. Of course, development and improvement of new components for instrumentation is critical in Green Analytical Chemistry. Using examples, we have illustrated the power and the versatility of modern analytical systems and their potential for minimizing the consumption of hazardous substances and the amounts of waste generated during assays.

GREEN ASPECTS IN ANALYTICAL METHODOLOGIES

Nowadays, the trend is to develop analytical methods enabling a broad spectrum of analytes to be determined in a single analytical run (multiresidue methods—MRM); but the problem here is that the compounds to be determined simultaneously, often present at low concentrations, have different physicochemical properties depending on their chemical structure [12]. Figure

1 presents the steps of a multiresidue method. Such a methodology, apart from being able to determine a large number of compounds in one run, should:

- Ensure maximum removal of interferents from extracts,
- Give large recoveries of target compounds, high sensitivity and good precision,
- Be environmentally-friendly, *i.e.*, require the smallest possible quantities of samples and chemical reagents, especially organic solvents,
- Be cheap, quick and easy to carry out.

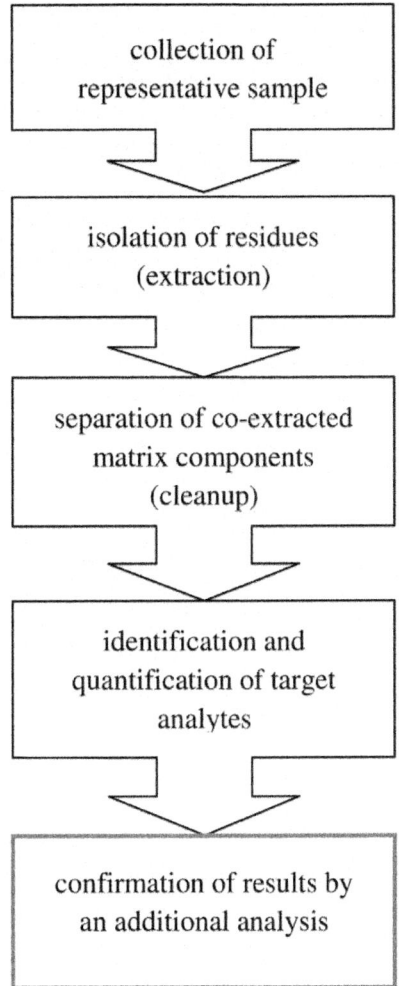

Figure 1. Steps in the determination of pesticide residues in samples characterized by complex composition of the matrix.

Generally the analytical procedure consists of numerous stages, the most important of which is the collection of a sample and its preparation for analysis. This stage is a complicated process, and its operations can be both a cause of analyte loss and a source of additional contamination. All errors at this stage will affect the final result of determination. A further difficulty is the fact that the collection and preparation of a sample takes up to *ca.* two thirds of the time required to perform the complete analysis. New techniques have been developed which eliminate many of these inconveniences and also increase the precision, throughput, reproducibility, and cost-effectiveness. In many cases the capability for smaller initial sample sizes, even for trace analyses, is also essential [16].

From an analytical point of view, environmental and food samples are highly diverse and complex: the factors affecting the nature of the sample are the sampling site, the type of matrix, the presence of interferents and the low concentration of target analytes. Whether or not the analysis yields reliable information about the sample content depends to a large extent on the proper sample preparation. The quality of sampling and sample pretreatment largely determine the success of an analysis from complex matrices. Ideally, sample preparation should be as simple as possible, because it not only reduces the time required, but also decreases the possibility of introducing contaminants. Figure 2 presents trends in the development of techniques of sample preparation.

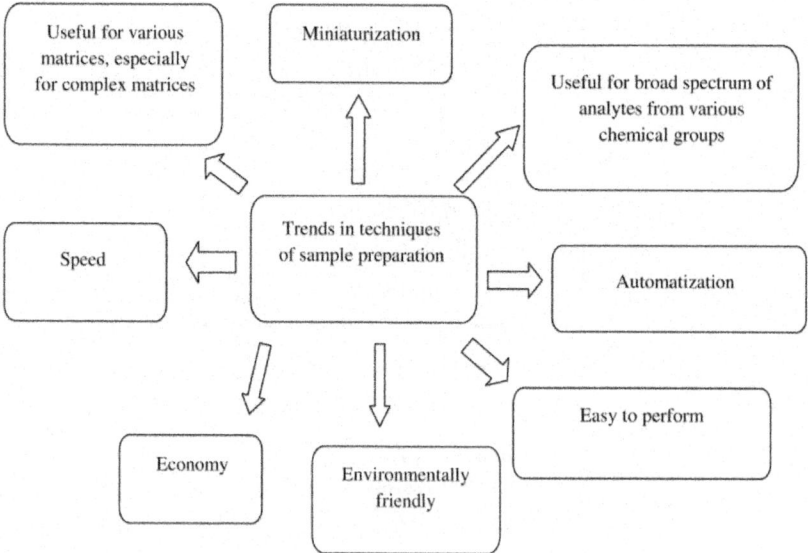

Figure 2. Trends in the development procedures for determination of trace constituents in samples characterized by complex composition of the matrix.

One of the oldest extraction techniques, and at the same time one of the most common in routine sample preparation, is liquid-liquid extraction (LLE). The solvents in LLE are usually dichloromethane [17–20], mixtures of petroleum ether and dichloromethane [21] or hexane and dichloromethane [18]. LLE is recognized as an attractive technique for screening tests of unknown pesticides [22,23] not only because of its simplicity, efficiency, minimal operator training, but also because of its wide acceptance in many standard methods. However, this technique has a number of drawbacks: it requires relatively large quantities of toxic solvents and multistage operation, there is a risk of emulsion forming during agitation, and there is the problem of disposal of the post-extraction solvents. To achieve the desired preconcentration coefficient, the excess solvent usually has to be evaporated. Also extract cleanup is often necessary. To minimize these disadvantages, numerous improvements have been made to this method, most of which have involved miniaturizing the process to reduce the amounts of solvents consumed.

Microextraction techniques, such as: liquid-liquid microextraction, dispersive liquid-liquid microextraction, single drop microextraction, solid-phase micro-extraction (SPME), stir-bar sorptive extraction (SBSE), liquid-phase micro-extraction (LPME), and on-line solid-phase extraction (SPE), have several advantages over the traditional approaches of liquid–liquid extraction (LLE) and conventional SPE [24–28].

The main advantages are minimal consumption of harmful solvents, and typically, the high enrichment factor. The improved sensitivity makes it possible to electron the amount of sample needed in the analysis. All these techniques are readily combined with GC, either off-line, at-line or sometimes even on-line [29]. Off-line procedures are good alternative when the number of samples is small, because there is usually no need for an automated method and the time-consuming development of such a method. Conventional methods will suffice. Setting up an automated method, either at-line or on-line, becomes more worthwhile when the number of analyzed samples increases. Automation typically improves the quality of the data, increases the sample throughput, decreases costs and improves the productivity of personnel and instruments. On-line systems are beneficial when the analytes are labile, the amount of sample is limited, or very high sensitivity is required. The selection of an extraction technique is made on the basis of several factors. Naturally, the sample preparation must be tailored to the final analysis. The sample matrix and the type and amount of analytes in the sample are of primary importance. Also crucial are speed of extraction, complexity of the instrumentation, simplicity and flexibility of the method development, and ruggedness of the method. Moreover, a method good for target-compound analysis may not be good for comprehensive chemical profiling of samples. Selectivity of the sample preparation is often a key factor for target-compound analysis while an exhaustive extraction is the better choice for profiling.

In practice, these novel developed techniques can be performed by following two general methodologies. These are solvent microextraction, where the extraction is performed by using a small amount (drop) of water-immiscible solvent suspended in a sample (MLLE, SDME, HS-SDME, CFME) and extraction via a membrane (HF(2)ME, SLME, MMLLE, MASE) which can be a selective barrier between two phases (see Table 2). The dispersion of very fine droplets of organic solvents into the aqueous phase in a ternary solvent component system (liquid samples) is another new option and called dispersive liquid-liquid microextraction (DLLME) [30,31]. Table 2 presents the most commonly used novel techniques for sample preparation in pesticides analysis.

Table 2. The most commonly used novel (green) techniques for sample preparation in pesticide analysis.

Technique of Sample Preparation	Volume of Organic Solvent	Description	Literature
MLLE (micro liquid-liquid extraction)	about 1 mL per 1 L of sample	It is possible to decrease the consumption of organic solvents by miniaturization and proper design of extraction vessel. The most commonly used solvents for microextraction are dichloromethane, toluene and methyl-tert-butyl ether.	[32,33]
SDME (single drop microextraction)	0.9–1.5 µL	The extraction phase is a drop of organic solvent (e.g., n-hexane, toluene, butyl acetate) suspended at the tip of microsyringe, so it is practically a solvent-free method. It can be carried out in two different ways by direct immersion (DI) or from the headspace (HS). Analyte isolation and preconcentration take place in a single step. The extraction process is assisted by mixing. When the extraction is complete, the microdroplet is directly injected into a gas chromatograph (GC) or high-performance liquid chromatograph (HPLC) for further analysis. The universality of SDME makes it widely applicable to the analysis of pesticides in samples with a complex composition containing target analytes in trace amounts.	[34–37]

CFME (continuousflow microextraction)	1–5 μL	This technique is similar to SDME. The drop of extraction solvent is injected by microsyringe into a glass chamber (0.5 mL) and held at the outlet tip of a polyetheretherketone (PEEK) connecting tube. The sample solution flows past the tube and through the glass extraction unit to waste. Extraction takes place continuously between the organic drop and the flowing sample solution. Because the drop of solvent makes full contact with the sample solution, the technique achieves higher concentration factor than static SDME.	[30]
DLLME (dispersive liquid-liquid microextraction)	disperser solvent 0.5–2 mL; extraction solvent 10–50 μL	The mixture of extraction solvent (e.g., chlorobenzene, carbon tetrachloride, tetrachloroethylene, carbon disulfide) and disperser solvent (e.g., acetone or methanol) is rapidly injected into an aqueous sample, resulting in the formation of a cloudy solution. The DLLME procedure is very convenient to operate and extraction could be completed in a few seconds. DLLME has advantages of simplicity of operation, rapidity and low cost. DLLME can be coupled with GC and HPLC. The non-selective characteristic of the extraction solvents can be sometimes a disadvantage. Recently He *et al.* used as extraction solvent ionic liquid 1-octyl-3- methylimidazolium hexafluorophosphate ([C_8MIM][PF_6]) for the determination of organophosphorus pesticides in water sample. Ionic liquids belong to non-molecular solvents with unique properties such as negligible vapor pressure associated to a high thermal stability. Hydrophobic ionic liquids incorporating the imidazolium cation and hexafluorophosphate anion have higher density than water. Compared with commonly used solvents they are more compatible with reversed-phase HPLC due to the non-harmfulness to column.	[38–45]

HF(2)ME (hollow fiber-protected two-phase solvent microextraction)	2–3 µL	The method is straightforward, quick, inexpensive and eliminates necessity of extract cleanup prior to final determination. Toluene, hexane or 1-octanol are usually used for the extraction of pesticides. It is based on the partition of analytes between the aqueous solution and the small quantity of organic solvent in a microporous tube (the rod configuration). The hollow fiber can be also in the U-shape configuration. The process is assisted by stirring. About 1–1.5 µL of extract is taken for further analysis using appropriate chromatographic techniques. For more complex matrices and moderately polar pesticides. Basheer *et al.* developed binary solvent based on HF(2)ME with GC-MS. The mixture (1:1) toluene: hexane was used as solvent. The limits of detection (LODs) were in the range of 0.3–11.4 ng L^{-1} and relative standard deviations (RSD) were 9–13%. This technique gave higher analytes enrichment, especially when applied to complex matrices (wastewater).	[40,46]
LPME-SFO (liquid-phase microextraction based on the solidification of a floating organic drop)	10 µL	The small volume of an extraction solvent (usually 1-undecanol) is floated on the surface of aqueous solution. The process is assisted by stirring. After the extraction, the floated extractant droplet can be collected easily by solidifying it at low temperature. The solidified organic solvent can be melted quickly at room temperature, which is then determined by either chromatographic or spectrometric methods. The technique is cheap, quick and sensitive, but the rate of extraction is slightly slow.	[47,48]

MMLLE (microporous membrane liquid-liquid extraction)	0.2 mL	Advantages of this technique compared to LLE are small sample volumes, the lack of emulsion formation, the clean extracts obtained and it can be coupled online to gas chromatography. The flat-sheet membrane extraction unit consisted of two blocks, one made of poly(tetrafluoroethylene) (PTFE) and the other of poly(etheretherketone) (PEEK). The membrane constitutes a barrier between two phases: acceptor (usually toluene) and the aqueous donor solution (sample). The donor solution is pumped to the donor channel of the membrane block, while the acceptor is stagnant during the extraction period.	[49,50]
LLSME (liquid-liquidsolid microextraction)	6–100 μL	This technique combines the advantages of solid-phase microextraction and liquid-phase microextraction. The molecularly imprinted polymer (MIP)—coated silica fiber is protected with a length of porous polypropylene hollow fiber membrane which is filled with water-immiscible organic phase (usually toluene). This technique is a three-phase microextraction approach. It is fast, selective and sensitive method for trace analysis of pesticides in complex aqueous samples.	[51,52]

These microextraction techniques eliminate the disadvantages of traditionally used extraction methods such as time-consuming operation and need for specialized apparatus. They are inexpensive and offer considerable freedom in selecting appropriate solvents for the extraction of different analytes. Moreover, they minimize exposure to toxic organic solvents.

Recently also increasing interest is observed in ad/absorption-based methods using beds of solid enrichment sorbents, which have gradually replaced conventional LLE for sample pretreatment and have gained wide acceptance because of their simplicity and economy in terms of time and solvent needs. Sorptive extraction techniques mainly include solid-phase extraction (SPE), solid-phase microextraction (SPME) and stir bar sorptive extraction (SBSE) (Table 3).

Table 3. The most commonly used novel techniques for sample preparation in pesticide analysis (minimization of toxic reagents).

Technique of Sample Preparation	Volume of Organic Solvent	Description	Literature
SPE (solid-phase extraction)	<15 mL	The advantages of this method are: requires a lower volume of solvent than traditional LLE, involves simple manipulations which are not time consuming, the SPE cartridges can be used for short-term storage of the species and provides high enhancement factors proportional to the volume of water passed through the SPE cartridge. Conventional sorbents such as C_{18} silica, graphitized carbon black and macroporous polystyrene divinylbenzene (PS-DVB), show low retention for polar compounds. In order to improve the extraction efficiency for polar compounds, the development of new adsorbents and modification of the adsorbents by introducing the polar groups become a major research direction. Nanomaterials are one kind of novel adsorbents. Carbon nanotubes (CNTs), including single-walled carbon nanotubes (SWCNTs) and multi-walled carbon nanotubes (MWCNTs), are a kind of carbonaceous nanomaterial and have received significant attention in many fields. In recent years, molecular imprinting polymer (MIP) technology with high selectivity evolves rapidly. MIP technology is now well established for the preparation of tailor-made polymers with cavities capable to extract or clean-up of OPPs.	[53–61]

SPME (solid phase microextraction)	solvent-free extraction	This technique uses polymer-coated fibers to extract analytes from aqueous or gaseous samples. After extraction, the analytes are either desorbed thermally by exposing the fiber in the injection port of a GC or chemically desorbed and analyzed by LC. SPME does not require the use of organic solvents. It is quick, universal, sensitive and convenient for use in the field and is simply applied in sample preparation. However the fiber is comparatively expensive, fragile and has limited lifetime. The materials used for coating fibers include: polydimethylsiloxane (PDMS), polyacrylate (PA), and also mixtures of: polydimethylsiloxane and polydivinylbenzene (PDMS-DVB), carbowax and polydivinylbenzene (CW-DVB), carbowax and molecularly imprinted resin (CW-TP). Depending on where the fiber is situated in relation to the sample, SPME can be carried out in two different ways by direct immersion (DI) or from the headspace (HS). The advantage of this method is that the limited capacity of the adsorbent precludes column overloading.	[62–68]

| SBSE (stir bar sorptive extraction) | solvent-free extraction | This techniques uses a 1.5 cm long glass magnetic stirrer coated with a thick layer of polydimethylsiloxane (PDMS) where sorption usually takes place. Its sorption capacity is a hundred times greater in comparison with sorption capacity of SPME fibers. Its main advantage is high sensitivity and a wide application range that includes volatile aromatics, halogenated solvents, polycyclic aromatic hydrocarbons, polychlorinated biphenyls, pesticides or organotion compounds. Because of the non-polar character of PDMS, the SBSE cannot be used to extract strong polar compounds unless derivatization was utilized. | [69–71] |

Approaches are being sought to develop pesticide determination techniques that are quick, easy, cheap, effective, rugged and safe. QuEChERS is a highly effective sample preparation technique for pesticide residue analysis. It is a combination of liquid-liquid extraction (LLE) and solid-phase extraction (SPE) and was developed by Anastassiades et al. [72]. The original Quechers method is based on a number of stages (see Figure 3). Samples are milled in frozen state (dry ice is added) to get the best recovery. Extraction is done in acetonitrile buffered at pH 5–5.5. After centrifuged, the organic phase is cleaned-up by dispersive SPE using primary secondary amine—PSA (and graphitized carbon black—GCB as necessary). Additional $MgSO_4$ is added to remove any residual water. The PSA treated extract is acidified with formic acid to improve the stability of base-sensitive pesticides. The extract is ready for GC and LC analysis. For samples with low water content (<80%), water is added before the initial extraction to get a total of ca. 10 mL water. Quality control is performed by adding ISTD to the acetonitrile extraction step.

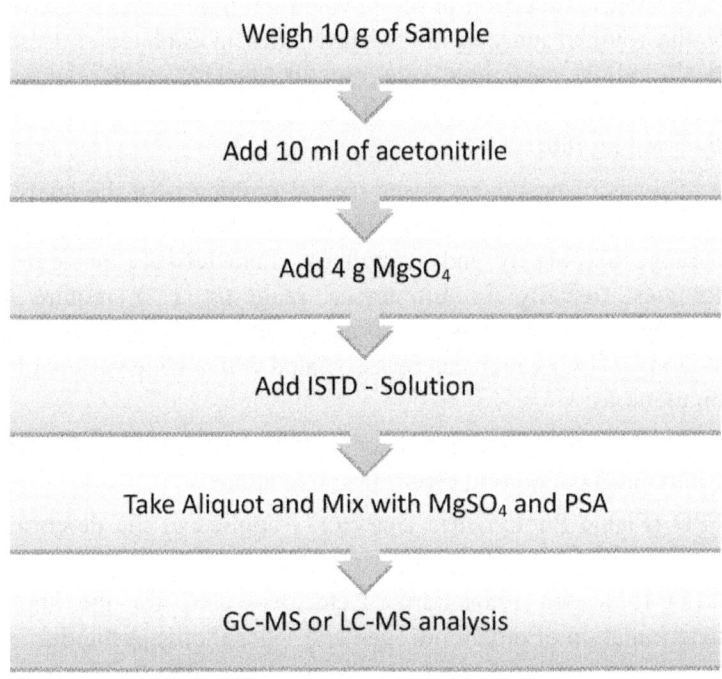

Figure 3. Steps in the QuEChERS procedure of sample preparation for the determination of pesticide residue in fruit and vegetables.

The consumption of sample and toxic solvents with the QuEChERS method is minimal. By applying QuEChERS to the determination of pesticides in fruit and vegetables, matrix effects are eliminated and high recoveries of target analytes are possible. The method can be modified depending on the type of sample and the target analytes. To improve the extraction of polar organophosphorus pesticides, the method is modified by the addition of acetic acid. When samples of citrus fruit are under investigation, protective wax coatings can be removed by freezing the samples for at least one hour. For samples, with a high content of carotenoides or chlorophyll, cleanup with PSA is not satisfying and there is a need to use GCB which is best in handling and effect. QuEChERS approach takes advantages of the wide analytical scope and high degree of selectivity and sensitivity provided by gas and liquid chromatography (GC and LC) coupled to mass spectrometry (MS) for detection. QuEChERS is a multi-residue method with fast sample preparation and low solvent consumption [73–79].

Nguyen *et al.* proposed a multiresidue method based on the QuEChERS sample preparation method and gas chromatography with the electron impact mass spectrometric detection in the selected ion monitoring mode (GC-SIM-MS) for the routine analysis of 107 pesticides in cabbage and radish. The recoveries for all the pesticides were from 80% to 115% with relative standard deviation lower than 15%. The limits of quantifications were in the range 0.002–0.05 mg/kg [80].

The analysis of pesticides poses special problems for the analysts, since the pesticides belong to different groups of chemical substances, having a broad range of polarity and acidic/base characteristic. Pesticide analysis methodologies (usually in ultratraces range-µg L^{-1}) require typically analytical separative techniques such as gas chromatography (GC) or liquid chromatography (LC) which can be associated with a wide variety of selective detection methods:

- ECD (Electron Capture Detector)—highly sensitive in relation to compounds containing electronegative atoms,
- FPD (Flame Photometric Detector)—applied in the determination of organophosphorus compounds,
- NPD (Nitrogen Phosphorus Detector)—used for the simultaneous determination of organonitrogen and -phosphorus pesticides.

Most pesticides are volatile and thermally stable, and therefore are amenable to GC. In contrast to GC, procedures based on application of LC technique have the advantage of being suitable for thermally unstable and polar/ionic pesticides, as these compounds require derivatization prior to GC analysis. The selective detectors are the most common used in routine residue analysis. Unfortunately, these do not allow confirmation of the analysis results without ambiguity [81]. The detection by mass spectrometry (MS) employing quadrupole, ion trap and/or time-of-flight analyzers offers simultaneously the confirmation and the quantification of numerous pesticides [82]. It has become very popular in laboratories performing monitoring of pesticide residues analysis [83]. As powerful as MS is, the low-resolution, scanning MS system has limits in data collection rate, avoidance of interferences, and spectral information provided for identification purposes. Currently, low-resolution (unit mass) MS detectors employing either single quadrupole or ion trap analyzers are most routinely used in applications [84–86]. Furthermore, innovations in chromatographic particle chemistry (from 5 to 3.5 or 1.8 µm packing in LC, as well as new bonding chemistries) have improved the separation of pesticides [87].

The confirmation of analysis results can also be performed by another independent method. The conditions of the process can be altered, by changing the temperature program or by using a different chromatographic column. It is crucial to obtain confirmation by another method as identification based solely on retention times is insufficient. Tandem mass spectrometry (MS/MS) improves sensitivity and selectivity of analytical methods. In this technique, ions that were separated in the first analyzer are again fragmented and the derivative ions analyzed in the second one. The chromatogram background is reduced, as a result of which the signal value is enhanced with respect to noise and the LOD (limit of detection) of the target analytes is lowered [88,89]. Better chromatographic peak resolution and a smaller influence of the matrix on the final result can also be achieved using two-dimensional (2D) gas chromatography (GCxGC). This uses two columns: the partially separated constituents from the first column are further separated in the second one by a different mechanism. The advantage of this method is that the separation mechanisms in the two columns are independent of each other, so that constituents that were co-eluted from the first column can be separated. Moreover, GCxGC can simplify the preparation of samples for determination of the presence of pesticides. This method is widely used because of its high resolving power, greater sensitivity and ordered nature of the chromatograms.

Fast GC is equally frequently used to shorten the time of analysis, which allows increasing sample throughput, and to obtain better peak resolution. Consequently the laboratory operating costs per sample can be reduced significantly [90,91]. Compared to classical GC, it requires shorter capillary columns with a smaller diameter and thinner film of stationary phase *ca*. 0.1 μm in thickness, as well as a faster flow rate and higher pressure of the carrier gas. These parameters yield determination results in higher precision [88,92,93].

Alder *et al.* applied multiresidue GC-MS method with electron impact ionization (EI) and the combination of LC with tandem mass spectrometers (LC-MS/MS) with electrospray ionization (ESI) for determination of 500 high priority pesticides. Only for one substance class, the organochlorine pesticides, GC-MS achieves better performance. For all other classes of pesticides a wider scope and better sensitivity were observed for LC-MS/MS. Table 4 lists the number of pesticides in each class that could not be detected by either of the two methods (GC-MS and LC-MS/MS).

Table 4. Pesticide classes and number of pesticides in each class that cannot be detected by GC-MS or LC-MS/MS.

Chemical Class	Number of Pesticides in That Class	Not Detected by GC-MS	Not Detected by LC-MS/MS
organophosphorus	81	0	1
carbamate	43	17	1

organochlorine	40	0	33
sulfonylurea	26	26	0
triazole	24	1	0
triazine	23	6	0
urea	22	16	0
pyrethroid	19	0	2
aryloxyphenoxy-propionate	12	4	0
aryloxyalkanoic acid	10	9	0
other	200	56	12
Total number	500	135	49

Based on the data from Table 2 it can be concluded, that more pesticides and their metabolites can be determined by using LC and ESI than by GC–MS. It is well known that sulfonyl or benzoyl ureas and many carbamates or triazines can be better or exclusively detected by LC–MS/MS techniques. Furthermore, a wider scope of LC–MS/MS was found for most of the other chemical classes too, for example, the organophosphorus pesticides. Only 49 compounds out of 500 exhibited no response, if LC–MS/MS in combination with positive and negative ESI was used. On the other hand, 135 pesticides/metabolites could not be analyzed by GC/MS using EI ionization, most often because of incompatibility with evaporation of the intact molecule in the GC injector. Both of these instruments have special merits, but neither of them can detect the full range of all pesticides. However, if the selection of the most appropriate techniques is focused on the enforcement of maximum residue levels, simultaneous identification, and quantification of a very large number of target analytes will be more important than the detection, identification, and quantification of non-regulated (non-target) pesticides and/or metabolites [94].

Research is continuing into the improvement of existing analytical methods and the development of new ones capable of supplying reliable results for a wide range of analytes in a short time and will be more economical and environmentally friendly.

CONCLUSIONS

Due to scientific and public concern about environment pollution, environmentally-friendly practices have been introduced in different areas of society and research. Investigation of green analytical methodologies encompasses a number of strategies to minimize or to eliminate the use of toxic substances and the generation of waste. The main focus has been on the development of new routes to minimize the amounts of side products and to replace toxic solvents. Progress in analytical methodologies has contributed to the development of new, greener options.

REFERENCES

1. *About Pesticides*; EPA-United States Environmental Protection Agency: Honolulu, HI, USA, 2011. Available online: http://www.epa.gov/pesticides/about/index.htm accessed on 7 August 2011.

2. Tankiewicz, M.; Fenik, J.; Biziuk, M. Determination of organophosphorus and organonitrogen pesticides in water samples. *Trends Anal. Chem* 2010, *29*, 1050–1063.

3. *Biuletyn Informacji Publicznej Ministerstwa Rolnictwa i Rozwoju Wsi, Tabela 1-Agregacja według rodzajów rodków ochrony ro lin*, Available online: http://www.bip.minrol.gov.pl accessed on 7 August 2011.

4. Ware, G.W.; Whitacre, D.M. *The Pesticide Book*, 6th ed; MeisterPro Information Resources: Willoughby, OH, USA, 2004.

5. Beyer, A.; Biziuk, M. Przegląd metod oznaczania pozostałości pestycydów i polichlorowanych bifenyli w próbkach żywności. *Ecol. Chem. Eng* 2007, *14*, 35–58.

6. Namieśnik, J.; Górecki, T. Preparation of environmental samples for the determination of trace constituents. *Pol. J. Environ. Stud* 2001, *10*, 77–84.

7. EPA-United States Environmental Protection Agency, *Standardized Analytical Methods for Environmental Restoration Following Homeland Security Events—Revision 5.0*; EPA-United States Environmental Protection Agency: Cincinnati, OH, USA, 2009.

8. Jakość wody. Oznaczanie wybranych chloroorganicznych insektycydów, polichlorowanych bifenyli i chlorobenzenów.*Metoda chromatografii gazowej po ekstrakcji ciecz-ciecz*, PN-EN ISO 6468:2002. Available online: http://enormy.pl/?m=doc&nid=PN-13.060.50-00246 access on 9 November 2011.

9. Jakość wody. Oznaczanie wybranych związków azoto- i fosforoorganicznych. *Metody chromatografii gazowej*, PN-EN ISO 10695:2004. Available online: http://enormy.pl/?m=doc&v=met&nid=PN-13.060.50-00228&key= access on 9 November 2011.

10. Jakość wody. *Oznaczanie parationu, parationu metylowego i kilku innych zwi zków fosforoorganicznych w wodzie metod chromatografii gazowej*

po ekstrakcji dichlorometanem, PN-EN 12918:2004. Available online: http://enormy.pl/?m=doc&v=met&nid=PN-13.060.50-00181&key= access on 9 November 2011.

11. Jakość wody. Oznaczanie wybranych środków ochrony roślin. *Metoda z zastosowaniem wysokosprawnej chromatografii cieczowej z detekcj UV po ekstrakcji ciało stałe-ciecz*, PN-EN ISO 11369:2002.Available online: http://enormy.pl/?m=doc&nid=PN-13.060.50-00248 access on 9 November 2011.

12. Greulich, K.; Alder, L. Fast multiresidue screening of 300 pesticides in water for human consumption by LC-MS/MS.*Anal. Bioanal. Chem* 2008, *391*, 183–197.

13. Tobiszewski, M.; Mechlinska, A.; Zygmunt, B.; Namiesnik, J. Green analytical chemistry in sample preparation for determination of trace organic pollutants. *Trends Anal. Chem* 2009, *28*, 943–951.

14. Armenta, S.; Garrigues, S.; Guardia, M. Green analytical chemistry. *Trends Anal. Chem* 2008, *27*, 497–511.

15. Molina-Diaz, A.; Garcia-Reyes, J.F.; Lopez, B.G. Solid-phase spectroscopy from the point of view of green analytical chemistry. *Trends Anal. Chem* 2010, *29*, 654–666.

16. Escuderos-Morenas, M.L.; Santos-Delgado, M.J.; Rubio-Barroso, S.; Polo-Diez, L.M. Direct determination of monolinuron, linuron and chlorbromuron residues in potato samples by gas chromatography with q nitrogen–phosphorus detection. *J. Chromatogr. A* 2003, *1011*, 143–153.

17. Sankararamakrishnan, N.; Sharma, A.K.; Sanghi, R. Organochlorine and organophosphorous pesticide residues in ground water and surface waters of Kanpur, Uttar Pradesh, India. *Environ. Int* 2005, *31*, 113–120.

18. Pirard, C.; Widart, J.; Nguyen, B.K.; Deleuze, C.; Heudt, L.; Haubruge, E.; De Pauw, E.; Focant, J.F. Development and validation of a multi-residue method for pesticide determination in honey using on-column liquid-liquid extraction and liquid chromatography-tandem mass spectrometry. *J. Chromatogr. A* 2007, *1152*, 116–123.

19. Tse, H.; Comba, M.; Alaee, M. Method for the determination of organophosphate insecticides in water, sediment and biota. *Chemosphere* 2004, *54*, 41–47.

20. Sabik, H.; Jeannot, R. Determination of organonitrogen pesticides in large volumes of surface water by liquid-liquid and solid-phase extraction using gas chromatography with nitrogen-phosphorus detection and liquid chromatography with atmospheric pressure chemical ionization mass spectrometry. *J. Chromatogr. A* 1998, *818*, 197–207.

21. Tahboub, Y.R.; Zaater, M.F.; Al-Talla, Z.A. Determination of the limits of identification and quantitation of selected organochlorine and organophosphorous pesticide residues in surface water by full-scan gas chromatography/mass spectrometry. *J. Chromatogr. A* 2005, *1098*, 150–155.

22. Fatoki, O.S.; Awofolu, R.O. Methods for selective determination of persistent organochlorine pesticide residues in water and sediments by capillary gas chromatography and electron-capture detection. *J. Chromatogr. A* 2003, *983*, 225–236.

23. Mahara, B.M.; Borossay, J.; Torkos, K. Liquid-liquid extraction for sample preparation prior to gas chromatography and gas chromatography-mass spectrometry determination of herbicide and pesticide compounds. *Microchem. J* 1998,*8*, 31–38.

24. Ridgway, K.; Lalljie, S.P.D.; Smith, R.M. Sample preparation techniques for the determination of trace residues and contaminants in foods. *J. Chromatogr. A* 2007, *1153*, 36–53.

25. David, F.; Sandra, P. Stir bar sorptive extraction for trace analysis. *J. Chromatogr. A* 2007, *1152*, 54–69.

26. Giordano, A.; Fernández-Franzón, M.; Ruiz, M.J.; Font, G.; Picó, Y. Pesticide residue determination in surface waters by stir bar sorptive extraction and liquid chromatography/tandem mass spectrometry. *Anal. Bioanal. Chem* 2009, *393*, 1733–1743.

27. David, F.; Hoeck, E.; Sandra, P. Stir bar sorptive extraction for the determination of pyrethroids in water samples. A comparison between thermal desorption in a dedicated thermal desorber, in a split/splitless inlet and by liquid desorption. *J. Chromatogr. A* 2007, *1157*, 1–9.

28. Quintana, J.B.; Rodriguez, I. Strategies for the microextraction of polar organic contaminants in water samples. *Anal. Bioanal. Chem* 2006, *384*, 1447–1461.

29. Hyötyläinen, T. Principles, developments and applications of on-line coupling of extraction with chromatography. *J. Chromatogr. A* 2007, *1153*, 14–28.

30. Lambropoulou, D.A.; Albanis, T.A. Liquid-phase micro-extraction techniques in pesticide residue analysis. *J. Biochem. Biophys. Methods* 2007, *70*, 195–228.

31. Pinto, M.I.; Sontag, G.; Bernardino, R.J.; Noronha, J.P. Pesticides in water and the performance of the liquid-phase microextraction based techniques. A review. *Microchem. J* 2010, *96*, 225–237.

32. Zhou, Q.; Liu, J.; Cai, Y.; Liu, G.; Jiang, G. Micro-porous membrane liquid-liquid extraction as an enrichment step prior to nonaqueous capillary electrophoresis determination of sulfonylurea herbicides. *Microchem. J* 2003, *74*, 157–163.

33. Zapf, A.; Heyer, R.; Stan, H.J. Rapid micro liquid-liquid extraction method for trace analysis of organic contaminants in drinking water. *J. Chromatogr. A* 1995, *694*, 453–461.

34. Xiao, Q.; Hu, B.; Yu, Ch.; Xia, L.; Jiang, Z. Optimization of a single-drop microextraction procedure for the determination of organophosphorus pesticides in water and fruit juice with gas chromatography-flame photometric detection. *Talanta* 2006, *69*, 848–855.

35. Ahmadi, F.; Assadi, Y.; Milani Hosseini, S.M.R.; Rezaee, M. Determination of organophosphorus pesticides in water samples by single drop microextraction and gas chromatography-flame photometric detector. *J. Chromatogr. A* 2006, *110*, 307–312.

36. Lambropoulou, D.A.; Psillakis, E.; Albanis, T.A.; Kalogerakis, N. Single-drop microextraction for the analysis of organophosphorous insecticides in water. *Anal. Chim. Acta* 2004, *516*, 205–211.

37. Bagheri, H.; Khalilian, F. Immersed solvent microextraction and gas chromatography-mass spectrometric detection of *s*-triazine herbicides in aquatic media. *Anal. Chim. Acta* 2005, *537*, 81–87.

38. *Solvent Microextraction, Theory and Practice*; Kokosa, J.M., Przyjazny, A., Jeannot, M.A., Eds.; Wiley: Horboken, NJ, USA, 2009.

39. Nagaraju, D.; Huang, S.D. Determination of triazine herbicides in aqueous samples by dispersive liquid-liquid microextraction with gas chromatography-ion trap mass spectrometry. *J. Chromatogr. A* 2007, *1161*, 89–97.

40. Rezaee, M.; Yamini, Y.; Faraji, M. Evolution of dispersive liquid-liquid microextraction method. *J. Chromatogr. A* 2009, *1217*, 2342–2357.

41. He, L.; Luo, X.; Xie, H.; Wang, C.; Lu, X.J.K. Ionic liquid-based dispersive liquid-liquid microextraction followed high-performance liquid chromatography for the determination of organophosphorus pesticides in water sample. *Anal. Chim. Acta* 2009, *655*, 52–59.

42. Huddleston, J.G.; Visser, A.E.; Reichert, W.M.; Willauer, H.D.; Broker, G.A.; Rogers, R.D. Characterization and comparison of hydrophilic and hydrophobic room temperature ionic liquids incorporating the imidazolium cation. *Green Chem* 2001, *3*, 156–164.

43. Visser, A.E.; Swatloski, R.P.; Reichert, W.M.; Griffin, S.T.; Rogers,

R.D. Traditional extractants in nontraditional solvents: Groups 1 and 2 extraction by crown ethers in room-temperature ionic liquids. *Ind. Eng. Chem. Res* 2000, *39*, 3596–3604.

44. He, L.J.; Zhang, W.Z.; Zhao, L.; Jiang, X.S. Effect of 1-alkyl-3-methylimidazolium-based ionic liquids as the eluent on the separation of ephedrines by liquid chromatography. *J. Chromatogr. A* 2003, *1007*, 39–45.

45. Xiao, X.H.; Zhao, L.; Liu, X.; Jiang, S.X. Ionic liquids as additives in high performance liquid chromatography. Analysis of amines and the interaction mechanism of ionic liquids. *Anal. Chim. Acta* 2004, *519*, 207–211.

46. Basheer, C.; Alnedhary, A.A.; Rao, B.S.M.; Lee, H.K. Determination of organophosphorous pesticides in wastewater samples using binary-solvent liquid-phase microextraction and solid-phase microextraction: A comparative study.*Anal. Chim. Acta* 2007, *605*, 147–152.

47. Khalili-Zanjani, M.R.; Yaminia, Y.; Yazdanfar, N.; Shariati, S. Extraction and determination of organophosphorus pesticides in water samples by a new liquid phase microextraction-gas chromatography-flame photometric detection.*Anal. Chim. Acta* 2008, *606*, 202–208.

48. Khalili-Zanjani, M.R.; Yaminia, Y.; Shariati, S.; Jönsson, J.Å. A new liquid-phase microextraction method based on solidification of floating organic drop. *Anal. Chim. Acta* 2007, *585*, 286–293.

49. Lüthje, K.; Hyötyläinen, T.; Riekkola, M.L. Comparison of different trapping methods for pressurised hot water extraction. *J. Chromatogr. A* 2004, *1025*, 41–49.

50. Hyötyläinen, T.; Lüthje, K.; Rautiainen-Rämä, M.; Riekkola, M.L. Determination of pesticides in red wines with on-line coupled microporous membrane liquid-liquid extraction-gas chromatography. *J. Chromatogr. A* 2004, *1056*, 267–271.

51. Hu, Y.; Wang, Y.; Hu, Y.; Li, G. Liquid-liquid-solid microextraction based on membrane-protected molecularly imprinted polymer fiber for trace analysis of triazines in complex aqueous. *J. Chromatogr. A* 2009, *1216*, 8304–8311.

52. Hu, X.; Ye, T.; Yu, Y.; Cao, Y.; Guo, C. Novel liquid-liquid-solid microextraction method with molecularly imprinted polymer-coated stainless steel fiber for aqueous sample pretreatment. *J. Chromatogr. A* 2011, *1218*, 3935–3939.

53. Chen, J.; Duan, C.; Guan, Y. Sorptive extraction techniques in sample preparation for organophosphorus pesticides in complex matrices. *J.*

Chromatogr. B 2010, *878*, 1216–1225.

54. Fontanals, N.; Marce, R.M.; Borrull, F. New hydrophilic materials for solid-phase extraction. *Trends Anal. Chem* 2005,*24*, 394–406.

55. Fontanals, N.; Marce, R.M.; Borrull, F. New materials in sorptive extraction techniques for polar compounds. *J. Chromatogr. A* 2007, *1152*, 14–31.

56. Yoshioka, N.; Ichihashi, K. Determination of 40 synthetic food colors in drinks and candies by high-performance liquid chromatography using a short column with photodiode array detection. *Talanta* 2008, *74*, 1408–1413.

57. Lv, Y.Q.; Lin, Z.X.; Feng, W.; Zhou, X.; Tan, T.W. Selective recognition and large enrichment of dimethoate from tea leaves by molecularly imprinted polymers. *Biochem. Eng* 2007, *36*, 221–229.

58. Yang, R.Z.; Wei, X.L.; Gao, F.F.; Wang, L.S.; Zhang, H.J.; Xu, Y.J.; Li, C.H.; Ge, Y.X.; Zhang, J.J.; Zhang, J. Simultaneous analysis of anthocyanins and flavonols in petals of lotus (*Nelumbo*) cultivars by high-performance liquid chromatography-photodiode array detection/electrospray ionization mass spectrometry. *J. Chromatogr. A* 2009, *1216*, 106–112.

59. Le Moullec, S.; Begos, A.; Pichon, V.; Bellier, B. Selective extraction of organophosphorus nerve agent degradation products by molecularly imprinted solid-phase extraction. *J. Chromatogr. A* 2006, *1108*, 7–13.

60. Le Moullec, S.; Truong, L.; Montauban, C.; Begos, A.; Pichon, V.; Bellier, B. Extraction of alkyl methylphosphonic acids from aqueous samples using a conventional polymeric solid-phase extraction sorbent and a molecularly imprinted polymer. *J. Chromatogr. A* 2007, *1139*, 171–177.

61. Kugimiya, A.; Takei, H. Selectivity and recovery performance of phosphate-selective molecularly imprinted polymer.*Anal. Chim. Acta* 2008, *606*, 252–256.

62. Gonçalves, C.; Alpendurada, M.F. M ultiresidue method for the simultaneous determination of four groups of pesticides in ground and drinking waters, using solid-phase microextraction-gas chromatography with electron-capture and thermionic specific detection. *J. Chromatogr. A* 2002, *968*, 177–190.

63. Yao, Z.; Jiang, G.; Liu, J.; Cheng, W. Application of solid-phase microextraction for the determination of organophosphorous pesticides in aqueous samples by gas chromatography with flame photometric detector. *Talanta*2001, *55*, 807–814.

64. Frías, S.; Rodríguez, M.A.; Conde, J.E.; Pérez-Trujillo, J.P. Optimisation of a solid-phase microextraction procedure for the determination of triazines in water with gas chromatography-mass spectrometry detection. *J. Chromatogr. A* 2003,*1007*, 127–135.

65. Su, P.; Huang, S.D. Determination of organophosphorus pesticides in water by solid-phase microextraction. *Talanta*1999, *49*, 393–402.

66. Rocha, C.; Pappas, E.A.; Huang, C. Determination of trace triazine and chloroacetamide herbicides in tile-fed drainage ditch water using solid-phase microextraction coupled with GC-MS. *Environ. Pollut* 2008, *152*, 239–244.

67. Magdic, S.; Boyd-Boland, A.; Jinno, K.; Pawliszyn, J. Analysis of organophosphorus insecticides from environmental samples using solid-phase microextraction. *J. Chromatogr. A* 1996, *736*, 219–228.

68. Basheer, Ch.; Jegadesan, S.; Valiyaveettil, S.; Kee Lee, H. Sol-gel-coated oligomers as novel stationary phases for solid-phase microextraction. *J. Chromatogr. A* 2005, *1087*, 252–258.

69. Bicchi, C.; Cordero, Ch.; Liberto, E.; Rubiolo, P.; Sgorbini, B.; David, F.; Sandra, P. Dual-phase twisters: A new approach to headspace sorptive extraction and stir bar sorptive extraction. *J. Chromatogr. A* 2005, *1094*, 9–16.

70. Chen, J.; Duan, Ch.; Guan, Y. Sorptive extraction techniques in sample preparation for organophosphorus pesticides in complex matrices. *J. Chromatogr. B* 2010, *878*, 1216–1225.

71. Kawaguchi, M.; Ito, R.; Sakui, N.; Okanouchi, N.; Saito, K.; Nakazawa, H. Dual derivatization-stir bar sorptive extraction-thermal desorption-gas chromatography-mass spectrometry for determination of 17β-estradiol in water sample. *J. Chromatogr. A* 2006, *1105*, 140–147.

72. Anastassiades, M.; Lehotay, S.J.; Štajnbaher, D.; Schenck, F.J. Fast and easy multiresidue method employing acetonitrile extraction/partitioning and dispersive solid-phase extraction for the determination of pesticide residues in produce. *J. AOAC Int* 2003, *86*, 412–431.

73. Albero, B.; Sanchez-Brunete, C.; Tadeo, J.L. Multiresidue determination of pesticides in juice by solid-phase extraction and gas chromatography-mass spectrometry. *Talanta* 2005, *66*, 917–924.

74. Gonzalez-Curbelo, M.A.; Hernandez-Borges, J.; Ravelo-Perez, L.M.; Rodriguez-Delgado, M.A. Insecticides extraction from banana leaves using a modified QuEChERS method. *Food Chem* 2011, *125*, 1083–1090.

75. Nguyen, T.D.; Yu, J.E.; Lee, D.M.; Lee, G.-H. A multiresidue method

for the determination of 107 pesticides in cabbage and radish using QuEChERS sample preparation method and gas chromatography mass spectrometry. *Food Chem*2008, *110*, 207–213.

76. Koesukwiwat, U.; Lehotay, S.J.; Miaoc, S.; Leepipatpiboon, N. High throughput analysis of 150 pesticides in fruits and vegetables using QuEChERS and low-pressure gas chromatography-time-of-flight mass spectrometry. *J. Chromatogr. A* 2010, *1217*, 6692–6703.

77. Lehotay, S.J.; Sonb, K.A.; Kwon, H.; Koesukwiwat, U.; Fud, W.; Mastovska, K.; Hoha, E.; Leepipatpiboon, N. Comparison of QuEChERS sample preparation methods for the analysis of pesticide residues in fruits and vegetables.*J. Chromatogr. A* 2010, *1217*, 2548–2560.

78. Park, J.-Y.; Choi, J.-H.; Abd El-Aty, A.M.; Kim, B.M.; Oh, J.-H.; Do, J.-A.; Kwon, K.S.; Shim, K.-H.; Choi, O.-J.; Shin, S.C.; Shim, J.-H. Simultaneous multiresidue analysis of 41 pesticide residues in cooked foodstuff using QuEChERS: Comparison with classical method. *Food Chem* 2011, *128*, 241–253.

79. Lehotay, S.J.; Mastovska, K.; Lightfield, A.R.; Gates, R.A. Multi-Analyst, multi-matrix performance of the QuEChERS approach for pesticide residues in foods and feeds using HPLC/MS/MS analysis with different calibration techniques.*J. AOAC Int* 2010, *93*, 355–367.

80. Nguyen, T.D.; Yu, J.E.; Lee, D.M.; Lee, G.H. A multiresidue method for the determination of 107 pesticides in cabbage and radish using QuEChERS sample preparation method and gas chromatography mass spectrometry. *Food Chem* 2008, *110*, 207–213.

81. Godula, M.; Hajslová, J.; Alterová, K. Pulsed splitless injection and the extent of matrix effects in the analysis of pesticides. *J. High Resolut. Chromatogr* 1999, *23*, 395–402.

82. Hernando, M.D.; Agüera, A.; Fernández-Alba, A.R.; Piedra, L.; Contreras, M. Gas chromatographic determination of pesticides in vegetable samples by sequential positive and negative chemical ionization and tandem mass spectrometric fragmentation using an ion trap analyser. *Analyst* 2001, *126*, 46–51.

83. Martínez Vidal, J.L.; Arrebola, F.J.; Mateu-Sánchez, M. Validation of a gas chromatographic-tandem mass spectrometric method for analysis of pesticide residues in six food commodities selection of a reference matrix for calibration. *Chromatographia* 2004, *59*, 321–327.

84. Stepan, R.; Ticha, J.; Hajslova, J.; Kovalczuk, T.; Kocourek, V. Baby food production chain: Pesticide residues in fresh apples and products. *Food Addit. Contam* 2005, *22*, 1231–1242.

85. Mastovska, K.; Hajslova, J.; Lehotay, S.J. Ruggedness and other performance characteristics of low-pressure gas chromatography-mass spectrometry for the fast analysis of multiple pesticide residues in food crops. *J. Chromatogr. A* 2004, *1054*, 335–349.

86. Garcia-Reyes, J.F.; Ferrer, C.; Gomez-Ramos, M.J.; Fernandez-Alba, A.R.; Molina-Diaz, A. Determination of pesticide residues in olive oil and olives. *Trends Anal. Chem* 2007, *26*, 239–251.

87. Leandro, C.C.; Hancock, P.; Fussell, R.J.; Keely, B.J. Comparison of ultra-performance liquid chromatography and high-performance liquid chromatography for the determination of priority pesticides in baby foods by tandem quadrupole mass spectrometry. *J. Chromatogr. A* 2006, *1103*, 94–101.

88. Sadowska-Rociek, A.; Cieślik, E. Stosowane techniki i najnowsze trendy w oznaczaniu pozostałości pestycydów w żywności metodś chromatografii gazowej. *Metrologia* 2008, *13*, 33–38.

89. Walorczyk, S. Różne możliwości wykorzystania chromatografii gazowej połączonej ze spektrometrią mas w analizie pozostałości środków ochrony roślin. *Progr. Plant Protect* 2007, *47*, 111–114.

90. Korytar, P.; Janssen, H.G.; Matisova, E.; Brinkman, U.A.T. Practical fast gas chromatography: Methods, instrumentation and applications. *Trends Anal. Chem* 2002, *21*, 558–572.

91. Mastovska, K.; Lehotay, S.J. Practical approaches to fast gas chromatography-mass spectrometry. *J. Chromatogr. A* 2003, *1000*, 153–180.

92. Namieśnik, J. Modern trends in monitoring and analysis of environmental pollutants. *Pol. J. Environ. Stud* 2001, *10*, 127–140.

93. Beyer, A.; Biziuk, M. Methods for determining pesticides and polychlorinated biphenyls in food samples-problems and challenges. *Crit. Rev. Food Sci* 2008, *48*, 888–904.

94. Alder, L.; Greulich, K.; Kempe, G.; Vieth, B. Residue analysis of 500 high priority pesticides: Better by GC-MS or LC-MS/MS? *Mass Spectrom. Rev* 2006, *25*, 838–865.

CITATION

CHAPTER 1

Renata Raina (2011). Chemical Analysis of Pesticides Using GC/MS, GC/MS/MS, and LC/MS/MS, Pesticides - Strategies for Pesticides Analysis, Prof. Margarita Stoytcheva (Ed.), ISBN: 978-953-307-460-3, InTech, DOI: 10.5772/13242.

CHAPTER 2

Mariana Furio Franco Bernardes, Murilo Pazin, Lilian Cristina Pereira, and Daniel Junqueira Dorta (2015). Impact of Pesticides on Environmental and Human Health, Toxicology Studies - Cells, Drugs and Environment, Dr. Ana Cristina Andreazza (Ed.), ISBN: 978-953-51-2140-4, InTech, DOI: 10.5772/59710.

CHAPTER 3

Alewu B. and Nosiri C. (2011). Pesticides and Human Health, Pesticides in the Modern World - Effects of Pesticides Exposure, Dr. Margarita Stoytcheva (Ed.), ISBN: 978-953-307-454-2, InTech, DOI: 10.5772/18734.

CHAPTER 4

Harsimran Kaur Gill and Harsh Garg (2014). Pesticides: Environmental Impacts and Management Strategies, Pesticides - Toxic Aspects, Dr. Sonia Soloneski (Ed.), ISBN: 978-953-51-1217-4, InTech, DOI: 10.5772/57399.

CHAPTER 5

Zhi-Jun Zhou (2011). Health Problem Caused by Long-Term Organophosphorus Pesticides Exposure - Study in China, Pesticides in the Modern World - Effects of Pesticides Exposure, Dr. Margarita Stoytcheva (Ed.), ISBN: 978-953-307-454-2, InTech, DOI: 10.5772/17519.

CHAPTER 6

Dipsikha Bora, Bulbuli Khanikor and Hiren Gogoi (2012). Plant Based Pesticides: Green Environment with Special Reference to Silk Worms, Pesticides - Advances in Chemical and Botanical Pesticides, Dr. R.P. Soundararajan (Ed.), ISBN: 978-953-51-0680-7, InTech, DOI: 10.5772/47832.

CHAPTER 7

Simon Koma Okwute (2012). Plants as Potential Sources of Pesticidal Agents: A Review, Pesticides - Advances in Chemical and Botanical Pesticides, Dr. R.P. Soundararajan (Ed.), ISBN: 978-953-51-0680-7, InTech, DOI: 10.5772/46225.

CHAPTER 8

Marcela Varona Uribe, Sonia Mireya Díaz, Andrés Monroy, Edwin Barbosa, Martha Isabel Páez and René A. Castro (2012). Exposure to Pesticides in Tomato Crop Farmers in Merced, Colombia: Effects on Health and the Environment, Pesticides - Recent Trends in Pesticide Residue Assay, Dr. R.P. Soundararajan (Ed.), ISBN: 978-953-51-0681-4, InTech, DOI: 10.5772/48640.

CHAPTER 9

Jolanta Stocka, Maciej Tankiewicz, Marek Biziuk, and Jacek Namieśnik (2011). Green aspects of techniques for the determination of currently used pesticides in environmental samples, Int. J. Mol. Sci. 2011, 12(11), 7785-7805; doi:10.3390/ijms12117785

INDEX